During the period from the Neoproterozoic to the early Paleozoic, numerous continental fragments in the Southern Hemisphere consolidated along a series of interconnected orogenic belts to form the supercontinent of Gondwanaland. The Ross orogen of the Transantarctic Mountains is the part of the orogenic system that formed at the Pacific continental margin of present-day Antarctica. According to one hypothesis, which is perhaps among the most significant recent developments in regional tectonics, this continental margin was created by the rifting and subsequent drift of Laurentia from Gondwanaland.

With an unparalleled breadth and depth of information, this book provides a detailed synthesis of the history of the Ross orogen, which commenced in the Neoproterozoic with passive margin sedimentation and progressed through a series of tectonic events that culminated in the Ross orogeny approximately 500 million years ago. In doing so, it incorporates classical studies with discussions of the most recent and controversial research from the international community. The book also includes a comprehensive bibliography and a historical chronology of all expeditions that have worked on the Ross orogen, from the first sightings by Ross in 1840 to the present.

This review will be valuable to all geologists interested in these episodes in the earth's history and to researchers of the geology of Antarctica.

# The Ross Orogen of the Transantarctic Mountains

The Tusk, a horn of coarsely crystalline marble on the west bank of Liv Glacier flowing from the right rear of the photo. Mt. Fridtjof-Nansen towers in the distance, displaying the Kukri peneplain, the physical and temporal boundary of the Ross orogen. Devonian to Triassic Beacon Supergroup and sills of Jurassic Ferrar dolerite overlie the erosion surface. Exposed beneath the surface are plutonic rocks of the Queen Maud–Wisconsin Range batholith. The Tusk is composed of Henson Marble, a member of the Early to Middle Cambrian Liv Group.

# THE ROSS OROGEN OF THE TRANSANTARCTIC MOUNTAINS

Edmund Stump
*Arizona State University*

CAMBRIDGE UNIVERSITY PRESS
Cambridge, New York, Melbourne, Madrid, Cape Town, Singapore, São Paulo

Cambridge University Press
The Edinburgh Building, Cambridge CB2 2RU, UK

Published in the United States of America by Cambridge University Press, New York

www.cambridge.org
Information on this title: www.cambridge.org/9780521433143

First published 1995
This digitally printed first paperback version 2005

*A catalogue record for this publication is available from the British Library*

*Library of Congress Cataloguing in Publication data*
Stump, Edmund.
The Ross orogen of the Transantarctic Mountains / Edmund Stump.
p.   cm.
Includes bibliographical references and index.
ISBN 0-521-43314-2
1. Orogen – Antarctica – Transantarctic Mountains.   2. Geology –
Antarctica – Transantarctic Mountains.   3. Transantarctic Mountains
(Antarctica)   I. Title.
QE621.5.A6S78      1995
551.8′2′09989 – dc20                                           94-19781
                                                              CIP

ISBN-13 978-0-521-43314-3 hardback
ISBN-10 0-521-43314-2 hardback

ISBN-13 978-0-521-01999-6 paperback
ISBN-10 0-521-01999-0 paperback

This book is dedicated to Mort D. Turner,
with great affection.

# Contents

# Preface

As an undergraduate in the autumn of 1966, I took John Haller's course on structural geology, taught in the classic, descriptive Billings style. Often with twenty minutes of the lecture left, Professor Haller would dim the lights and show his slides of East Greenland, incredible fjord walls thousands of meters high and kilometers long displaying natural cross sections in vivid detail from every sector of the orogenic belt. Haller's knowledge of that icy land was intimate; he was its master. And I fancied that when I grew up I too might have a mountain range I could call my own, whose every valley I could conjure up from memory and whose history I might some day set down. My fantasy was generic, except that the mountains were always big and remote; glaciers and ice caps were not part of the picture.

*Geology of the East Greenland Caledonides,* published by Haller in 1971, was an inspiration, and I believe that it still stands as the standard of monographs synthesizing knowledge about major mountain systems. Its great strength lies not only in the copious, annotated photographs of perhaps the best-exposed orogenic belt on earth, but in the many figures and maps, hand-drafted by Haller himself, that bring the details of rock distribution and structure so accurately to the page.

From 1968 to 1970 I spent a disappointing two years at Yale. However, during that time I had an opportunity to observe John Rodgers laboring on his own book, *The Tectonics of the Appalachians,* which came out in 1970. Like Haller's, it was a comprehensive statement from the master, but unlike Haller's it lacked graphics except for a couple of maps. I was critical of this omission at the time, but have come to appreciate the extraordinary time and effort that go into the preparation of such figures.

That I evoke the names of Haller and Rodgers is not to say that my own work is in any measure comparable to theirs. Rather it is to acknowledge the inspiration that these two men gave to a young student to become a "regional geologist," regardless of the fact that such a title never appears in job descriptions for academic positions.

In the spring of 1970 I was at loose ends, looking to get as far away as possible, when Dick Armstrong suggested that I try Antarctica. As fate would have it, a helicopter-supported party from Ohio State was bound for the Queen Maud Mountains the following season, and I fell into a spot with the group. This expedition rode the coattails of an extraordinarily successful venture the preceding year when researchers from Ohio State had found *Lystrosaurus* at Coal Sack Bluff, nailing the correlation of the Triassic throughout all of the Gondwanaland fragments. The party's stunning scientific success was matched by a stunning logistic failure, with only a few days of helicopter time during the entire season, so a second helicopter-supported party was fielded in 1970–71 to reach areas proposed but not gotten to the year before.

I had never previously considered going to the Antarctic, and so the impact of pristine, polar wilderness, its purity, its vastness, and its detail, was overpowering. When I got back to Ohio I had only one goal – to return to that icy fastness. The folks at the Institute of Polar Studies – David Elliot, Gunter Faure, John Splettstoesser, Colin Bull – were extremely supportive, and after three attempts at proposal writing I was rewarded in 1974–75 with my own ground party of four to go to the Queen Maud Mountains. That was in the days before peer review at the National Science Foundation, and I continue to be humbly grateful to the program manager, Mort Turner, for his faith in a long-haired kid who at the time had little more to show than a relentless enthusiasm for those desolate mountains at the end of the earth.

My first seasons on The Ice were in the area explored by the parties of Larry Gould and Quin Blackburn during Byrd's First and Second Antarctic Expeditions, and of Al Wade and Vic McGregor in the early 1960s. My own studies followed in their footsteps. Those footsteps were few and the coverage was thin, but color existed on most of the map, and their pioneering work was my guide. Because the area is so vast, the logistic costs are so high, and it is such a privilege to be sent there, I have always felt that in part a measure of one's contribution in Antarctica is the amount of ground one covers in a field season. In this my predecessors demonstrated a prowess that has always challenged me. I have a keen respect for those who have gone before, perhaps because I know firsthand the realities of the field conditions that they endured.

One of the goals of this book has been to document the field parties that have explored the various parts of the Transantarctic Mountains. While this may be dry reading for most, I feel that the chronology of exploration should be set down, lest younger generations lose sight of those who preceded them. Accordingly, this chronology is partitioned into sections that the reader can skip without losing the geological account.

I have attempted to be comprehensive in assembling the bibliography, although I am sure that there are some Russian and perhaps German citations that I have missed, as well as a few others recently published that slipped through my net. To all those authors whose work has not been cited, I sincerely apologize. Send me your reprints and if there ever is a second edition, God forbid, I will include them.

All isotopic dates quoted in the text that were determined using old decay constants have been recalculated using the constants of Steiger and Jäger (1977) following procedures outlined in Faure (1986).

The question of the absolute time boundaries on the Cambrian is critical to an accurate history of the Ross orogen, since there is possible overlap in the timing of isotopically dated plutonism and fossil-dated sedimentation. The age of the Precambrian–Cambrian boundary has been controversial for more than a decade. Harland et al. (1982) set it at 590 Ma, whereas Odin et al. (1983) suggested 530 Ma, an age considerably younger than other proposals up to that time. Recent dating of ash layers from localities scattered around the globe (Canada, Morocco, China, Australia) favors an age more in line with that of Odin et al. (Compston et al., 1992; Cooper et al., 1992). A few months before the completion of this

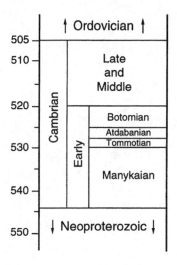

**Figure P.1.** Geological timescale for the Cambrian adopted for this work. After Bowring et al. (1993).

project, Bowring et al. (1993) published an age for the Precambrian–Cambrian boundary at about 544 Ma, based on U/Pb zircon dating of rocks from Siberia. Although it is still too soon to know whether the consensus will side with this number, I have used it in the places where the discussion centers on the timing of events in the Early Cambrian and have adopted the other age boundaries for stages in the Cambrian cited by Bowring et al. (1993) (Fig. P.1). The subdivision of the Proterozoic follows the recommendations of the Subcommission of Precambrian Stratigraphy (Plumb, 1991).

# Acknowledgments

This project was undertaken during a sabbatical leave at the British Antarctic Survey in 1990–91. I gratefully acknowledge the hospitality of my hosts, Mike Thomson and David Drewry. It was a luxury to spend a year in Cambridge and to have at my disposal not only the BAS library, but also the library of the Scott Polar Institute and the Sedgwick Library, whose librarian, Ruth Banger, was particularly helpful. The two areas of the Transantarctic Mountains that I have not visited are the Thiel Mountains and the Pensacola Mountains. Both of these have been sites of prior research by BAS geologists, and I found it particularly valuable to discuss their geology with Bryan Storey and Dave Macdonald, as well as to be given access to their unpublished material.

The project was funded by the Division of Polar Programs of the National Science Foundation (Grant DPP-8916057). Many thanks to Herman Zimmerman, program manager at the time. Various researchers have been generous through their discussions and supply of unpublished material. These include Bert Rowell, Peg Rees, and John Goodge, whose recent contributions to our understanding of the Ross orogen have been numerous and insightful. In Germany, Franz Tessensohn, Georg Kleinschmidt, and Norbert Roland were helpful at various stages. Carlo Alberto Ricci kindly supplied the volumes of the Italians' work. In New Zealand, David Skinner provided encouragement and unpublished manuscripts, and Malcolm Laird, really the godfather of the Ross orogen, has always had the welcome mat out for transient foreigners stopping in Christchurch on their way south.

The greatest reward for me in taking students to Antarctica has been in watching their awestruck reactions on first contact, thereby reliving my own. Pat Lowry, Jerry Smit, David Edgerton, and Scott Borg are as fine a group of fieldmates as one could have. Scott, who took two degrees from Arizona State, deserves special mention, for his power, finesse, and observational skills in the field were matched only by his meticulous technique in the lab. His contributions to our understanding of the chemical and isotopic evolution of plutonic rocks of the Ross orogen have reached far beyond the continent. Through my early seasons Phil Colbert served as field assistant, mountaineer, musician, philosopher, true friend, and all-around dirtbag. Together we learned to survive Antarctica with grace.

The graphics that figure prominently in this work owe their existence to Sue Selkirk. Her skill with both computer-assisted and old-fashioned, hand-drawn techniques is outstanding. All of the figures from other works have been scanned, stripped of extraneous information, and relettered for uniformity throughout. The geographic and location maps were produced by a combination of scanning and meticulous redrafting to give the composite result. The hand draft of the Dry Valleys area is a work of art. I offer my sincere thanks to Sue for the many hours she invested in this project.

Finally, I must acknowledge the support of my wife, Harriet Maccracken, and my children. Since my second trip to The Ice in 1974–75, Harriet has each time been at the airport to see me off and then to welcome me back home. Our periods of separation now run into years. Through all this time she has supported my work and encouraged its continuance. By the 1985–86 field season, Simon had been born, and by 1987–88, Molly. Harriet traded the solitude of our separations for the courageously hard work of a single mom. Nick was conceived in England during the writing of this book. It was all but finished when we returned to the United States, but the pressures of a departmental chairmanship have strung it out for the past two and a half years and pushed it into night work when I should have been home tucking the children into bed. Words are not adequate to express the appreciation I feel for all that Harriet has been through and done.

# Abbreviations

DV1, DV2    Dry Valley 1 and Dry Valley 2 Suites (Smillie, 1992)

DVDP    Dry Valley Drilling Project

$\varepsilon_{Sr}$    Epsilon strontium, a notation of measured $^{87}Sr/^{86}Sr$ relative to a uniform reservoir (Borg et al., 1990):
$$\{[(^{87}Sr/^{86}Sr_{measured})/(^{87}Sr/^{86}Sr_{UR})] - 1\} \times 10^4$$

$\varepsilon_{Nd}$    Epsilon neodymium, a notation of measured $^{143}Nd/^{144}Nd$ relative to a chondritic reservoir (Borg et al., 1990):
$$\{[(^{143}Nd/^{144}Nd_{measured})/(^{143}Nd/^{144}Nd_{CHUR})] - 1\} \times 10^4$$

Ga    Billions of years before present

IGY    International Geophysical Year

IR    Initial $^{87}Sr/^{86}Sr$ ratio

Ma    Millions of years before present

MSWD    Mean squares of weighted deviates, a measure of the statistical goodness of fit of Rb/Sr isochrons to the data

m.y.    Millions of years

$T_{DM}$    Neodymium model age; the time of separation of the Sm/Nd system from the mantle (Borg et al., 1990)

# 1 Introduction

Spanning the continent of Antarctica for 3,500 km and rising to heights in excess of 4,000 m, the Transantarctic Mountains are among the great mountain ranges on earth. They are the physiographic boundary between East and West Antarctica, holding back the East Antarctic Ice Sheet that issues through them in giant outlet glaciers, emptying into the Ross Sea and the Ross Ice Shelf (Fig. 1.1). Between the Horlick Mountains and the Patuxent Range, the West Antarctic Ice Sheet inundates the Transantarctic Mountains merging with its East Antarctic counterpart. Beyond this the Pensacola Mountains are drained on both sides by broad ice streams flowing into the Filchner and Ronne Ice Shelves.

Unlike most major mountain belts, which owe their origin to plate-margin tectonics, either subduction- or collision-related, the Transantarctic Mountains are intracontinental and probably best described as a rift shoulder whose boundary follows the front of the range along the Ross embayment (Fitzgerald et al., 1986; Stern and ten Brink, 1989). Outboard of this is a region of extended continental crust (Cooper and Davey, 1987). The rift margin itself may arc outward from the Horlick Mountains to the west of the Whitmore and Ellsworth Mountains (Behrendt and Cooper, 1991), but there is geological continuity in the Thiel and Pensacola Mountains with the rest of the Transantarctic Mountains. However, the latter areas do have geological features distinct from the rest of the Transantarctic Mountains, in particular a lack of intense deformation during the Ross orogeny, as will be discussed later in the book (Chapters 6 and 7).

In a number of ways the Transantarctic Mountains are also distinct from other rift systems. Their scale greatly exceeds the rifts of the Rio Grande, Oslo, and Rhine. They are of a length comparable to that of the East African rift system, but there the rifts would appear to be wholly intracratonic, whereas in Antarctica they occur at the boundary between the cratonic basement of East Antarctica and Phanerozoic accreted rocks of West Antarctica. Perhaps the most analogous feature is the Wasatch Range at the eastern boundary of the Basin and Range province in the Cordillera of the western United States. The Wasatch front marks the boundary between the Phanerozoic Cordilleran geocline to the west and the onlapped cratonic sequences to the east, as well as the front of the Sevier fold and thrust belt. By contrast, the Basin and Range forms a terrestrial topography, while much of West Antarctica is below sea level.

As indicated by fission-track analysis, the main phase of uplift of the Transantarctic Mountains commenced around 50 Ma (Gleadow and Fitzgerald, 1987; Fitzgerald, 1992), but earlier precursors are also indicated at around 80 and 110 Ma (Stump and Fitzgerald, 1992). It is probably no coincidence that the present-day elongation of the range is parallel to structural features in its basement

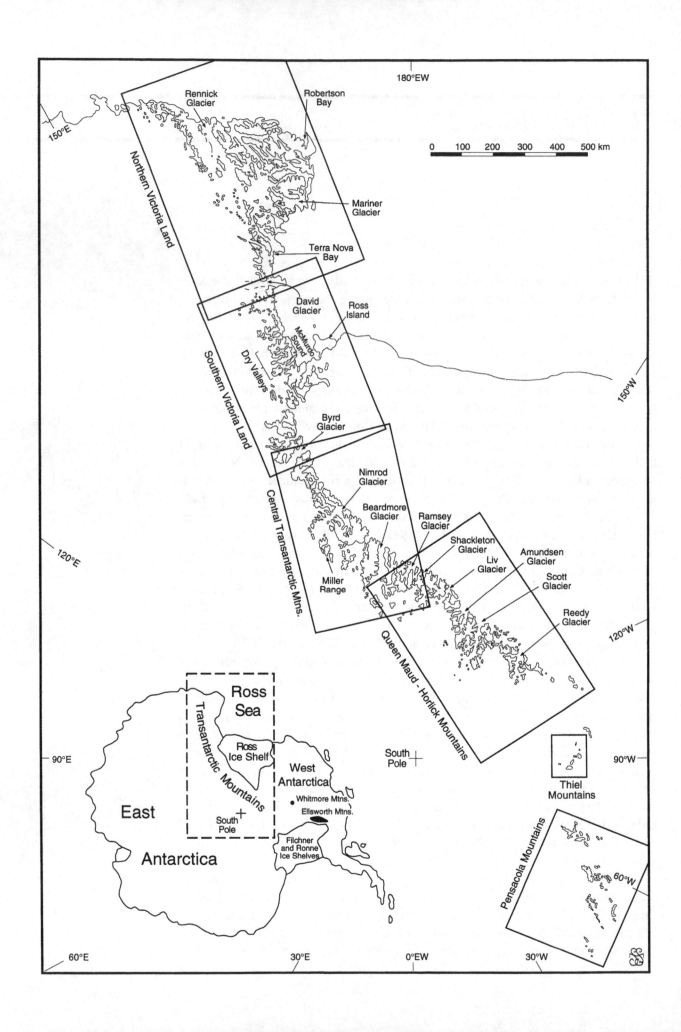

formed approximately half a billion years earlier. This basement is the Ross orogen. Its temporal and spatial boundary is a mid-Paleozoic erosion surface, often called the Kukri peneplain, that is overlain by the Beacon Supergroup, a Devonian to Triassic sequence largely of terrestrial origin (Barrett, 1981; Collinson, 1991), intruded and overlain by Jurassic tholeiites of the Ferrar Group (Tingey, 1991).

The history of the Ross orogen commences in the Neoproterozoic with passive, continental margin sedimentation, perhaps following the rifting of Laurentia from Gondwanaland (Stump, 1992b). Earlier Proterozoic tectonism is indicated in the Miller and Geologists Ranges at the craton margin. The passive margin was activated by compressional deformation and plutonism in the latter part of the Neoproterozoic, which evolved through a full-blown cycle of orogeny during the Cambrian, with exhumation and cooling of the belt during the Ordovician, making way for Beacon sedimentation.

The history of the Ross orogen is the subject of this volume. The chapters are divided by geographical area, as shown in Figure 1.1. Although a common history runs through all of the areas, each has distinct characteristics within the boundaries as they are defined, and, as can be seen, uncertainties remain regarding correlation of some units from one area to another.

**Figure 1.1** (*facing page*). Location map for the Transantarctic Mountains. The boxes outline areas covered by location maps in the succeeding chapters.

# 2 Northern Victoria Land

## Geological Summary

Northern Victoria Land boasts the widest exposure in all of the Transantarctic Mountains, spanning more than 350 km. Elements common to the Ross orogen occur on its western side, but to the east the area is underlain by rocks without kin throughout the rest of the Transantarctic Mountains. The concept of terranes provides the best framework for discussing the geological and tectonic development of the region. Their recognition by New Zealand geologists (Weaver et al., 1984a; Bradshaw et al., 1985b) closely followed the introduction of the notion to the discipline of geology as a whole (Coney et al., 1980; Howell, 1985). The original subdivision of Weaver et al. (1984a) of a western Wilson terrane, eastern Robertson Bay terrane, and medial, fault-bounded Bowers terrane (Fig. 2.4) is now widely accepted, although a consensus on the details of timing and kinematics of terrane assembly has not been reached. Because of differences among ranges within the Wilson terrane and the presence of likely major subglacial faults, this terrane may warrant further subdivision (Bradshaw, 1985).

The Wilson terrane is characterized by medium- to high-grade schists and gneisses and lesser low-grade metasediments, with nomenclature distinct to subareas. The original designation of "Wilson Group" for the metamorphics found throughout the Wilson terrane north of the Terra Nova Bay area has evolved to "Wilson metamorphics," confined to the western ranges, "Lanterman and Mountaineer metamorphics," occurring in those respective ranges, and several more local names discussed later. In the Terra Nova Bay area the rocks are called Priestley Formation. This unit has been interpreted to overlie unconformably an older suite called the Snowy Point Gneiss Complex. Although it seems likely that the metasedimentary rocks of the Wilson terrane are at least broadly related, the stratigraphic relationships between sequences in various areas are poorly constrained. Fragments of body fossils indicative of a Paleozoic age have been found in the Terra Nova Bay area (Lombardo et al., 1989), while a U/Pb zircon study from the Daniels Range has suggested that the metasediments there were involved in an early magmatic episode at around 544 Ma (Black and Sheraton, 1990). Elsewhere the depositional age of the metasedimentary rocks in the Wilson terrane is constrained only by crosscutting Granite Harbour Intrusives, and so may be considered either Proterozoic (?) or Cambrian (?) in age.

Deformation and metamorphism of the sedimentary component of the Wilson terrane are considered to be of the Ross orogeny. In the western area, plutonism appears to have been involved in the deformation and peak metamorphism at around 540–535 Ma, whereas plutonism in the Lanterman Range and Terra Nova

**Figure 2.1.** The heart of the Robertson Bay terrane, northern Victoria Land. View to east across Tucker Glacier (middle ground) and central Admiralty Mountains toward Mt. Minto and Mt. Adam, the high peaks on center horizon. Most of the area is underlain by Robertson Bay Group, with Mt. Adam composed of Admiralty Intrusives.

Bay area commenced after the main deformation. A major pulse of posttectonic Granite Harbour Intrusives occurred throughout the terrane around 480 Ma, cooling to argon retention temperatures within 10–15 m.y.

The Bowers terrane consists of a varied sequence of sedimentary and volcanic rocks primarily of Cambrian age, called the Bowers Supergroup. The lowermost Sledgers Group contains two interfingering formations, the volcanic Glasgow Formation and the sedimentary Molar Formation. The volcanics have affinities with primitive island arcs and, to a lesser extent, rifted or marginal basins (Weaver et al., 1984a). The sedimentary rocks are shelf or deeper basinal deposits containing abundant volcaniclastic fractions, but also conglomeratic horizons with granitic and metamorphic clasts. Fossils indicate that the Sledgers Group was probably deposited entirely within the Middle Cambrian.

Sledgers Group is followed conformably by Mariner Group, a sequence of fine-grained clastic and calcareous sedimentary rocks, which has been interpreted as representing a regressive marine environment. Abundant fossils within the group range from late Middle to Upper Cambrian. Leap Year Group lies above a subtle regional unconformity that cuts successively deeper into Sledgers Group toward the north. The group is characterized by coarse-grained sandstones and conglomerates, interpreted to have been deposited primarily in terrestrial settings. No body fossils have been found in these rocks.

The Bowers terrane was folded during a single event that produced a series of

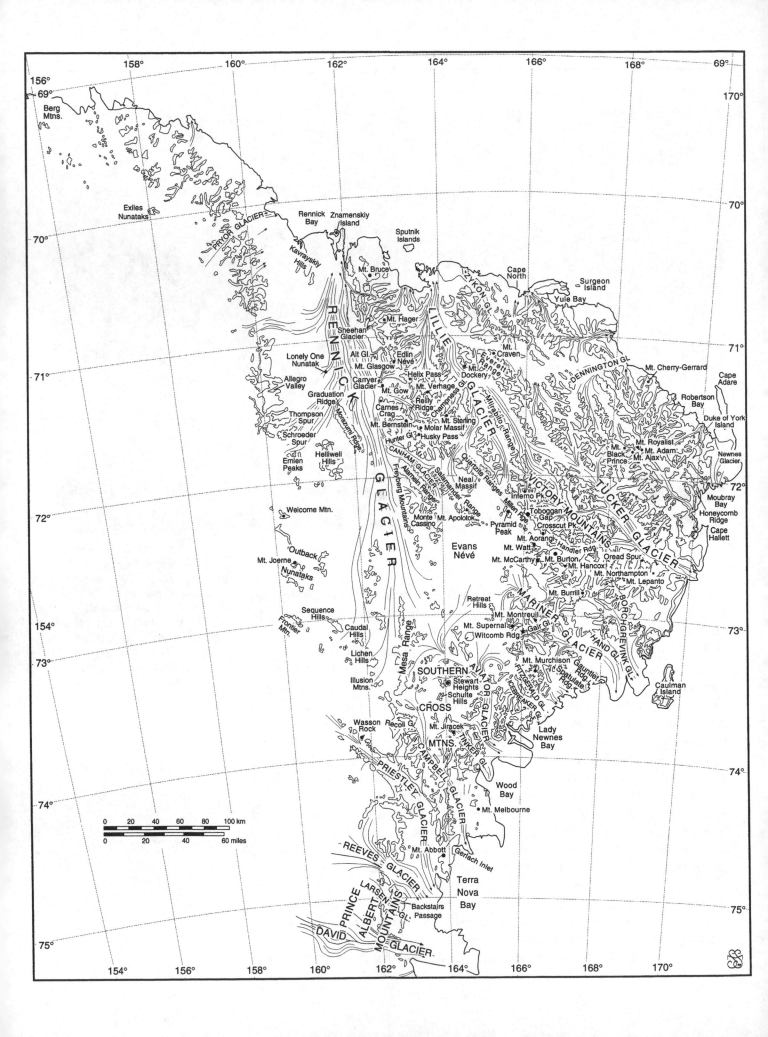

tight folds whose axes trend parallel to the elongation of the terrane. Accompanying metamorphism did not exceed prehnite–pumpellyite or lower greenschist grade, except along the terrane boundaries.

The Robertson Bay terrane is predominantly a sequence of graywackes and shales, probably Cambrian to Lower Ordovician in age, known as the Robertson Bay Group. Uppermost Cambrian to Tremadocian body fossils have been found in a slideblock of limestone within the Handler Formation, thought to be in the uppermost stratigraphic levels of the group. The Robertson Bay Group was deformed during a single episode that produced folds and cleavage under low-greenschist conditions. Ar/Ar dating of sericite indicates diachronous cleavage development between around 500 and 460 Ma from west to east across the terrane.

Both the Bowers and Robertson Bay terranes, and possibly the Wilson terrane, were intruded during the Devonian by high-level crosscutting plutons of the Admiralty Intrusives. Devonian to Carboniferous volcanics are scattered across all three terranes. Besides the history of the rocks within each terrane, the timing and mechanisms of assembly of the terranes are critical problems in the overall tectonics of the region.

Antecedent to the concept of terranes in northern Victoria Land, the subdivision of the region into discrete blocks or units dates to Klimov and Soloviev (1960), who recognized a block west of Matusevich Glacier containing phyllitic and limestone schists, which they named the "Berg series," a block east of Rennick Glacier with rocks equivalent to the graywacke and shale of Robertson Bay, and an intervening block composed of a complex of crystalline schists, migmatites, and granitoids, which they named the "Wilson series." Soloviev (1960) separated the blocks with meridional faults, one up the Matusevich Glacier, the other inland from Rennick Bay. He also noted that the one extending south from Rennick Bay is aligned on the same meridian with the boundary fault between the western Ross Sea and the Transantarctic Mountains south of northern Victoria Land.

Le Couteur and Leitch (1964) were the first to recognize the Camp Ridge Quartzite, which was incorporated into the "Bowers Group" of Sturm and Carryer (1970). By the time of production of the Antarctic Map Folio Series, the distribution of major rock units and their boundary faults were known in all areas except along the rugged coast between Borchgrevink and Aviator Glaciers (Gair et al., 1969). Mapping from helicopters in that area in 1981–82 demonstrated the continuance of the Bowers Supergroup from upper Mariner Glacier through to the coast (Stump et al., 1983b), and the application of terrane tectonics followed.

## Chronology of Exploration

Serendipity has always been the handmaiden of exploration, and no truer case can be found than the discovery of the Transantarctic Mountains by the British expedition led by Capt. James Clark Ross. With two vessels, HMS *Erebus* and HMS *Terror,* strengthened for ice conditions, the expedition was charged to record magnetic observations at high southerly latitudes and, if possible, to reach

**Figure 2.2** (*facing page*). Location map for northern Victoria Land.

the South Magnetic Pole. Ross was ably qualified for the undertaking, having made a study of terrestrial magnetism and having reached the North Magnetic Pole in 1831.

Upon arrival in Hobart in August 1839, Ross learned that his publicly announced southerly course along longitude 160°E had been preempted the previous season by a U.S. expedition, headed by Capt. Charles Wilkes. Piqued by this action and bound not to follow in the wake of such a ragtag flotilla, Ross chose instead to sail south along longitude 170°E, thereby making the most important discoveries of Antarctic exploration: the gateway to the icy interior of the continent, the world's largest ice shelf, and the last major mountain range to be discovered on earth.

On 31 December 1839 in heavy seas, the expedition encountered the ice pack. After three days, the weather became more moderate and the ships began their assault. Six days of pounding were rewarded when, on the morning of 9 January 1840, the ships broke free into open water. Ross reckoned that the South Magnetic Pole was at 76°S, 145°20′E and set a course in that direction. The excitement of possibly claiming both magnetic poles was cut short, however, when the following day land was sighted to the southwest.

As the ships moved in toward Cape Adare, the mountains rose up, filling the men with wonder. Robert McCormick, surgeon on the *Erebus,* described the view as "rising steeply from the ocean in a stupendous mountain range, peak above peak, enveloped in perpetual snow, and clustered together in countless groups resembling a vast mass of crystallisation, which, as the sun's rays were reflected on it, exhibited a scene of such unequaled magnificence and splendor as would baffle all power of language to portray or give the faintest conception of" (Ross, 1847, vol. 2, pp. 415–16).

Heavy seas prevented a landing on the mainland, but the following day a party put in at Possession Island, a small volcanic remnant, where the territory was claimed in the name of Queen Victoria and several rock specimens were collected. Following a half century of disinterest in things Antarctic, the specimens were finally described by Prior (1899) as consisting of basalt, palagonite–tuff, phonolite, and granite. The granite, containing muscovite, tourmaline, and garnet, apparently occurred as inclusions within the basalt.

From Possession Island the expedition headed southward parallel to the mountain front that continued to stand between themselves and the magnetic pole. On Franklin Island, another Tertiary volcano in the Ross Sea, the group made a second landing, before sailing on to the foot of Mt. Erebus, "covered with ice and snow from its base to its summit, from which a dense column of black smoke towered high above" (McCormick in Ross, 1847, vol. 2, p. 416).

The mountains appeared to continue south to beyond the horizon, but eastward from Mt. Erebus stretched the "Great Barrier" of the Ross Ice Shelf, dashing any lingering hopes of sailing to the magnetic pole. After charting the front of the ice shelf, Ross returned the ships to warmer waters.

The following year Ross again sailed south, this time along longitude 146°E. After 56 days beset in the pack, drifting eastward, the ships finally broke through, but were forced to continue their eastward course. They reentered the Ross Sea and returned to the barrier, but managed to extend the charting of the preceding year by only six miles before beating a prudent retreat. No landings were effected

**Figure 2.3** (*facing page*). Location map for figures in Chapter 2.

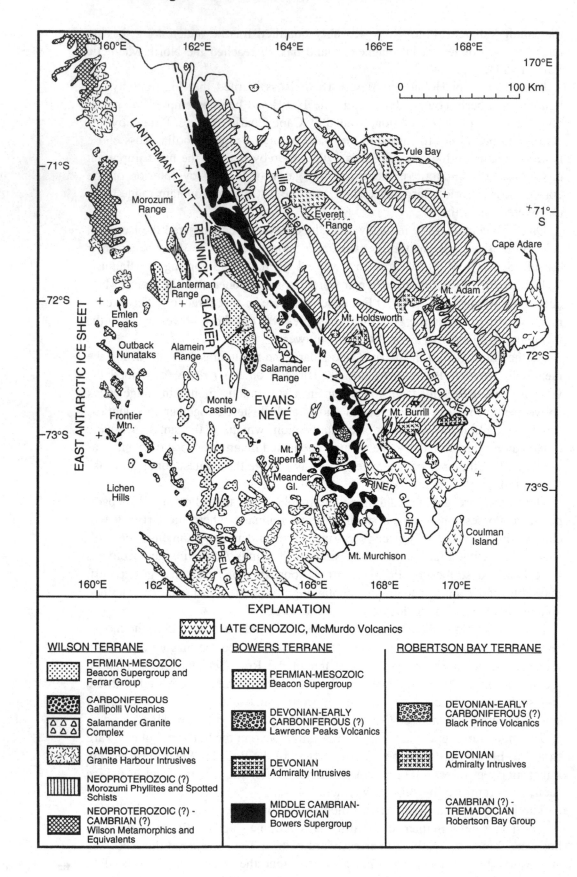

EXPLANATION

 LATE CENOZOIC, McMurdo Volcanics

**WILSON TERRANE**

PERMIAN-MESOZOIC
Beacon Supergroup and
Ferrar Group

CARBONIFEROUS
Gallipolli Volcanics

SALAMANDER GRANITE
Complex

CAMBRO-ORDOVICIAN
Granite Harbour Intrusives

NEOPROTEROZOIC (?)
Morozumi Phyllites and Spotted
Schists

NEOPROTEROZOIC (?) -
CAMBRIAN (?)
Wilson Metamorphics and
Equivalents

**BOWERS TERRANE**

PERMIAN-MESOZOIC
Beacon Supergroup

DEVONIAN-EARLY
CARBONIFEROUS (?)
Lawrence Peaks Volcanics

DEVONIAN
Admiralty Intrusives

MIDDLE CAMBRIAN-
ORDOVICIAN
Bowers Supergroup

**ROBERTSON BAY TERRANE**

DEVONIAN-EARLY
CARBONIFEROUS (?)
Black Prince Volcanics

DEVONIAN
Admiralty Intrusives

CAMBRIAN (?) -
TREMADOCIAN
Robertson Bay Group

that season, and the only rock specimens collected were pebbles from the bellies of penguins and seals (Prior, 1899).

Fifty-five years after Ross had discovered the Ross Sea and the Transantarctic Mountains, a whaling vessel, the enterprise of a Norwegian Australian, Henrik Johan Bull, returned to those icy waters. Although few whales and seals were taken, a party put in at Ridley Beach on Cape Adare, claiming the first landing on the mainland of the Antarctic continent. Rock specimens from the beach were collected by Carsten Borchgrevink, an Australian schoolteacher of Norwegian descent, who had been keen to accompany the voyage and acted as its de facto scientist. The rocks were mainly basalts, but mica schist and a tourmaline–garnet-bearing aplite were also returned (David et al., 1896). A landing at Possession Island recovered additional basalts. Petrographic descriptions of several of the basalts were made by Geikie (1897–99).

Though lacking scientific expertise, Borchgrevink was flush with enthusiasm, and the following year found him in London addressing the Sixth International Geographical Congress on the virtues of wintering a party at Cape Adare, with himself as leader, that would undertake meteorological, biological, and geological observations and a possible overland traverse to the South Magnetic Pole. The Royal Geographical Society turned a cold shoulder to this interloper from Australia, since it had been steadily building toward a full-blown national expedition. However, Borchgrevink found a patron in Sir George Newnes, who provided £40,000 for the venture.

The expedition sailed on the *Southern Cross,* a refitted Norwegian whaling vessel. On 17 February 1899, after 43 days in the pack, the ship sailed into an ice-free Robertson Bay, off-loading supplies and building a hut at Ridley Beach. Ten men and 75 Siberian huskies, the first used in Antarctic work, remained for the winter. Throughout the following year, short sledge trips were made on the sea ice around Robertson Bay and down the eastern side of the Adare Peninsula. Geological specimens were collected mainly from beaches and moraines, but localities were not recorded at the time. Returned samples were described by Prior (1902) and included phonolitic trachyte and kenyte, basalt, quartz felsite, gabbro, and granite. The granites were both pink and gray, and contained biotite, biotite + muscovite, and biotite + hornblende. The only in situ samples collected were "slates and quartz–grits" from Duke of York Island.

Following the party's pickup in early February 1900, the *Southern Cross* sailed southward with landings at Possession Island, Coulman Island, Mt. Melbourne, Franklin Island, and Ross Island. In all cases the bedrock was Tertiary volcanics.

Although in 1901 landings and collections were made at Cape Adare and Coulman Island during the southward passage of the *Discovery* carrying Scott's First Expedition (Ferrar, 1907), the substantive geological work in northern Victoria Land was done by the Northern Party of Scott's Second Expedition, 1910–13. This six-man party, including geologist R. E. Priestley, had aimed at exploration along the northern coast of northern Victoria Land as far west as Cape North, but due to ice conditions it was forced to land at Ridley Beach on Cape Adare, where it was confined, as Borchgrevink had been, to work around the perimeter of Robertson Bay. The "Slate–Greywacke Formation of Robertson Bay" was mapped and collected from Newnes Glacier to Cape Barrow (Rastall

**Figure 2.4** (*facing page*). Geological map of northern Victoria Land north of Terra Nova Bay.

and Priestley, 1921). Its quartz-rich clast content and matrix containing chlorite and calcite were described, and the presence of numerous anticlines and synclines with well-developed cleavage was noted.

The Northern Party was moved by the *Terra Nova* to Inexpressible Island in January 1912, from which it conducted geological investigations northward into the Northern Foothills and the southern end of the Deep Freeze Range, collecting both in situ and morainal specimens. Metamorphic rocks included porphyritic biotite gniess, biotite gneisses and granulites, and graphitic mica schists (Smith and Priestley, 1921). Plutonic rocks were granite, diorite, and enstatite–peridotite (Smith, 1924).

After the *Terra Nova* failed to pick up the party six weeks later as planned, the six men survived a horrific winter on minimal rations in an ice cave before man-hauling 350 km south the following spring to rejoin the expedition at Cape Evans, recovering a cache of rocks at Dunlop Island left by Professor David in 1908.

No further geological work was done in Victoria Land until the International Geophysical Year (IGY). For the decade that followed, much of northern Victoria Land was systematically mapped by New Zealand field parties. The first of these, in 1957–58, worked in the vicinity of Cape Hallett and the lower Tucker Glacier (Harrington et al., 1964, 1967).

During the 1958–59 season geologists on the Russian ship *Ob* surveyed the King George V and Oates Coasts (Klimov and Soloviev, 1958a, 1960; Soloviev, 1960; Ravich and Krylov, 1964; Ravich et al., 1965). In 1959–60, 1960–61, and 1961–62 the Australian ships *Thala Dan* and *Magga Dan* visited the Oates Coast in the vicinity of Rennick Bay. Observations were recorded by McLeod (1964) and McLeod and Gregory (1967).

Subsequent New Zealand ground parties and the areas in which they worked were as follows: 1962–63, area northeast of Evans Névé (Le Couteur and Leitch, 1964); 1963–64, lower Rennick Glacier area (Sturm and Carryer, 1970), upper Rennick Glacier area (Gair, 1964, 1967), and the area from Mawson to Priestley Glaciers (Ricker, 1964; Skinner and Ricker, 1968; Skinner, 1972); 1966–67, upper Mariner Glacier area (Riddols and Hancox, 1968) and upper Campbell and Aviator Glaciers area (Nathan, 1971a, b); 1967–68, lower Rennick Glacier area (Dow and Neall, 1972, 1974); 1969–70, southern Deep Freeze Range and Northern Foothills (Skinner, 1983a); 1971–72, northern Evans Névé (Laird et al., 1972, 1974; Andrews and Laird, 1976); and 1974–75, Bowers Mountains (Cooper et al., 1976, 1982; Laird et al., 1976, 1977, 1982; Bradshaw et al., 1982; Wodzicki et al., 1982).

In addition, in 1964–65 a U.S. helicopter-assisted party surveyed in reconnaissance fashion a broad area from Robertson Bay to Borchgrevink Glacier to the Lanterman Range (Crowder, 1968).

The 1980s saw the advent of the routine use of helicopters by several nations in northern Victoria Land and a burgeoning of geological research, whose extensive literature will be cited where appropriate throughout the remainder of the chapter. West Germany began work in northern Victoria Land in 1979–80 and continued with a series of expeditions under the acronym GANOVEX (German North Victoria Land Expedition) (Tessensohn, 1989). GANOVEX I was staged from a small base (Lillie Marleen) in the Everett Range, covering a strip of territory that extended from the Daniels Range to Robertson Bay (Kothe et al., 1981). In 1981–82 GANOVEX II was aborted when its ship sank to the north of Yule Bay. That same season the United States staged a remote helicopter camp at the northern end of Evans Névé that served U.S., New Zealand, and Australian parties in a radius covering most of northern Victoria Land north and west of the Aviator Ice Tongue and east and south of the Daniels Range (Stump et al., 1983c). In 1982–83 GANOVEX III built Gondwana Station on Gerlache Inlet and carried out fieldwork between there and Mariner Glacier (Kothe, 1984). They also reopened Lillie Marleen and worked areas as far-ranging as the Kavrayskiy Hills, Mt. Black Prince, and the Salamander Range. The most comprehensive contribution at this point was a geological map of northern Victoria Land at 1:500,000 (GANOVEX Team, 1987), incorporating the many new findings since the publication of the American Geographical Map Folio 12 in 1969 (Gair et al., 1969).

GANOVEX IV was conceived largely as a geophysical survey that based out of Gondwana Station in 1984–85 and covered a huge area of the western Ross Sea and a nearly 2° strip inland between 73°S and 75°S (Damaske et al., 1989; Dürbaum et al., 1989). Ground camps were set up at Frontier Mountain and upper Priestley Glacier primarily for ground control of the airborne surveys, but some bedrock geology was also undertaken in the vicinity. A party that included German, New Zealand, and U.S. geologists examined targeted areas on Handler Ridge and Reilly Ridge (Bradshaw et al., 1985a).

In 1988–89 GANOVEX V studied the western flank of the Ross Sea rift and aspects of the Ross orogen (Tessensohn, 1993). It operated with helicopters out of Gondwana Station and in the northern portion of northern Victoria Land. The latter effort was centered at a base camp in the Kavrayskiy Hills with airborne work from shipboard as far west as the Berg Mountains (Flöttman and Kleinschmidt, 1991a, b, 1993; Skinner, 1992; Schüssler et al., 1993).

In 1990–91 GANOVEX VI again operated out of Gondwana Station and along the northern coastal areas, with a base camp at Cape Williams. Preliminary results of the various parties are presented in the volume introduced by Tessensohn (1992).

Italian geologists began research in the Terra Nova Bay area in 1985–86 and continued work through subsequent seasons. In 1986–87 the Italians constructed Terra Nova Bay Station on Gerlache Inlet about 10 km southwest of Gondwana Station. A major contribution of the first three seasons of work was the production of a geological map at a scale of 1:500,000 between 73°S and 76°S incorporating new data, in particular the recognition of granulites in the Deep Freeze Range (Carmignani et al., 1988).

A U.S. party examined Robertson Bay Group in 1988–89 in the upper Tucker Glacier area, and then returned to work coastal areas by helicopter from an icebreaker in 1989–90 (Wright and Dallmeyer, 1991; Dallmeyer and Wright, 1992).

# Wilson Terrane

## Nomenclature

The Wilson terrane extends from the Oates Coast in the north to Terra Nova Bay in the south. To the east it contacts the Bowers terrane along a fault system exposed in the Lanterman and Mountaineer Ranges, and to the west it is covered by the East Antarctic Ice Sheet. A fairly continuous set of outcrops borders the western margin of the terrane along the ice sheet, but exposures of bedrock throughout the rest of the terrane are scattered among a number of widely separated ranges. Because a fair difference in lithology occurs between some of the areas, because the continuity of the underlying metamorphic rocks is uncertain, and because geologists have approached this region from different directions, the nomenclature of the metamorphic rocks that constitute the bulk of the Wilson terrane is diverse and has undergone numerous revisions.

The name "Wilson Series" was first applied by Klimov and Soloviev (1960) to the schists, migmatites, and granitoids in the Wilson Hills, and the name "Berg Series" for metamorphosed sandstones and shales in the Lev Berg Mountains. These were subsequently changed to "Wilson Group" and "Berg Group" by Grindley and Warren (1964) and Ravich et al. (1965). Gair (1967) introduced the name "Rennick Group" for schists and minor marble in the exposures between the Outback Nunataks and the Illusion Hills. On the map folio sheet for northern Victoria Land (Gair et al., 1969), all amphibolite-grade metamorphic rocks were designated "Wilson Group." The naming of the Wilson terrane stems directly from this widespread application of the term "Wilson Group." Notably, the greenschist-grade metasediments of the Morozumi Range were mapped as Robertson Bay Group (Gair et al., 1969).

A portion of the folio map was completed using data from Sturm and Carryer (1970), who had mapped a huge area from the Wilson Hills to the Emlen Peaks, and eastward to the upper Tucker Glacier. They chose to subdivide the Wilson Group informally into the "Wilson gneisses" for those rocks of gneissic and migmatitic affinity, and "Rennick schists" for amphibolite-grade metasediments. They also assigned those phases of the Wilson Group of Ravich et al. (1965) that were clearly plutonic in origin to the Granite Harbour Intrusives.

Also anticipated on the folio map, the study of the lower Rennick Glacier area by Dow and Neall (1972, 1974) continued the application of Wilson Group to the amphibolite-grade metasediments in the Lanterman Range and in the northernmost part of the Morozumi Range. These rocks were found to be distinct from occurrences along the polar plateau in that they lacked extensive migmatization and syntectonic intrusion. The remainder of the sedimentary rocks in the Morozumi Range, separated by a small pluton, were mapped as Robertson Bay Group.

Meanwhile, Riddolls and Hancox (1968), working in the area southeast of Evans Névé, had introduced the name "Retreat Hills Schist" for rocks at that locality, which they considered to be of lower metamorphic grade than the Rennick schists and proposed to be a formation of the Robertson Bay Group.

GANOVEX I (Tessensohn et al., 1981) continued the usage of Wilson Group and Robertson Bay Group in their investigation across central northern Victoria Land. They promoted the hypothesis that all Wilson and Robertson Bay Group rocks were of the same sedimentary succession, having suffered different degrees of metamorphism and deformation, and observed that in the Morozumi Range there was almost continuous section between what had previously been mapped separately as Wilson and Robertson Bay Groups by Dow and Neall (1972, 1974). Kleinschmidt and Skinner (1981) felt that the distinction between Rennick Schist and Wilson Gneiss was not applicable because on Thompson Spur in the southern Daniels Range there was a complete gradation between rocks representative of each type.

With the recognition of terranes in northern Victoria Land, the notion of continuity of sedimentary basins across the terranes has largely been abandoned; however, the relationship of the Morozumi Range metasediments continues to be troublesome (e.g., Kleinschmidt and Tessensohn, 1987; Bradshaw, 1989).

At the Antarctic Earth Science Symposium in Adelaide (1982), attending geologists, recognizing basic differences between the metamorphic rocks in the western and eastern portions of the Wilson terrane, agreed to use the terms "Wilson metamorphics" and "Lanterman metamorphics" as general descriptive terms for rocks in the respective areas, but without abandoning local designations. This was in anticipation of an influential paper by Grew et al. (1984) that identified two distinct metamorphic belts within the Wilson terrane, a high-temperature/low-pressure belt in the west and a medium-pressure belt in the east.

At the Adelaide symposium, Plummer et al. (1983) introduced the name "Daniels Range Metamorphic and Intrusive Complex" to emphasize the presence of a considerable portion of orthogneiss in the Daniels Range, but the same authors later dropped that name in favor of "Wilson Plutonic Complex" to categorize the same rocks (Babcock et al., 1986).

After new mapping in the rugged terrain between Mariner and Icebreaker Glaciers, Kleinschmidt et al. (1984) added "Mountaineer Metamorphics," "Murchison Formation," and "Dessent Formation" to the lexicon.

The 1:500,000-scale geological map of northern Victoria Land by the GANO-VEX Team (1987) is a benchmark in the formalization of terminology for the entire region. New terms to appear in that work were "Aviator Agmatites" for brecciated metamorphic rocks between Icebreaker and Tinker Glaciers, and "Morozumi phyllites and spotted schists."

In the southern extremity of northern Victoria Land, metasedimentary rocks in the Priestley Glacier area were named Priestley Formation by the early geologists (Ricker, 1964; Skinner and Ricker, 1968) and considered probably to be correlative with the Robertson Bay Group. Ricker (1964) also identified a suite of schists and gneisses around Snowy Point and east of Mt. Browning, which he informally identified as the "metamorphic complex." Nathan (1971b) extended Priestley Formation to occurrences in the upper Campbell Glacier area and also remapped the Retreat Hills Schist, agreeing tentatively that both were correlative with the Robertson Bay Group.

Skinner (1983a) later chose to restrict the name "Priestley Formation" to less metamorphosed portions of the previously mapped outcrops and introduced the informal term "Priestley Schist" for higher-grade, along-strike equivalents that crop out, in particular, south of Corner Glacier. He also proposed the term "Snowy Point Paragneiss" for the "metamorphic complex" of Ricker (1964), which because of an apparently older folding episode was suggested to be separated from the Priestley Schist by an unconformity. As with units to the north, the correlations of Priestley Formation and Retreat Hills Schist with Robertson Bay Group were abandoned with the recognition of terranes.

The map by the GANOVEX Team (1987) showed all of Skinner's (1983a) Priestley Formation and Priestley Schist occurrences as Priestley Schist and introduced the name "Terra Nova Formation" for Skinner's (1983a) Snowy Point Paragneiss.

Subsequently, Italian expeditions mapping in the Terra Nova Bay area continued the division between low-grade Priestley Formation and medium-grade Priestley Schist, but also distinguished a sizable area of high-grade metasediments, including relict granulites, that encompassed rocks of Skinner's (1983a) Snowy Point paragneiss in the area of Mt. Browning, though not at Snowy Point itself (e.g., Caramignani et al., 1989).

Recently, Skinner (1991b) has reemphasized that only two metasedimentary units should be recognized in the Terra Nova Bay area: (1) Priestley Formation, of which "Priestley Schist" should remain an informal descriptive term for its foliated portions, and (2) Snowy Point Gneiss Complex, with the slight change in nomenclature to characterize more accurately the variety of rocks in the unit.

The descriptions of metamorphic rocks in the Wilson terrane will follow the divisions formalized by the GANOVEX Team (1987), except in the Terra Nova Bay area, where we will follow Skinner (1991b).

## Berg Group

The Berg Mountains (69°11'S, 155°58'E) are an outlier of the rest of northern Victoria Land, located to the west of Matusevich Glacier along the George V Coast. One of this trio of small ridges was visited by a Russian ship in 1958–59, and a series of phyllites named the Berg Group were described mainly in terms of metamorphic assemblages (Ravich et al., 1965). However, the area was not included on the folio map of Gair et al. (1969), and it remained unvisited until GANOVEX V in 1988–89, which extended its outcrop area to isolated nunataks 20 km to the west and 15 km to the east. Skinner (1992) presents sedimentary descriptions and two measured sections from this work.

The Berg Group is of low metamorphic grade and has been tightly folded and cleaved. A pluton of granite intrudes it sharply on the western of the three ridges. The rocks are in general quartzofeldspathic graywackes and siltstones. A calcareous component is present in a fair portion of the section manifest as nodules, lenses, and discrete layers. A 50-m interval on the western ridge is predominantly carbonate beds. The rocks are identified as turbidites with partial or complete Bouma sequences. Sedimentary structures include graded bedding, stacked climbing ripples, flute casts, load casts, and laminations. The phyllites are composed primarily of quartz, biotite, sericite, and chlorite (Ravich et al., 1965). Calcareous units contain clinozoisite and actinolite. Tourmaline is also a common mineral in many horizons.

Acritarchs collected from the group have been identified as Riphean (middle Neoproterozoic) by Iltchenko (1972). Skinner (1991a) has suggested a correlation of Berg Group and Priestley Formation based on similarities of lithology, in particular carbonate content, and their similar positions on the western side of the Wilson terrane. The fossil data of Iltchenko (1972) would appear to be at odds with the identification of Paleozoic body fossils in the Priestley Formation (Lombardo et al., 1989; Casnedi and Pertusanti, 1991). It should be added that although apparently not as widespread as in the Berg Group and Priestley Formation, a calcareous fraction is present in portions of the Wilson metamorphics and marble has been mapped at one locality in the Sequence Hills (Gair, 1967). Since very little attention has been paid to the sedimentology of the Wilson metamorphics, their correlation with Berg Group and Priestley Formation should not be precluded.

## Wilson Metamorphics

The metamorphic rocks in the western portion of the Wilson terrane are characterized by a high-temperature/low-pressure paragenesis, with steep metamorphic gradients over relatively short distances (Grew et al., 1984). In the Helliwell Hills, for example, rocks change from very fine-grained phyllite to fibrolite-bearing migmatite across a distance of 4.5 km (Grew and Sandiford, 1982; Tessensohn et al., 1992a).

All gradations exist within the Wilson metamorphics from low-grade phyllites and schists retaining primary sedimentary attributes through migmatitic gneisses to anatectic granites. This is demonstrated in continuous outcrop on Thompson Spur in the southern Daniels Range (Kleinschmidt and Skinner, 1981) and in the Helliwell Hills (Tessensohn et al., 1992a). Elsewhere for mapping purposes, either Rennick Schist or Wilson Gneiss has been distinguished, apparently with reasonable assurance, although the assignment of transitional outcrops must be arbitrary. Diversity and complexity are characteristics of the Wilson metamorphics.

## Rennick Schist

At the less altered end of the spectrum is the Rennick Schist, which has been mapped in outcrops from as far north as Allegro Valley in the Daniels Range, southward through the Emlen Peaks, the Outback Nunataks, and the Sequence Hills, to the scattering of nunataks west of the Lichen Hills, as well as in the western portion of the Helliwell Hills (Gair, 1967; Plummer et al., 1983; Babcock et al., 1986; GANOVEX Team, 1987; Roland et al., 1989; Stump, 1989). The southernmost outcrops are approximately 50 km north of the northernmost occurrences of Priestley Formation.

The first observation of rocks to be called Rennick Schist was made at Welcome Mountain, as shown in Figure 2.5, during the completion of a geophysical traverse across the East Antarctic Ice Sheet. Weihaupt (1960) pictures the mountain and describes dark gray to black biotitic schists with distinct bedding 3.5–45 cm thick, intruded by numerous massive pegmatites.

Rennick Schist is generally described as fine-grained pelitic and psammitic schists and gneisses. Calc–silicate lenses and layers are found throughout much of the outcrop area, and minor marbles occur in the Sequence Hills and at Mt.

**Figure 2.5.** Recumbent folds in Rennick Schist on the east face of Welcome Mountain. Face approximately 100 m high.

Jøern (Gair, 1967; Roland et al., 1989). Metaconglomerate has been noted in the Emlen Peaks (Klobcar and Holloway, 1986).

Most workers have treated the Rennick Schist primarily as a metamorphic rock, so there is only meager description of its protolith in the literature. Although most authors record that the formation is well bedded, only Weihaupt (1960) gives any data on bedding thickness (3.5–45 cm). The only statement on the thickness of the formation is that of Gair (1967), who estimates a minimum of 1,000 m for schist and 200 m for marble in the Sequence Hills. Relict graded bedding is reported by Gair (1967), Kleinschmidt (1981), and Roland et al. (1989). Tessensohn et al. (1981) draw analogies between psammite–pelite alternations and graded bedding at Thompson Spur, and the less metamorphosed graywacke–shale turbidites of the Robertson Bay Group, implying that the Rennick Schist is of similar origin.

Diagnostic metamorphic minerals within the Rennick Schist include garnet, andalusite, sillimanite, cordierite, diopside, and amphibole, indicating amphibolite grade, and metamorphic conditions comparable to those experienced by the Wilson Gneiss (Gair, 1967; Grew et al., 1984; Babcock et al., 1986; Roland et al., 1989). Tourmaline is also an abundant accessory mineral in the schists, indicating the presence of an appreciable amount of boron during the metamorphism.

## Wilson Gneiss

Wilson Gneiss is at least in part a more segregated and mobilized equivalent of the Rennick Schist, but a considerable fraction of the gneissic rocks in the Daniels Range are orthogneisses intruded into the sedimentary country rock before or during the metamorphic interval (Plummer et al., 1983; Babcock et al., 1986; Sheraton et al., 1987). Wilson Gneiss has been mapped from the Wilson Hills in the north as far south as the southern Daniels Range along the polar plateau and also in the Kavrayskiy Hills west of the mouth of Rennick Glacier (Schubert et al., 1984).

In the Wilson Hills the rock is predominantly a biotite–quartz–feldspar paragneiss with segregated layers, more and less rich in biotite, at a scale of 1–5 mm (McLeod, 1964; Ravich et al., 1965; Sturm and Carryer, 1970). Plagioclase, generally andesine in composition, is the most common mineral. Garnet, sillimanite, and/or cordierite are found in small amounts in many rocks. McLeod (1964) pointed out that the rocks in the Wilson Hills were apparently lacking in any calcareous fraction in their protolith. The same can be said for the Kavrayskiy Hills (Schubert et al., 1984).

A whole range of migmatitic veins occurs within the gneisses of the Wilson Hills, concordant and crosscutting, simple and anastomosing, ptygmatic, and fine- to coarse-grained. The veins are quartzofeldspathic, and in many cases contain both biotite and muscovite, typically along with garnet (Ravich et al., 1965). In some places these veins grade into larger bodies of two-mica granite.

Gneisses with similar textures and paragenesis have been described from the Kavrayskiy Hills, but there apparently the rocks are not as migmatitic as in the Wilson Hills (Schubert et al., 1984). Chemical analyses of these gneisses indicate a protolith of graywacke and shale. Some of both the coarse- and fine-grained gneisses have very similar pelitic chemistry, with no apparent reason for the differences in texture.

The northern part of the USARP Mountains has been mapped only in broad reconnaissance fashion (Gair et al., 1969; Sturm and Carryer, 1970). Wilson Group is shown around Pryor Glacier, but no specific descriptions of this area have been made.

In the Daniels Range all gradations of the sedimentary protolith occur from schists retaining much of their sedimentary character and intruded only by quartz veins (Rennick Schist), through schists and gneisses intruded by multiple generations of granitic and pegmatitic dikes, to highly mobilized migmatites in an advanced state of anatexis (Wilson Gneiss). At the southern end of the range, the entire suite is continuously exposed on Andalusite Ridge and Thompson Spur with an increase in metamorphic grade toward the east (Kleinschmidt, 1981; Schubert, 1987; Ulitzka, 1987).

Mapping by Plummer et al. (1983), Babcock et al. (1986), and Sheraton et al. (1987) throughout the Daniels Range has tended to emphasize the igneous aspect of the Wilson Gneiss, as is reflected in their usage of the term "Wilson Plutonic Complex" for these rocks. Late syn- to posttectonic plutonic rocks that intrude the complex have been mapped as Granite Harbour Intrusives and will be described later.

Of note is a small (20-m-wide) breccia pipe located near Penseroso Bluff that contains quartzite and schist fragments set in a matrix ranging from nearly pure garnet composition to a hypersthene–cordierite–sillimanite–hercynite assemblage (Plummer et al., 1983; Babcock et al., 1986). The matrix is partially retrograded to amphibolite facies, but the granulite-facies mineralogy is interpreted possibly to be from rocks of cratonic affinity underlying the Daniels Range, though the authors also acknowledge that the rocks may represent depleted, partially melted metasedimentary rocks (Rennick Schist?) from a deeper level than is presently exposed in the Daniels Range.

## Deformation and Metamorphism of Wilson Metamorphics

Until the work of Kleinschmidt and Skinner (1981), structural observations in the western part of the Wilson terrane had been minimal. Undertaking a detailed study of the deformation on Thompson Spur, these authors determined that at least four episodes of folding had affected the rocks in that area. The first two episodes produced tight to isoclinal, moderately to steeply plunging folds, with associated penetrative cleavage or schistosity. The second episode folded the cleavage of the first, but in many locations it is not possible to distinguish between the two. The third deformation folds the previous cleavages around nearly horizontal axes, but is not associated with a penetrative cleavage of its own. The final deformation is an open, nearly horizontal folding of all previous structures.

Kleinschmidt (1981) has argued, principally from the presence of unaltered, undeformed andalusite in folded quartz veins from Andalusite Ridge, that the metamorphism of the Wilson metamorphics was of a single, prograde nature, with the exception of minor retrograde effects that have produced amphibole and chlorite in some rocks. Others have produced no evidence of more than one prograde event. This conclusion led Kleinschmidt and Skinner (1981) to postulate that the polydeformation seen on Thompson Spur was due to reorientations of stress during a single orogenic episode.

Modern thermobarometric techniques have been applied to Wilson Gneiss at two localities. At Thompson Spur, Ulitzka (1987) determined a peak temperature of metamorphism of 730°C at a $H_2O$ pressure of 3.8 kb. In the Kavrayskiy Hills, Schubert et al. (1984) determined a peak temperature of 600°–650°C at a $H_2O$ pressure of 4.5–5 kb.

Two mylonitic shear zones with oppositely directed thrust geometries have been recognized in the northern part of the Wilson terrane by Flöttmann and Kleinschmidt (1991a, b; Flöttmann et al., 1993a, b). Although the shear zones and their sense-of-shear indicators are well described by the authors, general bedrock descriptions of the surrounding areas have not been published. The eastern shear zone (named Wilson thrust) is traceable through the northern USARP Mountains, where a top-to-the-eastnortheast sense of movement is indicated (Fig. 2.6). The shearing is contained within Wilson Gneiss or Granite Harbour Intrusives at all but one locality. This is at McCain Bluff, where high-grade gneiss to the west is thrust over low-grade metasedimentary rocks to the east. The metasedimentary rocks are only briefly described as containing graded beds and some calc–silicate intercalations and showing similarities to the Moro-

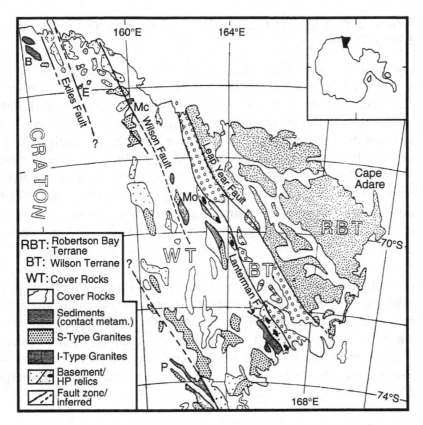

**Figure 2.6.** Generalized geological map of northern Victoria Land emphasizing thrust faults in the Wilson terrane. B, Berg Mountains; E, Exiles Nunataks; Mc, McCain Bluff; Mo, Morozumi Range; P, Priestley Glacier. After Flöttmann and Kleinschmidt (1991b). Used by permission.

zumi phyllites and spotted schists (Flöttmann and Kleinschmidt, 1993). The authors argue that the thrust separates a region of high-grade Wilson Gneiss from lower-grade metasedimentary rocks to the east. This may be an oversimplification, however, for they make no mention of the greenschist-grade metasediments in the Helliwell Hills to the west of the projection of their fault.

The western shear zone (named Exiles thrust) crops out in the Exiles Nunataks at the head of Matusevich Glacier and has been projected down the glacier. Sense of shear at this locality is top-to-the-westsouthwest. Shearing is apparently developed along the western portion of the Lazarev Mountains as well. Flöttmann and Kleinschmidt (1991a, b, 1993) suggest that this marks the western boundary of the Wilson Gneiss in fault contact with Berg Group exposed at Cape Horn to the west.

While migmatites of the Wilson Gneiss are cut by the Wilson and Exiles thrusts, indicating that shearing followed the peak of metamorphism and accompanying anatexis, the growth of fibrous sillimanite accompanied the movement of the Exiles thrust, indicating that amphibolite-grade metamorphism persisted. The Wilson thrust is itself cut by pegmatites, so the shearing did not outlast the plutonism in the area.

Flöttmann and Kleinschmidt (1991a, b) developed a model in which the mobile orogenic core is thrust outward over both fore-arc and back-arc sedimentary rocks resulting from subduction to the east of the Wilson terrane.

# Metamorphic Rocks of the Terra Nova Bay Area

Southward from the Wilson metamorphics along regional structural trends, medium- to high-grade metasedimentary rocks of the Priestley Formation crop out in the vicinity of Priestley Glacier, as well as around the central plateau of the Southern Cross Mountains (Fig. 2.7). These rocks bear considerable similarities to the Wilson metamorphics, both in their lithology and in their steep metamorphic gradients. However, they have been overprinted by a thermal metamorphic event producing hornfels textures, whereas the Wilson metamorphics largely have not. A sequence of high-grade metamorphic rocks known as the Snowy Point Gneiss Complex are a stratigraphically distinct suite according to Skinner (1991b, 1992). The mapping by Carmignani et al. (1988) subdivided units in the greater Terra Nova Bay area according to metamorphic grade. "Low-grade metasediments" (Priestley Formation) and "amphibolite facies metasediments" (Priestley Schist) largely correspond to the Priestley Formation of Skinner (1991b), and "high-grade metamorphics: gneiss and migmatite with granulite relics" largely correspond to Snowy Point Gneiss Complex, but there are areas where the stratigraphic and metamorphic designations are not coincident, as for example at Snowy Point, the locality for which the gneiss complex was named, where "amphibolite facies metasediments" are mapped by Carmignani et al. (1988).

# Priestley Formation

In its least metamorphosed occurrences Priestley Formation can be seen to be predominantly a pelitic sequence, with lesser quartzofeldspathic graywacke and minor limestone and quartzite (Skinner, 1983a). Mineralogy of the clasts in the graywackes includes quartz, minor plagioclase, lithic fragments of schistose metasediments, and mica (mainly muscovite, minor biotite) (Lombardo et al., 1989). Small, rounded organic fragments, possibly of echinoderms, have been found south of Foolsmate Glacier (Lombardo et al., 1989; Casnedi and Pertusanti, 1991). If this identification is correct, it constrains the age to be post-Precambrian.

Although Skinner and Ricker (1968) originally assigned an informal-type locality for Priestley Formation in the area from Ogden Heights to Timber Peak, Skinner (1991a) reestablished the type locality at O'Kane Canyon, and published three measured sections from there, as well as one from Lowry Bluff.

In O'Kane Canyon (Fig. 2.8) the lower 800 m are a fining-upward sequence of interbedded, laminated silty sandstone (graywacke), siltstone, and shale. Bedding thickness varies from 0.5 to 5 cm. Cross-bedding, defined by calcareous laminae, is present, as is graded bedding. In the upper portion of this part of the section, load casts, ripple marks, and "mudcracklike features" are present.

Above about 800 m the section changes to finer-grained, more pelitic lithologies, in which limestone beds (<1 m thick) appear. Shales are dark brown or black, containing some pyrite, and sandstones contain jarosite.

The Lowry Bluff section lacks the fining-upward succession of O'Kane Canyon, but it does contain an abrupt transition from siltstone and fine sandstone to

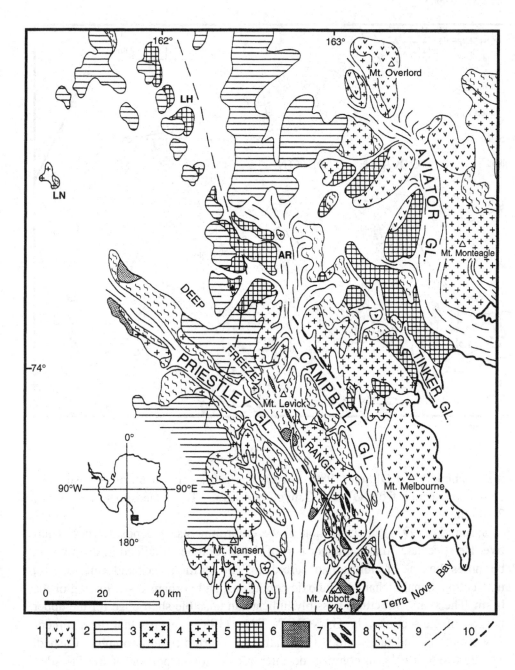

**Figure 2.7.** Schematic geological map of the area from Priestley to Aviator Glacier. 1, McMurdo Volcanics (Late Cenozoic–Quaternary); 2, Ferrar Supergroup and Beacon Supergroup (Permian–Mesozoic), Granite Harbour Intrusives; 3, calc–alkaline monzonitic association; 4, calc–alkaline granodioritic association; 5, postkinematic peraluminous granitoids; 6, mafic intrusives; 7, synkinematic peraluminous granitoids; 8, metamorphic complexes of Wilson terrane (Precambrian?); 9, main faults; 10, main shear zones. After Biagini et al. (1991a). Used by permission.

pyritic shales and limestones. These are overlain by varvelike alternations of red-brown calcareous sandstone and blue-gray calcareous shale that show evidence of considerable soft-sediment deformation. Above the varvelike portion of the section, carbonate disappears, there is an increasing amount of quartz, and slump features continue.

About 400 m from the top of the section is a limited interval of volcaniclastic detritus; then in the uppermost 60 m the rocks are predominantly volcanic, with pillow-appearing bodies interbedded with amphibolite, tuffaceous sandstone, and quartzite. The composition of the volcanics is ambiguous. Skinner (1991a) on the one hand reports amphibolites, but on the other recognizes high proportions of K-feldspar and muscovite, suggesting that the volcanics were silicic in composi-

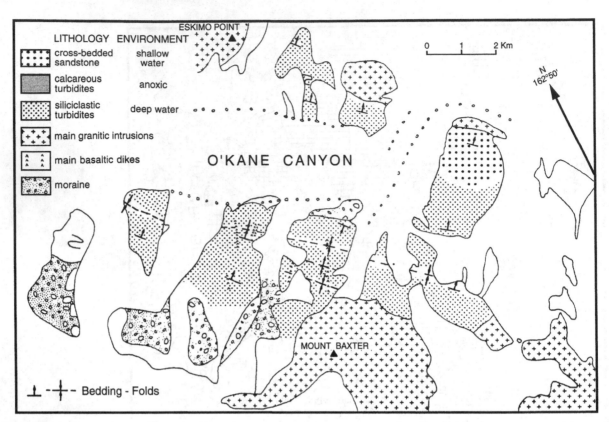

**Figure 2.8.** Geological map of O'Kane Glacier area emphasizing a lithological subdivision of Priestley Formation. After Casnedi and Pertusati (1991).

tion. Schubert and Olesch (1989) identify the tuffaceous material as basic. The composite section from O'Kane Canyon and Lowry Bluff measures in excess of 2,400 m.

Skinner's (1991a) interpretation of the environments of deposition are tentative. He speculates that the lower part of the section accumulated in deep water, but notes the lack of indicators of turbidites. The pyritic shales and limestones in midsection signal a change to anoxic conditions, and perhaps a shallowing of the basin. The slump structures possibly indicate an unstable shelf edge for the upper portion of the section. The GANOVEX Team (1987) concurs that the sedimentary structures and facies are not indicative of a turbiditic environment. Casnedi and Pertusati (1991), in contrast, do interpret the lower portion of the Priestley Formation as originating as turbidites.

All rocks assigned to the Priestley Formation have suffered some degree of regional metamorphism. In many of the lower-grade occurrences, a thermal overprint is also apparent. The areas of low-grade metasediments mapped as Priestley Formation by Carmignani et al. (1988) occur (1) in a zone on the west side of Priestley Glacier from Ogden Heights passing on the west side of Nash Ridge through O'Kane Canyon and Simpson Crags, (2) on the east side of Priestley Glacier from Wasson Rock to Mt. Burrows, and from there across upper Corner Glacier to Boomerang Glacier, and (3) at various localities in the southern portion of the Southern Cross Mountains.

Those rocks showing the lowest metamorphic effects are located south of

Foolsmate Glacier and in the vicinity of Mt. Jiracek (Carmignani et al., 1989). At both locations the metamorphism does not exceed the biotite zone of the greenschist facies, and thermal effects are minimal (Lombardo et al., 1989). At O'Kane Canyon thermal metamorphism has produced weakly schistose and hornfels textured rocks in which biotite and actinolite are characteristic minerals. Elsewhere, thermal effects are more pronounced, resulting in spotted rocks and prevasive hornfels textures. Mineralogy reflects the original compositions and degree of metamorphism. Cordierite, andalusite, sillimanite, corundum, and clinopyroxene have been noted at various localities.

Amphibolite-grade metasedimentary rocks, mapped as Priestley Schist by Carmignani et al. (1988), crop out in three areas: (1) in a belt from Browning Pass through Black Ridge crossing Priestley Glacier through Lowry Bluff to Mt. New Zealand, (2) east of Mt. Levick and upper Capsize Glacier, and (3) in the Southern Cross Mountains at Schulte Hills and Stewert Heights. The occurrences at Snowy Point and Cape Sastrugi north of Browning Pass have been mapped as Snowy Point Gneiss Complex by Skinner (1983a).

The rocks are fine-grained garnet–biotite gneisses and schists, biotite quartzites, and minor calc–silicate rocks and amphibolites. Garnet, cordierite, andalusite, sillimanite, and spinel are characteristic of the gneisses and schists; clinozoisite, spinel, brucite, olivine, and Ti-clinohumite are characteristic of the calc–silicates and marbles (Nathan, 1971b; Schubert and Olesch, 1989; Lombardo et al., 1989). Primary muscovite is absent in these rocks, although secondary sericite occurs at places. Lombardo et al. (1989) estimate metamorphic conditions in the range 600°–730°C and ~5 kb.

Overall, the metamorphic rocks of the Terra Nova Bay area are of a low-pressure/high-temperature paragenesis, as is found throughout the western Wilson terrane. Metamorphic gradients in places are steep, and whereas some plutonic rocks are pre- or syntectonic, many of the main intrusions in the area crosscut metamorphic zones and have caused local static recrystallization (Talarico et al., 1992).

## Snowy Point Gneiss Complex

The "metamorphic complex" of Ricker (1964) found around Cape Sastrugi, Snowy Point, and Mt. Browning was designated the Snowy Point Paragneiss by Skinner (1983a). Further mapping by Skinner (1991b) extended these rocks to a belt to the south and west of Mt. Dickason, and led to a renaming as Snowy Point Gneiss Complex. Further fieldwork by Skinner (1993) extended the Snowy Point Gneiss Complex along the west side of Campbell Glacier to Capsize Glacier. This belt had previously been mapped by Carmignani et al. (1988) as "high grade metamorphics: gneiss and migmatite with granulite relics."

The Snowy Point Gneiss Complex is primarily a quartz–plagioclase–biotite gneiss, locally containing garnet, cordierite, sillimanite, and/or andalusite (Skinner 1983a, 1991b). Marble, quartzite, calc–silicate, and amphibolite are developed locally. Migmatities of various stages of development are common in the gneisses, with well-developed leucosomes of two generations, an earlier concordant phase containing garnet, and a later crosscutting phase containing cordierite.

The pressure–temperature conditions for these rocks are estimated to have been 650°–750°C and 4–6.5 kb (Lombardo et al., 1989).

Granulite-facies rocks also occur in a number of outcrops along Campbell Glacier (Carmignani et al, 1989; Lombardo et al., 1989; Frezzotti and Talarico, 1990; Talarico, 1990). Quartzofeldspathic varieties are the most common, with a mineral association of plagioclase + quartz + orthopyroxene + biotite + K-feldspar ± garnet ± corundum ± spinel. Garnet + biotite + corundum ± sillimanite are found in aluminous granulites at a small locality at the head of Priestley Glacier. Talarico et al. (1989) recognized three stages of metamorphism within the granulites based on petrographic and thermobarometric observations. The oldest and highest grade is recorded in the cores of garnet and orthopyroxene in samples from the Boomerang Glacier area, estimated to have formed at 830°C and 7–10 kb. The main granulite metamorphism found throughout the extent of the outcrop area is estimated to have occurred at 700°–800°C and 5.5–7.5 kb. These rocks have been variably retrogressed to amphibolite facies.

## Deformation in the Terra Nova Bay Area

On the large scale, the metamorphic rocks of the Terra Nova Bay area are disposed in a series of anticlines and synclines, trending northwest to north-northwest parallel to the major glaciers in the area, with higher-grade rocks in the cores of anticlines and lower-grade rocks in the cores of synclines (Carmignani et al., 1989). Even in its lowest-grade occurrences (e.g., around O'Kane Canyon) the Priestley Formation apparently displays two episodes of folding (Skinner, 1991a). Schists are more widely developed in Priestley Formation in its southerly and easterly occurrences.

At least three episodes of deformation are developed in Snowy Point Gneiss Complex, with the earliest apparently predating the Priestley Formation. Although the contact between the two formations was not observed in outcrop around Snowy Point, the structural relations led Ricker (1964) and Skinner (1983a) to suggest that an unconformity exists between Snowy Point Gneiss Complex and Priestley Formation.

Further mapping by Skinner (1991b) to the south of Mt. Dickason revealed that the Snowy Point Gneiss Complex overthrusts the Priestley Formation in a southwesterly direction along the so-called Boomerang thrust (Fig. 2.9). The Snowy Point Gneiss Complex contains an intense flattening schistosity that overprints the gneissic foliation within 1.5–2.5 km of the fault. Adjacent to the fault a tourmaline-rich blastomylonite is developed. Very little apparent deformation occurs in the Priestley Formation adjacent to the fault. The contact makes a discordance of about 40° with the foliation of the Snowy Point Gneiss and 40°–13° with the foliation of the Priestley Schist.

Skinner (1991b) suggested that the fault zone and Priestley Formation were metamorphosed together after development of the Snowy Point Gneiss Complex. Talarico et al. (1992) concur. A younger bound is placed on the relative age of the thrust by dikes of cordierite–garnet-bearing granite.

Flöttman and Kleinschmidt (1991a, b), linked the west-vergent Exiles thrust with the Boomerang thrust, ending the Wilson thrust south of the Helliwell Hills. However, Kleinschmidt (1992a) carries the Wilson thrust beneath the Mesa

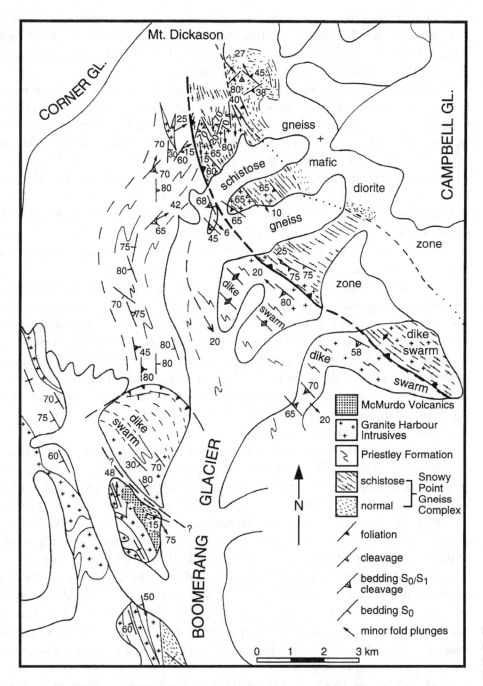

**Figure 2.9.** Geological map of Boomerang Glacier area showing distribution of Priestley Formation, Snowy Point Gneiss Complex, and the Boomerang thrust of Skinner (1991b). Used by permission.

Range and links it with east-vergent structures west of Campbell Glacier, at the same time reinterpreting the Boomerang thrust as conjugate to the east-vergent structures and not related to the Exiles thrust.

## Lanterman Metamorphics

A suite of amphibolite-grade gneisses and schists, designated the Lanterman metamorphics, crops out in the Lanterman and Salamander Ranges in the east-central portion of the Wilson terrane. Knowledge of these ranges has come in patchwork fashion from a succession of geologists who visited only limited areas (Crowder, 1968; Sturm and Carryer, 1970; Dow and Neall, 1972, 1974; Kleinschmidt and Skinner, 1981; Bradshaw et al., 1982; Grew and Sandiford, 1982, 1984, 1985; Wodzicki et al., 1982; Sandiford, 1985). The first comprehensive mapping was done by Roland et al. (1984).

Along the northeastern margin of the Lanterman Range, the Lanterman metamorphics contact the Bowers Supergroup along the highly tectonized Lanterman fault system, being the boundary between the Wilson and Bowers terranes. The western side of the range is marked by Jurassic or younger faults which juxtapose basement granites and Ferrar dolerites (Roland and Tessensohn, 1987). The western side of the range is intruded by several plutons of Granite Harbour Intrusives, whereas the eastern side is relatively pluton-free.

Lithologies of the Lanterman metamorphics include biotite–hornblende gneisses and schists, calc–silicate gneisses and schists, amphibolites, and ultrabasites. The biotite gneisses appear to be more prevalent in the northern and central parts of the Lanterman Range, while hornblende-bearing gneisses are more prevalent to the south (Roland et al., 1984). Except to say that they were derived from graywacke, shale, and marl (Roland et al., 1984), no treatment of the sedimentary nature of these rocks has been offered. The amphibolites are interbedded with the metasediments as pods and layers, and presumably are volcanic in origin (Sandiford, 1985). A controversial conglomeratic unit occurring only in the northeastern part of the Lanterman Range has previously been included in the Lanterman metamorphics, but more recently was interpreted as belonging in the Bowers Supergroup (Gibson, 1984). This will be discussed later in the section on Bowers Supergroup.

Metamorphism of the Lanterman metamorphics occurred in multiple stages, first at amphibolite grade, followed by retrogression at greenschist grade. The critical assemblages of sillimanite–muscovite–quartz for pelitic rocks and biotite–almandine–hornblende–cummingtonite for basic rocks indicates intermediate pressures at the sillimanite grade (Roland et al., 1984). Relict kyanite and staurolite from near Mt. Bernstein indicate higher pressures during early stages of the metamorphism (Grew and Sandiford, 1984, 1985; Grew et al., 1984).

Grew and Sandiford (1984, 1985) proposed that the initial metamorphism occurred at temperatures near 650°–750°C and pressures of 7–10 kb, based on the presence of kyanite in pelitic schist and a staurolite–talc association in associated ultramafic rock. Kleinschmidt et al. (1987) also estimated that the majority of the ultramafic bodies had initially been metamorphosed under granulite conditions (900°C, 8–9 kb).

Analyzing garnet, biotite, and plagioclase from a fibrolite-bearing schist, Grew and Sandiford (1984) estimated that the main phase of metamorphism occurred at 650°–700°C and 5–6 kb. Roland et al. (1984), using data from garnet zonations, postulated conditions of 550°–580°C at 5 kb. A similar estimate of 570°–630°C at 5 kb was deduced from the ultramafic rocks by Kleinschmidt et al. (1987).

Late-stage retrograde metamorphism, a feature not seen in the Wilson metamorphics in the western portion of the Wilson terrane, is thought to have occurred at 300°–370°C and 3–5 kb (Grew and Sandiford, 1984). Minerals characteristic of the retrograde metamorphism are chlorite, muscovite, epidote, actinolite, calcite, and pumpellyite.

Several deformational episodes are recognized in the Lanterman metamorphics. Isoclinal folding and the formation of axial-plane schistosity occurred once (Sandiford, 1985) or twice (Roland et al., 1984) during the amphibolite stage of metamorphism. The schistosity is northwest–southeast trending and dips at angles greater than 60°SW. Mesoscopic crenulation structures of two generations developed on the previous schistosity during the retrograde metamorphism (Roland et al., 1984; Sandiford, 1985).

Ultramafic and mafic rocks occur as small lenses in a narrow zone along the eastern margin of the Lanterman Range and on into the Salamander Range (Wodzicki et al., 1982; Grew and Sandiford, 1982; Roland et al., 1984; Kleinschmidt et al., 1987). The petrology of these rocks varies considerably due to metamorphic recrystallization. Rare spinel–olivine pyroxenite and hornblende–olivine–garnet pyroxenite are thought to contain appreciable relics of the original magmatic mineral assemblage (Kleinschmidt et al., 1987). More typically, olivine and pyroxene are minor phases or wholly absent in rocks in which combinations of anthophyllite, magnesite, chlorite, cummingtonite, tremolite, actinolite, and serpentine predominate.

The linear distribution of the ultramafic and mafic bodies at the eastern margin of the Wilson terrane and their discontinuous lensoid or phacoid shape suggest that they were tectonically emplaced. However, the bodies appear to fall into two groups according to their pressure–temperature paths of metamorphism, one prograding from an early serpentine stage, the other retrograding from granulite conditions, both converging at the amphibolite pressure–temperature conditions of the host Lanterman metamorphics. Kleinschmidt et al. (1987) have interpreted this to indicate a mixing of two different levels of oceanic lithosphere during incorporation at the terrane boundary.

## Mountaineer Metamorphics

Rising from sea level to more than 3,500 m, surrounded on three sides by steep-walled valleys with heavily crevassed glaciers and on the south in summer by open water of the Ross Sea, and unusually prone to avalanches, the Mountaineer Range is perhaps the most inaccessible in the Transantarctic Mountains. As such it was not visited until 1981–82, when helicopter landings were made at a number of sites and the continuation of rocks of the Wilson and Bowers terranes into the area was recognized (Stump et al., 1983b).

During the subsequent season, members of GANOVEX III, again with the use of helicopters, mapped throughout the Mountaineer Range (Gibson et al., 1984; Kleinschmidt et al., 1984; Mortimer et al., 1984; Tessensohn, 1984) (Fig. 2.10). The boundary between Wilson and Bowers terranes was traced from Whitcomb Ridge southward past the head of Argonaut Glacier, across Meander Glacier and east of Dessent Ridge. To the west of the boundary two distinct metamorphic sequences were recognized, the Murchison Formation and the Dessent Formation, together called the Mountaineer metamorphics (Kleinschmidt et al., 1987). The contact between the two formations in the vicinity of Wylde Glacier is not exposed, but it is thought to be a thrust or high-angle reverse fault because the Murchison Formation, which is of higher metamorphic grade than the Dessent

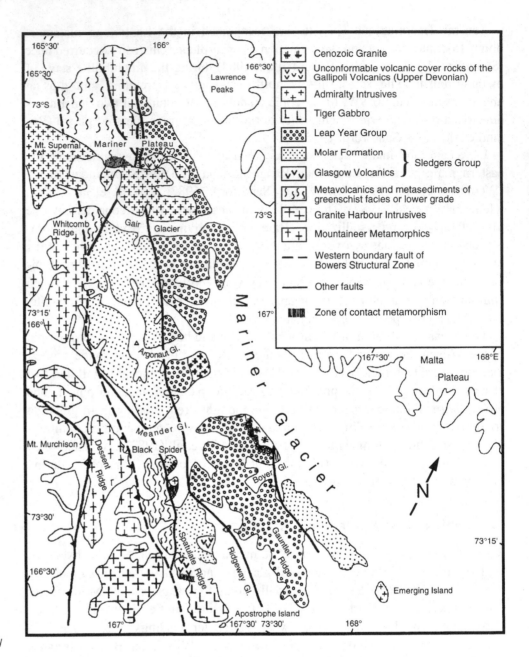

**Figure 2.10.** Geological sketch map of area west of Mariner Glacier. After Gibson et al. (1984). Used by permission.

Legend:

|  |  |
| --- | --- |
| ✛ ✛ | Cenozoic Granite |
| v v v | Unconformable volcanic cover rocks of the Gallipoli Volcanics (Upper Devonian) |
| + + + | Admiralty Intrusives |
| L L | Tiger Gabbro |
| ·°·° | Leap Year Group |
| :::: | Molar Formation } Sledgers Group |
| v v v | Glasgow Volcanics } Sledgers Group |
| ʃ ʃ ʃ | Metavolcanics and metasediments of greenschist facies or lower grade |
| + + | Granite Harbour Intrusives |
| + + | Mountaineer Metamorphics |
| – – – | Western boundary fault of Bowers Structural Zone |
| ——— | Other faults |
| ▥ | Zone of contact metamorphism |

Formation, is located topographically above it. Exposed high on the eastern side of Mt. Murchison, the western boundary of Murchison Formation with Granite Harbour Intrusives is also thought to be a thrust fault.

The Murchison Formation is composed of biotite schist and biotite–plagioclase gneiss, much of which is migmatitic (Kleinschmidt et al., 1984). The darker material often occurs as blocks or rafts, some of which are rotated. Relict garnet, diopside, and sillimanite, and the absence of primary muscovite, are indicative of the highest metamorphic grade. Highly mobilized textures indicate that anatectic conditions were probably reached. Retrograde metamorphic effects are manifest in diopside largely altered to hornblende, garnet replaced by biotite and plagioclase, and widespread development of chlorite and prehnite throughout the formation.

The Murchison Formation has undergone at least four episodes of deformation. Quartz veins injected parallel to the first schistosity ($S_1$) are isoclinally folded ($F_2$), with $S_2$ axial-planar to these folds and indistinguishable in most cases from $S_1$. Tight $F_3$ folds refold older structures and generally verge to the northeast, compatible with the hypothesis of thrusting of the formation. $F_4$ folds are open structures developed only locally.

Although the degree of mobility of the Murchison Formation is greater, it bears similarities in both type and uniformity of lithology to the Lanterman metamorphics. The Dessent Formation decidedly does not.

The Dessent Formation appears to be confined to Dessent Ridge. It is a highly varied sequence containing silicic, pelitic, calcareous, and mafic rock types. Sharp lithological changes indicate bedding planes that are otherwise largely indistinct in much of the sequence due to deformation. Some of the thin marble beds may be repeated by folding, and facing direction is not determinable in the metasediments. With those acknowledged difficulties, Kleinschmidt et al. (1984) have measured a 400-m section on Disthen Wand at the northern end of Dessent Ridge.

The (topographically) lower 250 m of the section are interbedded schists, gneisses, and marbles. The upper 150 m are a polymict metaconglomerate that has an amphibolitic matrix containing hornblende–cummingtonite–garnet–plagioclase–quartz. The clasts range in size from 1 to 40 cm, are matrix-supported, and include calc–silicate rocks, quartzites, carbonates, and gneisses.

In the lower part of the section, fairly pure marble, white to blue-gray, and fairly pure quartzite, in some cases containing muscovite or amphibole, are distinct. Amphibolites, too, are fairly common in the lower part of the section. Most are thought to be of sedimentary origin, with only one outcrop showing evidence of primary igneous textures. Pelitic and quartzofeldspathic rocks present a variety of mineralogies, with staurolite and kyanite being diagnostic of the highest metamorphic grade. Portions of the formation are intensely retrogressed with widespread development of chlorite and prehnite. A 3- to 7-m-thick quartz vein containing blue kyanite crystals to 15 cm in length crops out in the upper portion of the Disthen Wand, thus the place name, which is German for "Kyanite Cliff."

Drawing from various experimental phase boundaries Kleinschmidt et al. (1984) estimate that the main regional metamorphism of the Dessent Formation occurred at 600°–650°C and 6–7 kb, with slightly higher conditions indicated for the Murchison Formation due to the zonation of garnet, the presence of sillimanite, and the migmatitic structures. On the basis of amphibole–plagioclase and garnet–hornblende pairs in amphibolite of the Dessent Formation, Capponi et al. (1990) estimated similar pressure–temperature conditions of 600°–700°C at 6–8 kb.

Contacts of different rock types (i.e., bedding), as well as two generations of schistosity (usually not distinguishable), trend fairly uniformly at 130°/30°SW. Minor folding appears to have accompanied the second schistosity and to have folded the first. Very open mesoscopic folds are interpreted as $F_3$, since they distort isoclinally folded schistosity. One thrust plane, and possibly two, parallel to layering have cut the Dessent Formation on Disthan Wand.

## Retreat Hills Schist

About 20 km north of the Mountaineer Range at the southern end of Evans Névé, an isolated occurrence of metamorphic rocks crops out in the Retreat Hills. Named the Retreat Hills Schist by Riddols and Hancox (1968) and subsequently examined by Nathan (1971b) and Kleinschmidt et al. (1984), the unit is composed of carbonate-bearing quartz–mica schist and phyllite. The mineralogy includes quartz, biotite, muscovite, calcite, and minor plagioclase ($An_{38-43}$) and garnet (almandine–grossularite), indicating metamorphic conditions at the boundary between greenschist and amphibolite facies (Kleinschmidt et al., 1984).

The rocks are strongly foliated and contain numerous, concordant quartz veins and simple pegmatites. The attitude of the foliation is approximately 135°/30° SW. Apparently, the foliation was produced in two indistinguishable generations, since it can be seen to be isoclinally folded on a mesoscale in limited places (Kleinschmidt et al., 1984). A third deformation is represented by small-amplitude (5-cm) folds with their own penetrative foliation, associated with smaller drag folds and crenulations. Metamorphic mineral growth proceeded throughout all three phases of deformation.

## Morozumi Phyllites and Spotted Schists

Graywacke–shale sequences of low regional metamorphic grade crop out in the central portion of the Wilson terrane in the eastern portion of the Helliwell Hills, and the eastern portion of the Kavrayskiy Hills, at Lonely One Nunatak, and at the Morozumi Range. Except for a narrow zone along Graduation Ridge where higher-grade schists in contact with a pluton were designated Wilson Gneiss, the early workers mapped the sequence in the Morozumi Range as Robertson Bay Group (Gair et al., 1969; Sturm and Carryer, 1970; Dow and Neall, 1974). GANOVEX I and III followed a similar course (Kleinschmidt, 1981; Kleinschmidt and Skinner, 1981; Tessensohn et al., 1981; Engel, 1984), but after the recognition of terranes, the GANOVEX Team (1987) introduced the name "Morozumi phyllites and spotted schists" for these rocks.

Although contact metamorphic effects have overprinted the Morozumi phyllites and spotted schists, their sedimentary character is still fairly well preserved (Dow and Neall, 1974; Wright, 1981). Rythmically alternating beds of graywacke and pelite vary in thickness from 10 cm to 2 m. Sedimentary structures include graded bedding, laminations, cross laminations, convolute laminations, and possible flute casts. The tops of some of the pelitic units are calcareous. The sum of sedimentary characteristics earmark these rocks as turbidites.

The rocks in the Morozumi Range have been folded into tight, upright, chevron-style folds with well-developed slaty cleavage that strikes in a north-northwest direction (Kleinschmidt and Skinner, 1981). At one location near the southern end of the range (spot height 1,310 m), an isoclinal fold appears to be refolded and crosscut by cleavage. Therefore, the prevalent chevron folds may be second generation to an earlier, co-planar folding event. Very fine-grained muscovite parallel to the cleavage is widely developed in the phyllites, less so in the graywackes, in the southern portion of the range. In the northern portion, particularly adjacent to the plutonic rocks, foliated biotite is present (Engel, 1984).

A pervasive thermal overprint caused the Morozumi phyllites and spotted schists, producing widespread hornfelses and spotted rocks. The northern quarter of the range is composed of a granitic pluton, except for a thin outcrop of contact schists along the northeastern margin (the Wilson Group of Dow and Neall, 1974). Farther south three small stocks of granite intruding the phyllites and schists are likely apophyses of a large body not far beneath the surface.

In the southern portion of the range the metagraywackes are characterized by randomly oriented biotite flakes. The metapelites contain light-colored spots, about 1 cm in diameter, composed of fine-grained muscovite, biotite, quartz, and K-feldspar, possibly retrograded from spots largely of andalusite (Engel, 1984).

As the pluton in the northern Morozumi Range is approached, andalusite, then corundum and fibrolite, appear in the metapelites. In calc–silicate rocks, porphyroblasts of grossularite, hornblende, and pyroxene are present. Engel (1984) concluded that peak contact metamorphic conditions in the northern part of the Morozumi Range were 600°C or greater at 2 kb.

## Granite Harbour Intrusives

Plutonic rocks are found throughout the three terranes of northern Victoria Land. As with the other rock groups, the primary distribution of the plutonic rocks was mapped by the early reconnaissance parties, contacts were noted, and petrography was described, at times exhaustively. Local lithology-based names were applied to many of the occurrences (Gair, 1964, 1967; Le Couteur and Leitch, 1964; McLeod, 1964; Ricker, 1964; Harrington et al., 1967; McLeod and Gregory, 1967; Crowder, 1968; Nathan and Schulte, 1968; Riddols and Hancox, 1968; Skinner and Ricker, 1968; Sturm and Carryer, 1970; Adamson, 1971; Nathan 1971a, b; Dow and Neall, 1972, 1974; Skinner, 1972). Harrington (1958) had proposed the name "Admiralty Intrusives" to encompass all of the plutonic rocks of the Ross Sea sector of the Transantarctic Mountains, based on his studies in the Admiralty Mountains (Harrington et al., 1967). However, in their monograph on the geology of southern Victoria Land, Gunn and Warren (1962) proposed "Granite Harbour Intrusives" as the all-encompassing designator, following the early descriptions by Ferrar (1907) from that area. The first K/Ar dating of plutonic rocks from northern Victoria Land indicated cooling slightly older than 300 Ma (Starik et al., 1959), whereas plutonic rocks in southern Victoria Land were dated at around 500 Ma by the same technique (Goldich et al., 1958; Angino et al., 1962; Pearn et al., 1963). This led Grindley and Warren (1964) in their synthesis of nomenclature for the western Ross Sea area to restrict the Admiralty Intrusives to plutonics younger than the Ross orogeny found in the eastern portion of northern Victoria Land, and to apply the name "Granite Harbour Intrusives" to the widespread suite of plutonic rocks associated with the Ross orogeny, a practice that has continued to the present.

As summarized by Nathan (1971b), until 1970 only nine chemical analyses, as well as a small handful of isotopic dates, had been obtained on plutonic rocks of northern Victoria Land. The 1980s, however, saw a burgeoning of chemical and isotopic analyses from German, U.S., New Zealand, Australian, and Italian geologists working in the area. These studies have led to important conclusions regarding the genesis, timing, and tectonic setting of the plutonic rocks in northern Victoria Land. The Granite Harbour Intrusives will be covered in the remainder of this section; the Admiralty Intrusives will be visited at the end of the section on the Robertson Bay terrane. Many recent studies have considered the plutonic rocks of northern Victoria Land in terms of S- and I-types, following the classification of Chappell and White (1974). It is not surprising that the classification applies well to these rocks since it was in southeastern Australia, adjacent to northern Victoria Land before breakup, that the scheme was developed.

The Granite Harbour Intrusives in the Wilson terrane are a highly varied group of rocks intruded both syn- and posttectonically. North of Terra Nova Bay they are predominantly S-types, but I-types are found to various degrees throughout much of the region, and exclusively so from Terra Nova Bay southward (Wyborn, 1981; Borg et al., 1987b; Vetter and Tessensohn, 1987; Biagini et al., 1991a, b). Throughout the Wilson terrane north of Terra Nova Bay, the pattern is one of increased alkalinity toward the west. As has been observed in an east–west transect of the region, the intrusive suite in the Lanterman Range is dominated by I-type tonalites and quartz diorites, although the intrusives in the Freyberg Mountains are mainly S-types. The plutons in the Morozumi Range are mildly peraluminous. In the Daniels Range the suite is predominantly two-mica granite, highly contaminated by inclusions of Wilson metamorphics, but a few small tonalite bodies reside there as well. A similar east-to-west pattern exists between Mt. Murchison and Tinker Glacier in the southeast coastal region.

The S-type intrusive rocks occur in a belt through the series of ranges and nunataks bordering the East Antarctic Ice Sheet from Oates Coast southward into upper Campbell Glacier and Wood Bay. Along the southeasterly coast, the generalization of a westerly increase in alkalinity breaks down. West of Campbell Glacier the S-types give way to a preponderance of I-types that continue southward along the coast to the McMurdo Sound region and beyond. In the Deep Freeze Range between Campbell and Priestley Glaciers there is an unusually large fraction of mafic plutonic rocks. This coincides with the belt of granulite-facies metamorphic rocks discussed earlier.

## Chemical and Isotopic Variations

The distribution of S- and I-type granitoids throughout the Wilson terrane is underscored by the geochemistry of the Granite Harbour Intrusives. Borg et al. (1986b, 1987b) undertook a geochemical and isotopic reconnaissance of plutonic rocks throughout northern Victoria Land, except in the vicinity of Terra Nova Bay. They then plotted results by sample locality and contoured the data using trend-surface analysis. Sixty-two samples of the Granite Harbour Intrusives showed a distinct trend in most of the major and trace elements that had been analyzed. A statistically significant increase toward the westsouthwest was shown for $SiO_2$, Al-number, $K_2O/Na_2O$, Rb, Rb/Sr, and especially $K_2O$ and $K_2O$-index. An increasing trend toward the eastnortheast was observed for $TiO_2$, $Al_2O_3$,

Schematic geologic cross section

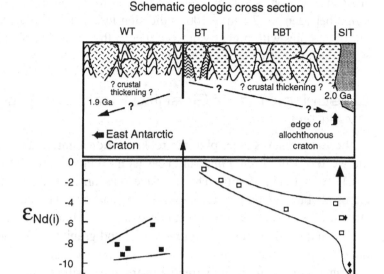

**Figure 2.11.** Initial $\varepsilon_{Nd}$ and $\varepsilon_{Sr}$ versus position along the northern Victoria Land transect of the Granite Harbour Intrusives (calculated at 500 Ma), the Admiralty Intrusives (calculated at 400 Ma), and the pre-Admiralty granites at Surgeon Island (calculated at 600 Ma) and Cooper Spur (calculated at 525 Ma). This diagram emphasizes the temporal and spatial separation of the Granite Harbour and Admiralty Intrusive Complexes and their correlation with metamorphic terranes. Breaks in the isotopic patterns indicate changes in lower crustal structure. □, Admiralty granites; ◆, Cooper Spur and Surgeon Island granites (pre-Admiralty); ■, Granite Harbour granites. WT, Wilson terrane; BT, Bowers terrane; RBT, Robertson Bay terrane; ST, Surgeon Island terrane. After Borg and DePaolo (1991). Used by permission.

$Fe_2O_3$, MnO, MgO, CaO, Sr, Y, and Zr. The trends were interpreted to indicate an increasing involvement of continental crustal material in the production of the Granite Harbour Intrusives in a westsouthwesterly direction across the Wilson terrane.

Rb/Sr and Sm/Nd isotopic data from a more limited sample set also reinforced this interpretation, as shown in Figure 2.11 (Borg et al., 1987b). $\varepsilon_{Sr}$ values vary in a broad band across the area with extremes of $-39.1$ and $+215.8$. All values indicate the involvement of crustal material, but there is a clear increase in a westsouthwesterly direction. Although all samples analyzed for neodymium indicated involvement of isotopically evolved material, $\varepsilon_{Nd}$ values ($-5.8$ to $-8.6$) showed no particular trend (the data set was small), except that the easternmost sample from a tonalite in the Aviator Glacier area showed less crustal involvement.

In the area between Campbell and David Glaciers, the Granite Harbour Intrusives are all I-types and the trends observed to the north break down. $\varepsilon_{Sr}$ values range between $+37$ and $+106$, indicating less crustal involvement than a number of samples with higher values farther to the north (Armienti et al., 1990a, b). $\varepsilon_{Nd}$ values from this area range from $-4.7$ to $-11.5$.

## Descriptions of Granite Harbour Intrusives by Area

In the Lanterman Range, plutonic rocks intrude Lanterman metamorphics along the western side of the range and near its northern end at Carnes Crag (Wyborn, 1981; Borg et al., 1986b). The intrusive relationships are complex and not well exposed, but the bodies are elongate in a northwest direction parallel to the structural grain of the metamorphic rocks. The primary compositions are quartz diorite and tonalite, but lesser granodiorite and granite are found. All are equigranular I-types containing biotite, hornblende, sphene, and allanite. Based on cross-cutting relations, there is a trend toward more silicic intrusions through time. The earliest magmatic activity was dike emplacement of mafic to intermediate composition. Some of the dike rock occurs as partly assimilated material within tonalite.

Wyborn (1981) noted well-developed secondary foliation in the most southerly tonalite in the Lanterman Range, which increases in intensity accompanied by grain-size reduction near its contact. He suggested that the emplacement of this pluton may predate the metamorphism of the Lanterman metamorphics. Borg et al. (1986b), in contrast, described equigranular rocks generally without foliation.

In the Freyberg Mountains plutonic rocks crop out in the southern portion of the Alamein Range and on the ridgelines centering on Monte Cassino. Dow and Neall (1974) broadly lumped these rocks as one suite, which they named the Freyberg Adamellite. However, later work by Borg et al. (1986b) showed that the plutonic rocks in the Alamein Range are markedly different from those in the Monte Cassino area, the former being I-type, the latter S-type. No contact between these two groups is exposed.

The intrusive rocks in the Alamein Range are metaluminous to slightly peraluminous granites and granodiorites. They are uniformly homogenous and equigranular to mildly porphyritic. Hornblende and sphene are common, and allanite is plentiful. The major- and trace-element geochemistry of this body of rock shows very uniform trends. Because of the similarity of petrography and geochemistry, Borg et al. (1986b) concluded that a small body of granodiorite north of Hunter Glacier in the Lanterman Range is genetically related to the pluton in the Alamein Range.

The intrusive rocks around Monte Cassino are a heterogeneous mixture of magma phases and metasedimentary inclusions in various states of assimilation (Borg et al., 1986b) (Fig. 2.12). The suite spans a range of compositions from granodiorite to syenogranite, with most phases being monzogranite. All analyzed samples are strongly peraluminous, with normative corundum 1.1–2.6%. Normative quartz is $>30\%$ in all cases. The geochemical and field characteristics indicate that these are S-types. Lithologies of the magmatic phases vary from ones rich in K-feldspar phenocrysts to biotite-rich or quartz/feldspar-rich.

**Figure 2.12.** Inhomogeneous, layered, S-type granite north of Monte Cassino. Note numerous metasedimentary inclusions and both concordant and discordant felsic dikes.

Inclusions within the Monte Cassino granites vary in size from less than a centimeter to 10 m. The most common type is biotite schist, but pieces of milky quartz are also abundant. The magmatic rocks are permeated with a distinct, undulatory, subhorizontal flow foliation, defined by mineral schlieren, inclusion trains, and the boundaries between magmatic phases. In many ways the characteristics of the plutonic rocks at Monte Cassino are similar to those in the Daniels Range, but the lack of primary muscovite, garnet, or cordierite in the former is a distinguishing feature.

Cropping out at the northern end of the Morozumi Range is a biotite granodiorite (Morozumi Adamellite of Dow and Neall, 1974) that intrudes Morozumi phyllite and spotted schist (Wyborn, 1981; Borg et al., 1986b). A thin band of the metamorphic rocks along the eastern edge of Graduation Ridge marks the eastern boundary of the pluton. The rock is a medium-grained porphyry with K-feldspar phenocrysts up to 5 cm in length. Much of the rock shows effects of strain and recrystallization in the form of quartz being either polygonized or with undulatory extinction, and having sutured grain boundaries. This main phase of porphyritic granodiorite intrudes a slightly more mafic, equigranular granodiorite toward the northern end of Graduation Ridge (Borg et al., 1986b). Two small cupolas of granodiorite also intrude the metasediments farther to the south on the eastern flank of the Morozumi Range (Tessensohn et al., 1981).

From the Kavrayskiy Hills, Schubert et al. (1984) report granodiorites and quartz diorites intercalated in the metasedimentary rocks, with mineral alignment roughly that of the foliation in the surrounding gneisses. At the southern end of the hills is a coarse-grained, posttectonic granodiorite containing biotite and hornblende, which is in turn intruded by a granite with red K-feldspar crystals. The final intrusive event in the area was the emplacement of aplitic dikes, up to 50 cm wide, in the metasediments.

In the western mountains bordering the East Antarctic Ice Sheet, the Granite Harbour Intrusives are extremely heterogeneous, in part showing evidence of anatexis and syntectonic deformation. In the western Wilson Hills, dikes and small stocks of undeformed granodiorite, quartz monzonite, granite, and pegmatite sharply crosscut paragneiss and migmatite. Biotite, apatite, garnet, and in some places muscovite are accessories in the more felsic phases. Increasingly to the east toward Manna Glacier, the intrusive contacts become more gradational (McLeod, 1964; McLeod and Gregory, 1967; Sturm and Carryer, 1970).

The only work to date in the northern USARP Mountains was undertaken during the reconnaissance study of Sturm and Carryer (1970), who describe a batholith (Mono batholith) more than 100 km long composed primarily of gray, massive, porphyritic, biotite-bearing granodiorite and tonalite. Feldspar phenocrysts are commonly 5 cm in length and may reach 11 cm. The presence of sphene as an accessory mineral would appear to indicate that this is an I-type intrusion. If the generalization is correct for the entire body, it is a much larger occurrence of I-type plutonics than is seen elsewhere along the western flank of the Wilson terrane. More detailed mapping is necessary to bear this out. At places along its western margin tonalite intrudes a pink coarse-grained granite containing biotite and muscovite. Near the borders of the batholith, pegmatites also are found, some containing biotite, apatite, and garnet.

Nowhere in northern Victoria Land is the boundary between igneous and metamorphic rocks more blurred than in the Daniels Range. Although some relatively homogeneous bodies intruded the area late syn- or posttectonically, a considerable portion of the range is underlain by heterogeneous gneisses with igneous characteristics. Because of the uncertainty in distinguishing between what is nearly melted Wilson Gneiss and true magmatic rock, Plummer et al. (1983), Babcock et al. (1986), and Sheraton et al. (1987) mapped Wilson Gneiss and all deformed igneous rock as one unit, which they called the Wilson Plutonic Complex.

In addition to migmatites (defined as interlayered rock with dark-colored metasedimentary material and light-colored melted or segregated material) two main types of pre- to syntectonic igneous rocks are described by Babcock et al. (1986): xenolithic granitoid and layered granitoid. The more common xenolithic granitoids, occurring in small plutons, sills, and dikes, are extremely heterogeneous, with abundant metasedimentary xenoliths and xenocrysts. Xenoliths range from several meters in length to small polycrystalline clots. The magmatic portion of these rocks ranges from muscovite–biotite granite to tonalite. The xenolithic granite both intrudes and is intruded by the layered granite. It crosscuts migmatitic wall rocks that are polydeformed and is generally well foliated itself. Portions with abundant clots and xenoliths appear not to have moved appreciably, while other portions appear to have been highly mobilized.

The layered granitoids occur in numerous small plutons and several sills. The layers are defined by irregular concentrations of biotite, or less commonly garnet, with spacing at a few millimeters to a few centimeters. Compositions are similar to the xenocrystic granitoids. Some layered granite intrudes and contains rafts of xenocrystic granite, while gradational boundaries between the two types also exist. Since elongate xenoliths that are found in the layered granitoids may have any orientation with respect to the layering, Babcock et al. (1986) suggested that the layering is not the result of magmatic flow, and that it represents some sort of preferred direction of nucleation within the magmas.

Scattered throughout the Daniels Range are dikes, sills, and plutons of homogeneous granite, leucogranite, and lesser granodiorite and quartz monzonite (Babcock et al., 1986). Small bodies of tonalite and diorite have also been mapped. The largest pluton underlies Schroeder Spur at the southern end of the range. North of this on Thompson Spur a complete anatectic gradation is preserved from Rennick Schist on the western end to homogeneous Granite Harbour Intrusives on the eastern, with heavily veined migmatites in the transition (Wyborn, 1981).

Nearly all of both the homogeneous and gneissic plutonic rocks in the Daniels Range are S-types. Most contain biotite and muscovite; garnet and altered cordierite are present in some parts; and in nearly all analyzed samples $Al_2O_3/(Na_2O + K_2O + CaO) > 1.1$ (Sheraton et al., 1987). Geochemical studies by Sheraton et al. (1987) and Ulitzka (1987) revealed that partial melting of sedimentary country rocks was responsible for the formation of the S-type granitoids in the Daniels Range. Although compositions of both the homogeneous and inhomogeneous suites are similar in geochemistry, Sheraton et al. (1987) note several differences, including relative abundances of MgO/(MgO + total Fe) and several trace elements, leading them to the conclusion that they were generated from different crustal source regions.

Isotopic data reinforce this conclusion. Adams (1986) determined an initial $^{87}Sr/^{86}Sr$ ratio (IR) of approximately 0.7205 for the plutonic Wilson Gneiss, and IRs in the range 0.709–0.715 for posttectonic Granite Harbour Intrusives.

Several small occurrences of hornblende-bearing tonalite and diorite are scattered throughout the Daniels Range as dikes, sills, and small plutons (Wyborn, 1981; Babcock et al., 1986). An older set is foliated and crosscut by layered and xenocrystic granite and granodiorite. A younger set intrudes Rennick Schist at several places and is crosscut by dikes of posttectonic Granite Harbour Intrusives.

**Figure 2.13.** Tourmaline-bearing pegmatites intruding Rennick Schist at Mt. Blair, Outback Nunataks.

On the basis of Rb/Sr analyses on the older set of tonalites, Vetter et al. (1983) and Kreuzer et al. (1987) calculated IRs of 0.707–0.708 assuming magma generation at 480 Ma. Borg et al. (1986b) assumed an older age of magma generation (550 Ma) and estimated an IR of 0.705–0.706, concluding that in either case a crustal prehistory must be assumed for at least some of the source material. Black and Sheraton (1990), in their ion microprobe study of zircons from a sample of xenolithic orthogneiss (Wilson Gneiss), suggested both Proterozoic and Archean source rocks for a fraction of the zircons.

Nonfoliated, two-mica S-type granites intrude Wilson metamorphics in the Emlen Peaks immediately south of the Daniels Range. Garnet and tourmaline occur as accessory minerals in some samples, and sillimanite is common as fibrolite, typically intergrown with biotite. A study of major and rare earth elements in the granites and the host rock by Klobcar and Holloway (1986) indicates that Wilson metamorphics could have been the source rock for the granites, if garnet was a residual phase following partial melting.

Farther to the south, the Outback Nunataks are characterized by Wilson metamorphics intruded by extensive aplites and pegmatites (Olesch and Schubert, 1989; Schubert et al., 1989; Stump, 1989) (Fig. 2.13). The pegmatites, both zoned and unzoned, contain a simple mineralogy consisting of quartz, albite,

**Figure 2.14.** Highly contorted Rennick Schist intruded by several generations of leucocratic Granite Harbour Intrusives in the Lichen Hills. Kukri peneplain is overlain by thin horizons of Beacon Supergroup, intruded by sills of Ferrar dolerite. Face approximately 200 m high.

and microcline. Black tourmaline, mainly schorl, is ubiquitous. Garnet and/or muscovite also occur in some instances.

South from the Outback Nunataks, S-type granites continue through the Caudal Hills, Lichen Hills, and Illusion Hills (Gair, 1967; Borg et al., 1986b). The intrusions vary from aplites to coarse-grained leucogranites, occurring as numerous sills and dikes, as well as small tabular plutons, within migmatitic Wilson metamorphics (Fig. 2.14).

As the narrowing line of granitic outcrops clears the south end of the Mesa Range (with its layers of Jurassic Ferrar basalts), it widens into a region of continuous (albeit broken) outcrop of Granite Harbour Intrusives that extends from close to the eastern boundary of the Wilson terrane in the vicinity of Fitzgerald Glacier, around through Terra Nova Bay, southward to David Glacier, and beyond, into southern Victoria Land. Two-mica granites continue south from their exposures in the Vantage Hills to the vicinity of Recoil Glacier, where they were named the Recoil Granite by Adamson (1971).

The highlands between Aviator and Campbell Glaciers were mapped by Nathan and Schulte (1968; Nathan, 1971b), who named four different phases of Granite Harbour Intrusives mainly on the basis of petrographic considerations. The coastline between Aviator and Mariner Glaciers remained unvisited until the arrival of helicopters in the 1980s.

The summaries by Ghezzo et al. (1989) and Biagini et al. (1991a, b) show that peraluminous two-mica granites and granodiorites, separated from the exposures in the vicinity of Recoil Glacier by a zone of metamorphic rocks, continue in a roughly north–south-trending belt that follows the west side of Aviator Glacier and straddles the Tinker Glacier (the Cosmonaut Granite and parts of the Northern Granite Complex of Nathan, 1971b) (Fig. 2.7). Two small outliers occur east of Campbell Glacier at Wood Ridge and Random Hills.

To the east in the scattering of steep exposures between Aviator Glacier and the crestline of Mt. Murchison, the rocks are primarily hornblende–biotite I-types of granitic to granodioritic composition intruded in places by tonalites. The western face of Mt. Murchison is composed of multiple intrusions of hornblende–biotite tonalite (Borg et al., 1986b; Ghezzo et al., 1989).

The spatial relations of predominantly I-type plutonic rocks to the east giving way to predominantly S-types in the west are the same along the southeastern coast of northern Victoria Land, as has been described from the Lanterman Range to the USARP Mountains. However, west of Tinker Glacier the belt of peraluminous plutonic rocks gives way to metaluminous rocks that continue down to southern Victoria Land (Biagini et al., 1991a, b). Distinct from these is a suite of high-potassium plutonics that crops out in the Northern Foothills and at Inexpressible Island (Mt. Abbott Intrusives of Armienti et al., 1990a).

The main suite of I-types ranges from gabbro through biotite–hornblende tonolite and granodiorite to biotite monzogranite and leucogranite (Biagini et al., 1991b). The bulk of these rocks were intruded late syn- or posttectonically, and crosscut the metamorphic rocks in the area, although some earlier deformed phases occur. The Larsen Granodiorite versus Irizar Granite (syntectonic vs. posttectonic, gray vs. pink) distinction made in southern Victoria Land by Gunn and Warren (1962) was carried to the Terra Nova Bay area by Skinner and Ricker (1968). As discussed in the following chapter, this division is no longer considered meaningful.

The plutonic rocks in the southern part of the Southern Cross Mountains (east of Campbell Glacier), informally designated the Southern Granite Complex by Nathan (1971b), have been visited only in reconnaissance fashion. The rocks are described mainly as fine-grained biotite granite and granodiorite.

Metamorphic rocks in the Deep Freeze Range are intruded by a number of fairly small plutons ranging in composition from diorite to monzogranite (Armienti et al., 1990a, b). The central portion of the range is underlain by a pluton of very coarse-grained porphyritic granite (Dickason Granite of Adamson, 1971). Phenocrysts of orthoclase up to 6 cm in length are aligned in portions of the pluton, particularly along the margins, due to flow during emplacement (Adamson, 1971; Ghezzo et al., 1989). Of note are numerous inclusions of biotite–amphibole microtonalite.

In the same area small stocks and sheets of quartz diorite and tonalite intrude the Priestley Schist (Corner Tonalite of Adamson, 1971). Although Adamson (1971) thought that these rocks intruded the Dickason Granite, Skinner (1983a) found that at Black Ridge tonalite is intruded by and enclosed as rafts within the granite. Biotite and hornblende are common to these bodies, but grain size varies considerably from one to another.

To the south and west of Priestley Glacier in the Eisenhower Range and Prince Albert Mountains, the mainly low-grade Priestley Formation is intruded by several large plutons of monzogranitic to granodioritic composition and by minor tonalites. These rocks also intrude the gabbro–tonalite rocks described later. The initial mapping and description were done by Skinner and Ricker (1968), who interpreted them as syntectonic and named them Larsen Granodiorite following the nomenclature in southern Victoria Land (Gunn and Warren, 1962). Ghezzo et al. (1989), however, report that many of these intrusive rocks are massive and homogeneous, but that some are foliated.

Much of the rock is medium-grained and sparsely porphyritic with phenocrysts of perthitic microcline or less commonly orthoclase (Skinner and Ricker, 1968; Armienti et al., 1990a, b; Biagini et al., 1991b). Plagioclase is strongly zoned with calcic cores of labradorite–bytownite. Myrmakitic and graphic textures are common. The main mafic mineral is biotite; hornblende in lesser quantity is present in some portions. Oxide mineralogy varies from magnetite alone, to magnetite + ilmanite, to ilmanite alone. Fine-grained, granular, mafic inclusions ("I-type") are widespread and at places abundant in these plutonic rocks.

A number of stocks and small plutons of mafic rock (gabbro–tonalite) are scattered throughout the Prince Albert Mountains. They commonly show strong compositional and textural variations over short distances, and generally are older than the more silicic rocks that encompass them.

At the northernmost contact with the mountains on their traverse to the South Magnetic Pole, David, Mackay, and Mawson ascended Backstairs Passage to Larsen Glacier, and thence to the polar plateau. Their collection of an in situ sample at the base of Backstairs Passage is described in a section titled "Biotite Granite from near Mount Larsen" by Mawson (1916). The rock is an even, fine-grained, light gray granite composed primarily of plagioclase, biotite, and strained quartz, with only rare orthoclase.

The Northern Party of the *Terra Nova* expedition collected and described similar specimens from moraines in the area of the Northern Foothills (Smith, 1924), distinguishing a "Mt. Larsen type" of gray granite with a paucity of K-feldspar from a "Granite Harbour type" as described by Prior (1907) with appreciable orthoclase. The Mt. Larsen type was generally foliated, the Granite Harbour type was not. However, Smith (1924) also acknowledged that all of the gray granites were to various degrees similar.

Skinner and Ricker (1968), working with specimens from the area of Backstairs Passage, found considerably more K-feldspar than the previous workers, either as an interstitial phase in equigranular rocks or as phenocrysts in porphyries, and on this basis expanded the definition to include what had previously been designated Granite Harbour type by Smith (1924). Skinner and Ricker (1968) also recognized zones of deformation marked by strongly foliated augen gneiss within these rocks.

The plutonic rocks cropping out in the Northern Foothills, Vegetation Island, and Inexpressible Island are distinct by virture of their petrology and their abundance of mafic (and ultramafic) types. They range in composition from gabbro through monzodiorite, quartz monzodiorite, and quartz monzonite, to syeno- and monzogranites (Biagini et al., 1991b). In a naming frenzy, Skinner

(1983a) identified seven divisions of this complex: Browning Mafites, Canwe Granodiorite, Russell Granite, Hells Gate Granite, Seaview Granite, Inexpressible Granite, and Abbott Granite. Armienti et al. (1990a, b) gave the name "Mt. Abbott Intrusives" to the entire complex. These rocks locally intrude older metamorphic rocks (Snowy Point Gneiss Complex).

The mafic–ultramafic bodies, consisting of gabbro, diorite, and ultramafites, are numerous and are found throughout the area (Ghezzo et al., 1989; Armienti et al., 1990a). These rocks typically have a highly varied, layered texture that is fragmented and mobilized across contacts with more silicic plutonics. Skinner (1972, 1983a) has evoked fractional crystallization to account for the petrological variations seen throughout outcrops.

Strongly foliated granodiorite and granite (Canwe Granodiorite) intrude the mafic rocks at the southern end of the Northern Foothills and Inexpressible Island. All are intruded by the Abbott Granite, the most abundant phase in the Northern Foothills, composed of coarse-grained, more and less porphyritic syenogranite containing both biotite and hornblende. The rocks are moderately foliated due either to syntectonic deformation (Skinner, 1983a) or to magmatic flow (Ghezzo et al., 1989).

Mineralogical characteristics that distinguish the silicic to intermediate phases of the Mt. Abbott Intrusives from other such plutonics in the Terra Nova Bay area include only moderately zoned plagioclase lacking calcic cores and, for the oxide phase, ilmanite alone (Biagini et al., 1991b). Geochemically the rocks show a greater abundance of potassium by silica content than do other plutonics in the area (Armienti et al., 1990a, b).

## Isotopic Dating of Wilson Terrane

Since the IGY a steadily accumulating set of isotopic ages has revealed the thermotectonic evolution of the Wilson terrane with increased clarity, although considerable uncertainty still exists regarding the precise chronology of the region. Ages derived from both the metamorphic rocks and the Granite Harbour Intrusives span a range from approximately 545 to 450 Ma. A summary of ages is shown in Figure 2.15.

Several Rb/Sr whole-rock isochrons have been determined from various parts of the Wilson terrane. As is typical of this type of dating, some of the isochrons are tight arrays of data, others are scattered, but several determinations from widely separated localities center around 535 Ma. A five-point isochron from plutonic rocks in the Terra Nova Bay area is $535 \pm 26$ Ma (IR = 0.70954, MSWD = 2.2) (Vetter et al., 1984), and a three-point isochron from plutonic rocks in the Lanterman and Alamein Ranges is $533 \pm 18$ Ma (IR = 0.7103, MSWD = 0.42) (Stump et al., in press). Two whole-rock isochrons on Rennick Schist from the Daniels Range, one on pelites, the other on psammites, are $531 \pm 17$ Ma (IR = 0.7243, MSWD = 1.8) and $534 \pm 24$ Ma (IR = 0.7187, MSWD = 7.8), respectively (Adams, 1986). These are interpreted as the time of peak metamorphism in that area.

Sheraton et al. (1987) reported conventional U/Pb zircon ages of approximately 652 and 634 Ma for migmatitic gneiss and xenolithic granite of the

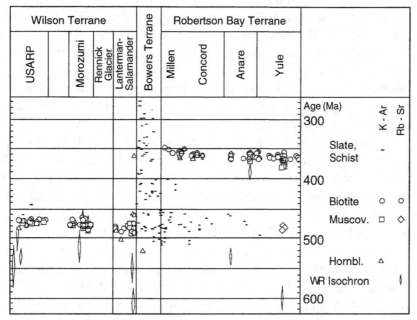

Figure 2.15. Variations of K/Ar and Rb/Sr dates from throughout northern Victoria Land projected onto a WSW–ENE profile. Dates are located according to nine zones, which correspond to prominent NW–SE- to NNW–SSE-trending physiographic features for which names are arbitrarily chosen. After Kreuzer et al. (1987). Used by permission.

Wilson Gneiss in the Daniels Range. These numbers were subsequently discounted by Black and Sheraton (1990), who presented a detailed ion microprobe study of zircons from a single sample of xenolithic plutonic rock from the "Wilson Gneiss," which showed a complex history of zircon growth. The main phase of zircon crystallization interpreted to have occurred during the anatectic event that produced the granite from the Rennick Schist was $544 \pm 4$ Ma. This age is consistent with the Rb/Sr isochrons on Rennick Schist, indicating that the time of metamorphism and initial plutonism in the western portion of the Wilson terrane was synchronous within error. No comparable metamorphic ages have been produced from the Lanterman metamorphics, but it is notable that the plutonic rocks dated at approximately 533 Ma in the Lanterman and Alamein Ranges are massive, posttectonic phases (Borg et al., 1986b). According to Skinner (1983a) the Dickason Granite, from which some of the samples came for the 535-Ma isochron reported by Vetter et al. (1984), is of a late syntectonic vintage, although Adamson (1971) considered the rock to be posttectonic.

Black and Sheraton (1990) found that a portion of the zircons in their study had cores that dated at $1130 \pm 50$ Ma, which they interpreted to be the crystallization age of the primary source of the sedimentary precursor to their sample of orthogneiss. A few zircon grains also gave U/Pb ages of around 2,000, 2,500, and 2,800 Ma, indicating a minor component of the source as old as Archean.

At the other end of the spectrum, Black and Sheraton (1990) found that overgrowths on the main magmatic phase of the zircons dated at $469 \pm 4$ Ma, which they interpreted as the time of intrusion of the main, posttectonic phase of the Granite Harbour Intrusives in the Daniels Range. This age is consistent with many of the K/Ar and Rb/Sr mineral dates on both the Granite Harbour Intrusives and metamorphic rocks from throughout the Wilson terrane. Before discussing these dates, however, we should mention several other Rb/Sr whole-rock isochrons that have been determined.

Adams (1986) produced a whole-rock isochron on 10 samples of Wilson Gneiss from the Daniels Range with an age of $490 + 33/-5$ Ma (IR $= 0.7205$, MSWD $= 5.8$). This was bolstered by two U/Pb dates on monazite of 484–488 Ma. However, given the conclusions of Black and Sheraton (1990), these dates may be considered to indicate open-system behavior. A five-point, whole-rock isochron on massive Granite Harbour Intrusives from the southern portion of the Daniels Range has an age of $495 \pm 12$ Ma (IR $= 0.7125$, MSWD $= 1.2$) (Vetter et al., 1983; Kreuzer et al., 1987).

In the Morozumi Range the main phase of the granodiorite at the northern end of the range has yielded a five-point, whole-rock isochron of $515 \pm 28$ Ma (IR $= 0.7136$, MSWD $= 0.9$) (Vetter et al., 1983; Kreuzer et al., 1987). A four-point isochron from the plutons at Jupiter Amphitheatre is $470 \pm 18$ Ma (IR $= 0.7114$, MSWD $= 2.0$) (Kreuzer et al., 1987). And from Unconformity Valley a four-point isochron is $475 \pm 12$ Ma (IR $= 0.7097$, MSWD $= 0.15$).

At the southern end of the Wilson terrane in the vicinity of Mt. Gerlache, Borsi et al. (1989) have determined a six-point, whole-rock isochron on Larsen Granodiorite of $461 \pm 24$ Ma (IR $= 0.71106$, MSWD $= 0.11$). This age is younger than three mineral isochrons ranging from 503 to 475 Ma on some of the same samples, leading the authors to interpret open-system behavior following emplacement of the Larsen Granodiorite, possibly due to fluid circulation during later emplacement of Irizar-type granite.

Rb/Sr and K/Ar dates on minerals, primarily from plutonic rocks but also from some metamorphics, show a fairly consistent pattern throughout the whole of northern Victoria Land. A scattering of dates comes from the older literature (Starik et al., 1961; Ravich and Krylov, 1964; Webb et al., 1964; Faure and Gair, 1970; Nathan, 1971a). More recent, expanded data sets are presented in Adams et al. (1982b), Kreuzer et al. (1987), Borsi et al. (1989), and Stump et al. (in press).

Throughout much of northern Victoria Land the oldest mineral ages are about 480 Ma and range down to not much less than 470 Ma (Kreuzer et al., 1987). Concordant mineral pairs, but somewhat different ages on relatively nearby localities, suggest rapid cooling from successive local pulses of Granite Harbour Intrusives (Kreuzer et al., 1987). An exception to this age distribution is in the Lanterman and Alamein Ranges, where some of the dates are older. K/Ar ages on hornblende and muscovite spread back to about 500 Ma (Adams et al., 1982b; Kreuzer et al., 1987), and several Rb/Sr mineral isochrons and whole-rock–mineral pairs are somewhat in excess of 500 Ma (Stump et al., in press). A K/Ar hornblende age of $355 \pm 3$ Ma from the south side of Husky Pass in the Lanterman Range (Adams et al., 1982b) is most likely due to thermal effects associated with movement on the Lanterman fault at the boundary with the Bowers terrane. In addition, some ages in the southern portion of the Wilson terrane are as young as 450 Ma, indicating that rocks there took longer to cool than elsewhere in the terrane (Nathan, 1971a; Borsi et al., 1989).

The thermotectonic evolution of the Wilson terrane in northern Victoria Land may be generalized as follows. Peak metamorphism and anatectic conditions had been reached by around 540–535 Ma with the early plutonism in the western area intimately involved with ongoing deformation, but with early plutonism in the Lanterman Range and Terra Nova Bay area occurring after the main deformation. Plutonism may have continued at scattered localities for the ensuing interval of time. Then beginning around 480 Ma a major pulse of posttectonic Granite Harbour Intrusives invaded the region, affecting K/Ar systems in the host metamorphic rocks. Cooling from this episode occurred within 10–15 m.y., except in the southern part of the terrane, where it lingered to around 450 Ma.

## Bowers Terrane

### Introduction

The Bowers terrane is an elongate, fault-bounded succession of rocks sandwiched between the Wilson and Robertson Bay terranes. It extends from Rennick Bay on the northern coast to Lady Newnes Bay in the southeast, reaching a maximum width of approximately 50 km at it southern end. The Bowers terrane is composed of the Bowers Supergroup, a varied sequence of sedimentary and volcanic rocks, mainly of Cambrian age. Understanding of the stratigraphy developed gradually through a succession of investigations over two decades.

The rocks that belong to the Bowers Supergroup were first mapped by Le Couteur and Leitch (1964), who discovered coarse-grained, cross-bedded quartzites and conglomerates in the East and West Quartzite Ranges, naming them the "Camp Ridge Quartzite." Although Crowder (1968), Riddols and Hancox (1968), and Gair et al. (1969) had already used the name in print, Sturm and Carryer (1970) are credited with having proposed "Bowers Group" for the predominantly shallow-water sedimentary sequence in the Bowers Mountains, based on their fieldwork in 1963–64. The group included the Camp Ridge Quartzite exposed to the south. To the west of the Camp Ridge Quartzite, Crowder (1968) identified a belt of calcareous sandstone and siltstone, with limestone, conglomerate, and considerable volcanic material, which he named the "Sledgers Formation."

The next addition to the Bowers Group was the "Carryer Conglomerate" of Dow and Neall (1972, 1974), named for massive polymict conglomerate, graywacke, and argillite exposed around the mouth of Carryer Glacier. These authors thought that the Carryer Conglomerate was stratigraphically the lowest formation of the Bowers Group, successively overlain by the Sledgers Formation and the Camp Ridge Quartzite. The discovery of the trace fossil *Rusophycus* in the Camp Ridge Quartzite indicated a Phanerozoic and probably Cambrian age for that formation.

The discovery of Late Cambrian body fossils (the trilobite *Pseudoagnothus* and the brachiopod *Billingsella*) in rocks at the head of Mariner Glacier provided the first sure age data for the Bowers Group (Laird et al., 1972). These data were difficult to resolve with what was known of the regional geology, for they indicated marine deposition coincident with the interval of time during which the Ross orogeny, with its deformation, metamorphism, and abundant plutonism, was occurring elsewhere throughout the Transantarctic Mountains.

A detailed study of these rocks at the head of Mariner Glacier resulted in an addition to the Bowers Group, the "Mariner Formation," a sequence of fissle mudstones and very fine-grained sandstones, with some limestone lenses (Andrews and Laird, 1976). The top of the Mariner Formation is scoured and conformably overlain at Eureka Spurs by Camp Ridge Quartzite. It was thought

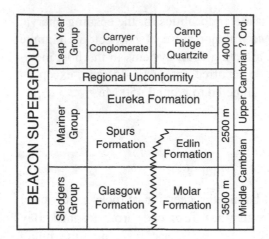

**Figure 2.16.** Nomenclature of Bowers Supergroup. After Laird et al. (1982).

to be a lateral equivalent of part of the Sledgers Group, but to be distinct enough to warrant separate formational status.

Further mapping by Laird et al. (1976) throughout a considerable portion of the Bowers Mountains and Quartzite Ranges resulted in a revision of the stratigraphy of the Bowers Group. Sledgers Formation was recognized to be the lowest unit, overlain conformably by equivalents of the Mariner Formation exposed to the south. These were in turn unconformably overlain with minor discordance by Camp Ridge Quartzite and Carryer Conglomerate, which crop out in two separate, parallel belts in the eastern and western Bowers Mountains, respectively. Additional fossil discoveries indicated that the Sledgers Formation is Middle Cambrian; the Mariner Formation, late Middle Cambrian; and the Camp Ridge Quartzite, middle or late Late Cambrian, with unfossiliferous upper portions reaching perhaps into the Ordovician (Cooper et al., 1976, 1982).

The benchmark in studies of the Bowers Supergroup is the paper by Laird et al. (1982) in which the authors elevate the Bowers Group to supergroup status, subdivide it into three groups, the Sledgers, Mariner, and Leap Year, and further subdivide the groups into a number of formations (Fig. 2.16). Of note is the Husky Conglomerate, a unit that has generated more controversy since its introduction than any other geological feature in northern Victoria Land.

Extension of mapping to the northernmost portion of the Bowers Mountains by Jordan et al. (1984) completed the areal coverage of the Bowers terrane and demonstrated that Sledgers Group crops out throughout the northern area.

## Bowers Supergroup

The Bowers Supergroup is now recognized to span a time period from Middle Cambrian through Upper Cambrian, and perhaps into the Lower Ordovician, with the uncertainty of the upper age being due to the lack of body fossils in the uppermost formations. The lower Sledgers Group contains abundant volcanic rocks erupted in a marine setting. The middle Mariner Group lacks volcanics and represents a regressive, shallow-marine environment. A regional unconformity, which truncates underlying Mariner Group with little or no angular discordance, is followed by the Leap Year Group, a succession of predominantly fluvial deposits.

The whole of the Bowers Supergroup was deformed during a single folding event that produced a series of tight anticlines and synclines parallel to the elongation of the Bowers terrane. Metamorphism during this deformation did not exceed prehnite–pumpellyite or lowermost greenschist conditions. Deformation and metamorphism along the terrane boundaries were, however, more intense.

In the following sections the stratigraphy of the Bowers Supergroup will be reviewed systematically, followed by discussions of the rocks associated with the western boundary and the overall deformation of the terrane.

## Sledgers Group

Sledgers Group is the most widely exposed group of the Bowers Supergroup. In the northern Bowers Mountains north of Alt Glacier, it comprises all of the known outcrops (Fig. 2.17). Throughout the central and southern Bowers Mountains it is widely exposed in the medial portion of the terrane. Farther to the south it crops out at an isolated occurrence at Jago Nunataks and then picks up again in the Barker Range and continues in a belt southward to the Malta Plateau. On the west side of Mariner Glacier it also crops out in a belt from central Gair Glacier southward to Spatulate Ridge.

Laird et al. (1982) designated a type area at the northern end of the Molar Massif where more than 2,500 m of nearly complete section occur. The total thickness of the group is probably in excess of 3.5 km. The Sledgers Group is divided into two interfingering formations, the primarily volcanic Glasgow Formation and the sedimentary Molar Formation. The most detailed mapping of the Sledgers Group is that by Wodzicki and Robert (1986) in the central Bowers Mountains, where five lithological units of Glasgow Formation and three of Molar Formation were recognized.

## Glasgow Formation

The Glasgow Formation consists of a variety of volcanic flows and breccias, which occur as thick sheets and lenses within the sedimentary rocks of the Molar Formation. Compositions are primarily basalt with lesser andesite, but range all the way to rhyolite. The type locality is designated as Mt. Glasgow, where at least 2,500 m of massive, basaltic breccia are exposed (Laird et al., 1982). At that locality the volcanics are sharply overlain by basal beds of Mariner Group. Elsewhere Glasgow Formation is both overlain and underlain by Molar Formation. For example, at the Molar Massif the volcanics are a massive sheet approximately 1,000 m thick with Molar Formation both above and below. In the lower Carryer Glacier area a 2,200-m section of Glasgow Formation overlies Molar Formation and is itself overlain by a thin section of Molar followed by Glasgow. This in turn is overlain by Leap Year Group above an erosion surface (Wodzicki and Robert, 1986). Around Mt. Soza a 2,700-m-thick section of Glasgow Formation is sandwiched between Molar Formation (Wodzicki and Robert, 1986). At Frolov Ridge two volcanic units, each about 1,000 m thick, are separated by about 300 m of mudstone (Jordan, 1981).

The most common lithology of the Glasgow Formation is volcanic breccia, with angular to subangular fragments up to 20 cm in length, set in a fine-grained

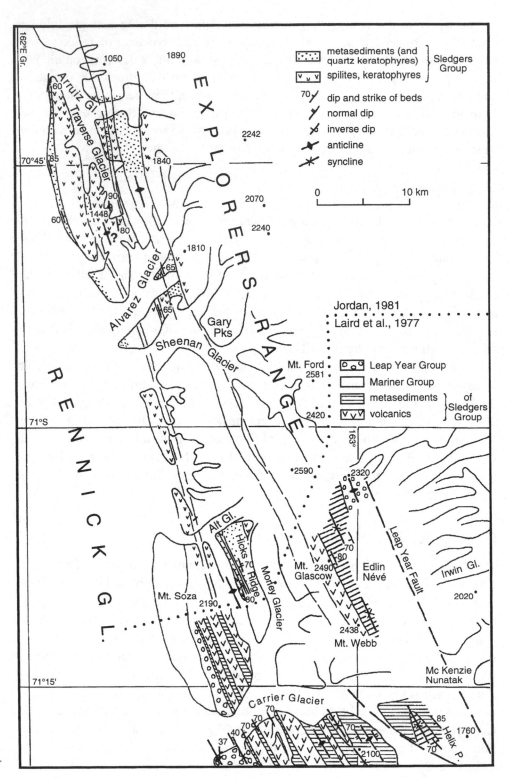

**Figure 2.17.** Geological sketch map from Frolov Ridge to Carryer Glacier. After Jordan (1981). Used by permission.

chloritic matrix (Jordan et al., 1984; Wodzicki and Robert, 1986). Both clast-supported and matrix-supported types occur. The majority are autobreccias composed entirely of volcanic clasts but some contain granitic and sedimentary clasts as well (Wodzicki and Robert, 1986). The breccias grade into both debris flows and tuffaceous sediments of the Molar Formation.

Lava flows are also fairly common in the Glasgow Formation. Some are massive, some banded. Vesicles are found in some. Pillow structures occur at numerous localities and have been pictured by Jordan (1981), Gibson et al. (1984), and Wodzicki and Robert (1986). Tuffaceous rocks are another important component, with variable amounts of crystal fragments, lapilli, and bombs (Jordan, 1981). Wodzicki and Robert (1986) report both crystal tuff with crystal fragments and well-preserved shard structures, and welded tuff with partially welded and flattened relict pumice fragments. The latter is indicative of subaerial eruption in what is otherwise taken to be a predominantly marine volcanic sequence. The only other suggestion of a nonmarine eruption is that of a 100-m-thick red lava flow, lacking pillows, in the lower Carryer Glacier area (Laird and Bradshaw, 1983). Dikes and sills of basaltic and andesitic composition occur locally.

Low-grade metamorphism and possible interaction with seawater have considerably altered the original mineralogy of the Glasgow volcanics, producing spilites and keratophyres. The most common mineralogy of the mafic to intermediate rocks is quartz + albite + chlorite + epidote + actinolite ± calcite ± sphene (leucoxene) (Jordan et al., 1984). Olivine, clinopyroxene, and orthopyroxene megacrysts are almost always pseudomorphed by combinations of chlorite, epidote, tremolite, calcite, prehnite, pumpellyite, albite, and/or quartz. Plagioclase is pseudomorphed by albite, calcite, chlorite ± sericite (Gibson et al., 1984; Jordan et al., 1984; Wodzicki and Robert, 1986).

## Solidarity Formation

Wodzicki and Robert (1986) have proposed the name "Solidarity Formation" for a unit of volcanic flows and breccias that crops out in the core of an anticline to the east of Mt. Gow crossing the Solidarity Range. They distinguish the unit from the rest of the Glasgow Formation because of its basal position, beneath about 1 km of Molar Formation, and by virtue of some differences in geochemistry from the rest of the Glasgow volcanics.

## Geochemistry

The composition of the Glasgow Formation ranges from basalt to rhyolite. Chemical analyses have been published by Weaver et al. (1984a), Gibson et al. (1984), and Wodzicki and Robert (1986). The silicic to intermediate rocks fall on a calc–alkaline trend, except for some low magnesium rhyolites from north of Alt Glacier (Wodzicki and Robert, 1986), whereas the basalts are mainly tholeiites. The basalts appear to fall into two groups, one low in titanium, the other high. The low-titanium basalts, which are the more numerous, have been interpreted on the basis of their trace-element geochemistry to have originated in a primitive island arc, whereas the high-titanium basalts suggest incipient rifting or marginal basin development before the cessation of volcanism (Weaver et al., 1984a). On a Ti/Zr discrimination diagram of Pearce (1982) most of the analyzed samples of Glasgow Formation fall within the arc lavas field, with a few falling in the overlapping part of the MORB field. Solidarity Formation plots within the

MORB field close to the boundary between arc lavas and within plate lavas (Wodzicki and Robert, 1986).

## Molar Formation

Molar Formation is a sequence of interbedded sandstone, mudstone, and minor conglomerate. The type area has been informally designated as the western Molar Massif, where at least 1,350 m (Laird et al., 1982), and perhaps more than 2,000 m (Wodzicki and Robert, 1986), of sedimentary rocks underlie volcanics of the Glasgow Formation. In the lower Carryer Glacier area approximately 1,500 m of Molar Formation are overlain by Glasgow Formation, while in the upper Carryer Glacier area approximately 800 m of Molar Formation interfinger with Glasgow Formation (Wodzicki and Robert, 1986). Around Mt. Soza, Glasgow Formation is underlain by about 1,000 m and overlain by an additional 450 m of Molar Formation. Approximately 900 m of Molar Formation are exposed to the west of Mt. Bruce (Jordan et al., 1984). The lower contact of the Molar Formation is not exposed, unless the Solidarity Formation is taken as basal to it (Wodzicki and Robert, 1986).

The predominant lithological association of the Molar Formation is dark mudstone interbedded with thin, fine-grained sandstone of graywacke type, typically laminated but sometimes containing cross laminations, ripples, or wavy bedding (Jordan, 1981; Laird et al., 1982). Flute casts and sole marks are fairly common. The sandstones are feldspathic to lithic wackes, with lithic fragments that include redeposited Molar sandstone, mudstone, and limestone, and Glasgow volcanics, as well as minor plutonic and metamorphic rock fragments (Wodzicki and Robert, 1986).

Intercalated within these rocks is a more coarse-grained association characterized by thinning and fining-upward sequences to 50 m thickness with conglomerates at the base, passing upward through sandstone and then mudstone-dominated units (Laird et al., 1983). Some of the sandstones are graded. The conglomerates contain clasts primarily of basalt, andesite, and graywacke, with lesser dacite, rhyolite, limestone, quartzite, granitoids, and muscovite schist (Wodzicki et al., 1982). Although some of the beds may have been deposited by turbidites, this is not characteristic of the sequence as a whole, which lacks complete Bouma sequences and cyclic bedding (Jordan et al., 1984).

A laterally persistent bed of limestone, 10 m thick, crops out in the lower Carryer Glacier area (Laird et al., 1982). Elsewhere, thin limestone intercalations or lenses occur within the mudstones. On Spatulate Ridge, at the southern end of the Bowers terrane, the transition from Glasgow volcanics to the more typical graywacke and mudstone of the Molar Formation is marked by red dolomitic limestone, limestone, and conglomerate (Gibson et al. 1984). Intercalated in the upper part of the formation in the Bowers Mountains are debris flows up to 100 m thick with blocks up to 5 m in length set in a muddy matrix. Clasts include limestone and volcanics, as well as the suite of lithologies found in the conglomerates already listed (Wodzicki and Robert, 1986; Laird and Bradshaw, 1983).

Although the basalt geochemistry of the Glasgow Formation indicates a primitive magmatic arc, the presence of granitoids and metamorphic rocks among the clasts of the Molar conglomerates indicates that a continental source was also nearby during deposition of the Sledgers Group.

Throughout most of the area the fine-grained association indicates transport from the northwest toward the southeast, except in the vicinity of Mt. McCarthy, where the opposite direction is indicated (Wodzicki and Robert, 1986; Laird, 1989). This is interpreted to indicate transport of muds by traction currents or thin turbidites parallel to the axis of the depositional basin (Laird, 1989) or parallel to a marginal shelf (Wodzicki and Robert, 1986). In the coarser-grained association the predominant transport is to the southeast, according to Laird (1989), with northeast and southwest directions also indicated. Wodzicki and Robert (1986) found southwest transport directions to predominate. These are interpreted to be mass flow deposits laid in channels cut across the slope and floor of the basin.

Both microfossils and macrofossils have been found in the Molar Formation (Cooper et al., 1976, 1982, 1983). Although there is uncertainty about the stratigraphic assignment of some fossil-bearing localities, as well as the age range of the microfossils, the age of the Molar Formation probably falls entirely within the Middle Cambrian. Polymerid trilobites from a limestone block within the upper Molar Formation from east of Neall Massif are likely late Middle Cambrian.

## Tiger Gabbro

At the southern end of Spatulate Ridge at Lady Newnes Bay, a layered gabbro intrudes Sledgers Formation. The body was defined by the GANOVEX Team (1987), though mention had been made of it by Gibson et al. (1984), Kleinschmidt et al. (1984), and Tessensohn (1984). It consists of cumulate layers of clinopyroxene, basic plagioclase, orthopyroxene, and accessory olivine. Pegmatites of plagioclase and hornblende cut through the body. The layers dip 60° to the southwest, and numerous shear zones cut through the body.

The gabbro intrudes both sedimentary and volcanic rocks thought to be of the Sledgers Group (Engel, 1987). The contact aureole is approximately 4.5 km wide. Adjacent to the gabbro, the metasediments are intensely disharmonically folded, but intensity drops off away from the contact. Folding is not demonstrable in the volcanics. Contact metamorphic conditions are estimated to have been 690°–730°C at 2–4 kb. K/Ar dating of hornblende gives an age of 521 ± 10 Ma (Kreuzer et al., 1987).

Kleinschmidt and Tessensohn (1987) have suggested that the Tiger Gabbro represents the deeper parts of an ophiolitic suite, whereas Bradshaw (1989) suggests that it may be a subvolcanic pluton related to the Glasgow arc.

## Mariner Group

Mariner Group, the least widespread of the Bowers Supergroup, crops out at the Eureka Spurs near the head of Mariner Glacier, in fault slivers on Reilly Ridge, and in a narrow band from the northern Neall Massif northward through the upper Carryer Glacier area to Mt. Glasgow. The type locality is designated as

Eureka Spurs (Andrews and Laird, 1976), where approximately 1,600 m of section are exposed. No complete section exists, but judging from other isolated occurrences the group is probably in excess of 2,500 m (Laird and Bradshaw, 1983).

The lower contact of the group is exposed at two localities, north of Mt. Glasgow and in the northern portion of the Molar Massif, where in both cases volcanics of the Glasgow Formation are conformably overlain (Laird and Bradshaw, 1983). The Mariner Group is conformably overlain by the Leap Year Group across a regional erosion surface that appears to cut more deeply into the group farther to the north (Laird et al., 1982).

A number of fossil localities have been reported in the Spurs and Eureka Formations (Laird et al., 1972; Cooper et al., 1976, 1982, 1983, 1990; Shergold et al., 1976, 1982; Rowell et al., 1983; Shergold and Cooper, 1985). Fossils include a variety of trilobites, plus inarticulate and articulate brachiopods, gastropods, and hyolithids. The age range of the fossils is late Middle Cambrian (Undillan or Boomerangian) to Upper Cambrian (late Idamean or early post-Idamean).

## Spurs Formation

Laird et al. (1982) named the Spurs Formation for the lower 900 m of section exposed at Eureka Spurs. From scattered outcrops in the area they suggest that it may be more than 1,500 m thick. Basal Spurs Formation overlies Glasgow Formation at the Molar Massif. The formation is characterized by medium gray fissle mudstones, slightly calcareous in places, with small lenses and thin beds of fine-grained sandstones (Andrews and Laird, 1976). Thin lenses of fine pebble conglomerate also occur in the lower portion. Ripples, cross lamination, and flute marks are fairly common sedimentary structures. Trilobite and brachiopod shells, either whole or fragmented, are concentrated in scattered beds of siltstone or sandstone.

A 12-m-deep channel was found at the type locality. The channel fill is bedded and contains slump-folded masses of sandstone, as well as blocks of mudstone and limestone.

On Reilly Ridge, Spurs Formation crops out in a series of fault slivers, which also contain Sledgers and Leap Year Group rocks. The predominant lithologies are typical fine-grained sandstone and mudstone, but in the upper part of the section coarse-sand and conglomeratic units are interspersed (Bradshaw et al., 1985a; Laird, 1989; Cooper et al., 1990). Some of the conglomerates are composed primarily of limestone blocks. Limestone units up to 500 m in length have been interpreted as allochthonous and emplaced by mass flow or gravitational sliding.

Other conglomerates of the Spurs Formation on Reilly Ridge are polymict and were previously mapped as Carryer Conglomerate (Laird et al., 1982). The reassignment was based on the similarity of interbeds of sandstone and mudstone to typical Spurs Formation, in contrast to interbeds in Carryer Conglomerate elsewhere, which are generally coarse-grained, cross-bedded sandstones (Bradshaw et al., 1985a; Cooper et al., 1990). These conglomerates are also interpreted as mass flow deposits, but with an input of rounded clasts in addition to breccia blocks. Clast lithologies include sandstone, limestone, basic and intermediate volcanics, and notably rare, two-mica, S-type granites. The polymict conglomerates of the Spurs Formation have been given the name "Southend Conglomerate" and assigned a member status (Bradshaw et al., 1985a).

## Edlin Formation

Edlin Formation is thought to be at least in part a lateral equivalent of the Spurs Formation, cropping out in the northern part of the Bowers terrane in the vicinity of Sheehan Glacier and at Mt. Glasgow where it overlies Glasgow Formation (Laird and Bradshaw, 1983). The type locality is 1 km north of Mt. Glasgow at the northern end of Edlin Névé, where 110 m of composite section are exposed (Laird et al., 1982). There are 260 m of section exposed at Sheehan Glacier. The formation contains quartz arenites, tuffs, conglomerate, and red and brown mudstone and, at both Mt. Glasgow and Sheehan Glacier, passes upward into limestone and mudstone of the Spurs Formation.

## Eureka Formation

The type locality of the Eureka Formation is at Eureka Spurs, where 700 m of section follow conformably on Spurs Formation (Laird et al., 1982). The lower 250 m are muddy, very fine-grained to fine-grained sandstones with characteristic wavy bedding. Slump balls and slump folds occur toward the top of the unit. These are followed by a unit (100 m thick) with similar sandstones and mudstones, but with lenticular bodies of limestone from 1 to 5 m thick and 5–50 m long. Most of the limestones are massive oosparites; some are massive sparites. These have been interpreted as shoals (Andrews and Laird, 1976). The upper 350 m of the formation are fine-grained sandstones and muddy sandstones with brownish limonite staining. Numerous trace fossils and mudcracks characterize this unit.

## Environmental Interpretation

Overall, the Mariner Group has been interpreted as a regressive shallow marine deposit (Andrews and Laird, 1976) with the fine sands and muds of the Spurs Formation representing shallow shelf conditions, giving way in the Eureka Formation to more near-shore sandstones, in part interspersed with oolite shoals, followed by deposits in a tidal estuary.

From slump folds beneath channels, the paleoslope for the basin is inferred to have been to the east or northeast (Laird, 1989). Currents are indicated to the northwest parallel to the basin axis. The basin geometry is inferred to have been similar to that of the Sledgers basin, except lacking a margin to the northeast.

## Leap Year Group

The Leap Year Group crops out in two elongate strips on opposite sides of the Bowers terrane. The eastern strip contains the Camp Ridge Quartzite, which is found from Sheehan Glacier in the north to Gauntlet Ridge in the south (Laird and Bradshaw, 1983; Stump et al., 1983b). The western strip, containing the Carryer and Reilly Conglomerates, is less widespread, cropping out between Mt. Soza and Reilly Ridge.

The Leap Year Group lies above a regional unconformity that from the head of Mariner Glacier cuts successively deeper toward the north. At Eureka Spurs, Camp Ridge Quartzite overlies Eureka Formation; 5 km north of Helix Pass it rests on Spurs Formation. On Reilly Ridge, Carryer Conglomerate appears to infill deep channels (up to 200 m) in Spurs Formation, whereas from lower Carryer Glacier to Mt. Soza it covers successively older horizons of Sledgers Group (Laird et al., 1982; Laird and Bradshaw, 1983). Relationships in the lower Mariner Glacier area are uncertain. Leap Year quartzites are in fault contact with Molar Formation, and only one isolated outcrop of possible Mariner Group rocks is exposed (Mortimer et al., 1984).

No body fossils have been found in the Leap Year Group, but several trace fossils have been recorded, including *Skolithos* (Laird et al., 1974; Cooper et al., 1982). The age is constrained by the underlying Upper Cambrian (Idamean) Eureka Formation and by a K/Ar date of 482 ± 4 Ma (Adams and Kreuzer, 1984).

## Carryer Conglomerate

The Carryer Conglomerate was named by Dow and Neall (1972, 1974), who designated a type locality along lower Carryer Glacier north of Mt. Gow. The formation is characterized by massive polymict conglomerate units, interspersed with units of sandstone and mudstone containing ripple marks and cross-bedding (Dow and Neall, 1974; Laird et al., 1982). At its type locality the formation reaches its greatest thickness of approximately 800 m. The lower 150–400 m of the formation are red and green wackes with minor conglomerate and red mudstone (Wodzicki and Robert, 1986).

The conglomerates are well rounded with boulders up to 50 cm in diameter. They contain a variety of clasts, including sandstones and mudstones, volcanics, and plutonics (Wodzicki et al., 1982; Wodzicki and Robert, 1986). The sedimentary clasts are similar to lithologies of the Molar Formation. Volcanics are pyroxene-bearing andesites, andesite breccias, and dacite porphyries. The plutonic rocks include two-mica S-type granites, biotite and hornblende granodiorites, quartz monzonites, and alaskites.

## Camp Ridge Quartzite and Reilly Conglomerate

The Camp Ridge Quartzite and Reilly Conglomerate are formations containing quartzose sandstones and conglomerates. The former crops out in the eastern strip of Leap Year Group, the latter in the western. Reilly Conglomerate was named by Laird et al. (1982) for monomict conglomerates at Reilly Ridge and nunataks to the south that overlie Carryer Conglomerate and that were thought to be equivalent to the basal Camp Ridge Quartzite. Cooper et al. (1990) suggest that the separate nomencalture may not be warranted, and Laird (1989) uses only the name "Camp Ridge Quartzite" for these rocks.

The Camp Ridge Quartzite was the first formation of the Bowers Supergroup to be recognized. Le Couteur and Leitch (1964) designated a type locality at Camp Ridge in the East Quartzite Range and published a generalized 1,800-m section. Laird et al. (1974) published a 2,900-m section from the type area. A succession of other workers have extended the original mapping and descriptions (Riddols and Hancox, 1968; Sturm and Carryer, 1970; Laird et al., 1974, 1976, 1982; Andrews and Laird, 1976; Laird and Bradshaw, 1983; Mortimer et al., 1984).

The Camp Ridge Quartzite is composed of reddish brown to buff-colored quartzose sandstone, quartzose conglomerate, and minor mudstone. Clasts in the conglomerate, well rounded and up to 8 cm in diameter, are almost exclusively quartz, although rare quartzose schist and sandstone have also been noted. The sandstones typically contain unimodal, trough cross-beds. The maximum thickness of the formation, 4,000 m, has been recorded by Laird and Bradshaw (1983) at the Leitch Massif, although Dow and Neall (1974) estimate a possible 7,000 m at that locality.

In the lower Mariner Glacier area, Mortimer et al. (1984) recognized three units of Leap Year rocks on the basis of color with a composite thickness of 3,300 m. Their lowest unit is brown (1,000 m), middle unit rose-colored (1,400 m), and upper unit white (900 m). There appears to be a general fining of grain size upward through the formation, with pebble and cobble conglomerates characterizing the lower part of the section, and quartzites and siltstones the upper portion.

Although the authors do not use a formational name, the rock descriptions are similar to those of Camp Ridge Quartzite. The one exception is that in the lowest part of the section on Gauntlet Ridge are 20-m-thick units of conglomerate containing volcanics that Mortimer et al. (1984) say resemble Carryer Conglomerate.

## Environmental Interpretation

The Carryer Conglomerate is interpreted as having been deposited in an alluvial basin or the proximal portion of a braided river close to the basin margin (Laird, 1989). Transport directions within the Carryer are wholly to the northeast. The Camp Ridge Quartzite is generally interpreted as fluvial in origin; however, the trace fossils in the lower portion of the formation may indicate transitional marine facies at the onset of deposition. Current directions are toward the northwest in the southern part of the area, swinging to northeast in the northern part. A quartzose source would appear to be indicated to the southeast of the Bowers terrane during deposition of the Camp Ridge Quartzite.

## Deformed Conglomerates

**Husky and Lanterman Conglomerates.** Probably the most problematic feature of the entire northern Victoria Land is the deformed conglomerates that crop out along the boundary between the Bowers and Wilson terranes, particularly in the eastern Lanterman Range. Crowder (1968) described a polymict conglomerate that he assigned to Sledgers Formation in contact with gneiss and schist on a spur to the west of Sledgers Glacier, but the exact location is uncertain because its coordinates were misprinted. Gair et al. (1969) and Dow and Neall (1972, 1974) mapped the contact between Bowers and Wilson Groups in the Lanterman Range but did not describe the rocks from that area.

Laird, Bradshaw, and Wodzicki were the first to appreciate the complexity of the geology at this locality (Laird et al., 1982; Bradshaw et al., 1982; Wodzicki et al., 1982). They mapped a series of fault slivers on Reilly Ridge that juxtaposed portions of Sledgers, Mariner, and Leap Year Groups. They also recognized a unit of dark green amphibolitic conglomerate and minor sandstone, which they examined in spurs from south of Husky Pass to Reilly Ridge and inferred from binocular observation to continue in spurs of the Lanterman Range to the west of Reilly Ridge (Laird et al., 1982).

The unit was called the Husky Conglomerate, and a type locality was designated west of the juncture of Reilly Ridge with the main Lanterman Range where a 370-m eastward-younging section is exposed. The upper contact was identified as a fault and no lower contact was observed, but the unit was inferred to be basal to the Bowers Supergroup. The majority of the clasts were amphibolitic, "indistinguishable from Wilson Group" (Laird et al., 1982, p. 538), but about 1% were rounded pebbles of quartz and granite. The amphibolitic clasts were subangular to subrounded and up to 3 m in diameter.

A second conglomerate with flattened and folded clasts cropping out between Reilly Ridge and Husky Pass was pictured and briefly mentioned by Bradshaw et al. (1982), and mapped with felsic and mafic units by Wodzicki et al. (1982); both sets of authors placed the conglomerate in the Wilson Group. The vague interpretations were misunderstood by Kleinschmidt and Skinner (1981), who were working from preprints of the others' work. Throughout the eastern Lanterman Range they recognized only one conglomeratic unit, Husky Conglomerate, with various degrees of deformation, and assigned it to the Wilson Group. Bradshaw and Laird (1983) later clarified their earlier work.

Further mapping by Laird and Bradshaw (1983) led to the conclusion that many of the clasts of the Husky Conglomerate were basic volcanics and that the rocks were similar to breccias found elsewhere in the Sledgers Group. They suggested that the Husky Conglomerate be considered a basal facies of the Sledgers Group and that the name be retained only on an informal basis. At "Fingernail Spur" (71°31′S, 163°12′E) a sharp contact between undeformed Husky Conglomerate and highly deformed conglomerate of the Lanterman metamorphics was interpreted as an unconformity.

Focusing only on Husky Conglomerate, Wodzicki and Robert (1986) interpreted it as a cataclastic rock formed along a major fault zone and suggested that the name be abandoned altogether as a stratigraphic term. They examined the contact between Husky Conglomerate and Wilson Group at two places (71°31′S, 163°03′E and 71°32′S, 163°07′E), both apparently different from the locality reported by Laird and Bradshaw (1983). At the northwest locality the contact is a fault, but at the southeast locality it appears to be gradational.

Following a study of all the conglomerates as far south as Husky Pass (Fig. 2.18), Gibson (1984) proposed a reclassification, including all mafic conglomerates regardless of degree of flattening as Husky Conglomerate and introducing the name "Lanterman Conglomerate" for the felsic conglomerates in the eastern

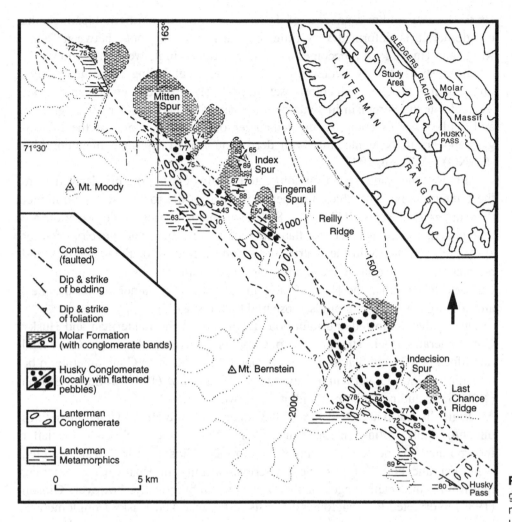

**Figure 2.18.** Geological map of conglomerates of the eastern Lanterman Range. After Gibson (1984). Used by permission.

Lanterman Range previously regarded as Wilson Group or Lanterman metamorphics.

The Lanterman Conglomerate everywhere separates Lanterman metamorphics on the west from Husky Conglomerate on the east. It consists of light gray polymict metaconglomerate with minor interstratified gneiss and schist. Conglomeratic clasts are flattened to various degrees, extremely so in many cases, with increased flattening toward the west. The clasts appear to be predominantly of metamorphic origin and include biotite, biotite–muscovite, chlorite, and quartzitic schists, quartzites, amphibolites, and lesser granitic and calc–silicate rocks and vein quartz (GANOVEX Team, 1987). Some of the clasts contain a foliation or crenulation cleavage not shared by the matrix, and many of the clasts could have been derived from the adjacent Lanterman metamorphics (Kleinschmidt and Skinner, 1981; Gibson, 1984). The contact with Lanterman metamorphics is nowhere seen, but Gibson (1984) inferred it to be a fault, allowing that it previously may have been an unconformity.

The Husky Conglomerate as redefined by Gibson (1984) is a dark green, predominantly mafic conglomerate and breccia with minor interbedded green sandstone and mudstone, cropping out east of the Lanterman Conglomerate and west of Molar Formation from Mitten Spur to north of Husky Pass and at a single locality 8 km southeast of Husky Pass. In its northern localities the conglomerate is deformed very little, but farther to the south between Reilly Ridge and Husky Pass it becomes extremely flattened in its western portions.

The clasts are mainly metamorphosed mafic rocks including actinolite amphibolites, actinolite schist, and chlorite schist, probably derived from volcanic rocks. Vesicular textures remain in some clasts. Psammites and minor marble, granite, and acid volcanics comprise the remainder of the suite. None of the lithologies appears to be compatible with derivation from Lanterman metamorphics (GANOVEX Team, 1987).

Although the Lanterman Conglomerate contains clasts possibly derived from the Lanterman metamorphics, the metamorphism of the rocks itself is no higher than upper greenschist (Gibson, 1984). Two major phases of deformation are recognized, the first a general flattening of the pebbles and development of associated schistosity in the matrix, the second a folding of the schistosity and pebbles and formation of crenulation cleavage, with various trends throughout the area. By contrast, the Husky Conglomerate is cut by numerous shear zones, and folding is uncommon or else obscured by the shearing.

Due to the tectonized and fault-bounded nature of the Lanterman and Husky Conglomerates, their stratigraphic position is uncertain. Although Grew and Sandiford (1982) and Sandiford (1985) found the Lanterman Conglomerate to be transitional into Lanterman metamorphics, Bradshaw (1987) follows Gibson (1984) in including both units within the Bowers terrane.

Originally Laird and Bradshaw (1983) interpreted the Husky Conglomerate as unconformably resting on Lanterman Conglomerate, citing clasts of the latter incorporated in the lowest 2 m of the former. Gibson (1984) also reported conglomeratic clasts of possible Lanterman affinity in Husky Conglomerate, reaffirming the relative ages of the two units. Subsequently, however, Gibson (1987) interpreted the conglomerate clasts to be reworked Husky Conglomerate, based on thin-section studies of Lanterman Conglomerate and the conglomerate clasts in question. Gibson (1987) went further to state that the previous interpretations of contact relations might need to be reconsidered and that a possible transition between the two units might exist, citing light green conglomerates with both Lanterman and Husky characteristics on Indecision Spur south of Reilly Ridge.

Both Gibson (1987) and Bradshaw (1987) agree that the Husky Conglomerate has affinities with Sledgers Group, but they disagree about its derivation. Gibson (1985, 1987) suggests that the Husky Conglomerate resulted from erosion of obducted Sledgers Group, and therefore postdates it. He suggests a correlation with Carryer Conglomerate, or perhaps lateral equivalency with Mariner Group. Bradshaw (1987) follows prior correlations of Husky Conglomerate with the Sledgers Group (e.g., Bradshaw and Laird, 1983; Laird and Bradshaw, 1983), but also notes that the clast chemistry is more boninitic than the main part of the Glasgow volcanics, suggesting that the source was more "outboard" of the main part of the arc presently exposed in the Bowers terrane.

Based on similarities in clast composition, Bradshaw (1987) suggests that Lanterman Conglomerate is contemporaneous with either Carryer Conglomerate or conglomerates of the Mariner Group. Gibson (1987) also suggests affinities between Lanterman Conglomerate and Leap Year Group. Both workers now regard Lanterman Conglomerate as younger than Husky Conglomerate.

**Black Spider Greenschists.** The initial reconnaissance of the Meander and lower Mariner Glaciers area included the observation of flattened conglomerates, which were correlated with Wilson Group (Stump et al., 1983b). Subsequent mapping revealed a belt of greenschist-grade metamorphic rocks east of the Dessent and Murchison Formations and west of typical Bowers Supergroup (Gibson et al., 1984). The rocks occur in two separate areas, one on the northern half of Spatulate Ridge, the other beneath the Mariner Plateau north of Gair Glacier. Rocks in the two areas have somewhat different characteristics, but both were included in the informal Blackspider Greenschists by the GANOVEX Team (1987).

Conglomerates are exposed only in the southern area, where they are somewhat flattened and become increasingly interbedded with green slates upward in the section (Gibson et al., 1984). Units are 30–100 m thick, with individual beds of 0.2–2 m. The conglomerates are crudely graded and pass upward into coarse sandstones. Both metamorphic and sedimentary clasts are represented, including chloritic schists, slate, fine-grained psammite, limestone, pink marble, and minor amphibolite and mafic, intermediate, and silicic metavolcanic rocks.

Localities north of Gair Glacier are isolated and were visited only briefly. There the rocks are primarily actinolite and actinolite–chlorite schists, with lesser amphibolite. Ultramafic schists also occur at several localities. The volcanic origin of these rocks is apparent in many of the rocks. Relict doleritic and/or porphyritic textures are preserved in some cases, as are vesicles. Relict pillow structure occurs at one locality. Some of the rocks give the appearance of metamorphosed pyroclastic rocks. Small pods of amphibolite in parts of the section probably represent metamorphosed dikes of gabbro.

Although the metamorphic and deformational aspects are distinct, Gibson et al. (1984) consider the Black Spider Greenschists to be correlative with the Sledgers Group.

## Deformation and Metamorphism of Bowers Supergroup

Deformation of the Bowers Supergroup occurred under prehnite–pumpellyite or pumpellyite–actinolite grades of regional metamorphism (Wodzicki et al., 1982), producing gently plunging, generally open folds oriented in a northwesterly direction parallel to the elongation of the Bowers terrane. Subvertical axial-plane cleavage is well developed in the more shaley units. In the central Bowers Mountains, where exposure of Bowers Supergroup is widest and best studied, three major synclines separated by anticlines have been mapped (Bradshaw et al., 1982; Wodzicki et al., 1986). The eastern syncline, which is cored by Camp Ridge Quartzite for most of its length, can be traced for more than 150 km from Edlin Névé to the vicinity of Pyramid Peak. The deformational style carries to the northernmost portion of the Bowers terrane; however, there the folding appears to be of a tight to isoclinal style (Jordan et al., 1984). In the Mountaineer

Range at the southern end of the Bowers terrane, folded Bowers Supergroup occurs in several fault-bounded slivers (Gibson et al., 1984).

Toward the western boundary of the Bowers terrane the rocks become more tightly folded and have variably plunging attitudes (Jordan, 1981; Gibson et al., 1984; Jordan et al., 1984). Metamorphism increases to greenschist grade in a 2- to 5-km-wide zone adjacent to Wilson terrane. On Reilly Ridge the rocks are also cut by numerous fault splays (Bradshaw et al., 1982; Cooper et al., 1990) (Fig. 2.19). As mentioned in the earlier section on Lanterman and Husky Conglomerates, in the eastern Lanterman Range the conglomerates were effected by two episodes of deformation, the first causing tight to isoclinal folds with penetrative schistosity and considerable flattening of the clasts, and the second producing open crenulations and kink folds (Gibson, 1987). Also, numerous shear zones, some containing mylonites, cut through these rocks. More recently, a preliminary report by Matzer (1992) identified three episodes of deformation in the Carryer Glacier area with $D_2$ shears indicating tops to the west and $D_3$ thrusts with tops to the east.

## Geochronology of Bowers Supergroup

More than 50 whole-rock K/Ar dates have been determined on slates and phyllites of the Bowers Supergroup, collected from throughout its stratigraphic and areal extent (Adams et al., 1982b; Adams and Kreuzer, 1984). The ages spread rather evenly over a period from 510 to 275 Ma (Fig. 2.15). While there is a minor but significant grouping of ages in the range 510–470 Ma, similar to cooling ages found in both the Wilson and Robertson Bay terranes, the majority of the ages are younger, completely spanning the period of emplacement of the Admiralty Intrusives and showing no marked signal from them.

Adams and Kreuzer (1984) note that there is a rough correlation of age with elevation and conclude that either (1) the Bowers terrane remained at a greater depth during the early to middle Paleozoic than the bounding terranes, or (2) that it cooled during the same Cambro-Ordovician time frame as the Wilson and Robertson Bay terranes, but was subsequently heated with the partial or total loss of radiogenic argon. In either case it is clear that the deeper parts of the Bowers terrane experienced temperatures higher than those required for argon retention after the period of cooling generally associated with the Ross orogeny. In their compilation of age and elevation data, Adams and Kreuzer (1984) also show that ages in the northern part of the Bowers terrane are somewhat younger than ages at equivalent elevations in the southern part, suggesting that there has been a differential cooling history along the length of the terrane.

## Robertson Bay Terrane

### Introduction

The Robertson Bay terrane extends from its western border with the Bowers terrane northeastward to the coastlines of the Southern Ocean and the Ross Sea. The terrane is composed mainly of a broad expanse of folded turbidites, the Robertson Bay Group, which is punctuated by a series of Devonian posttectonic

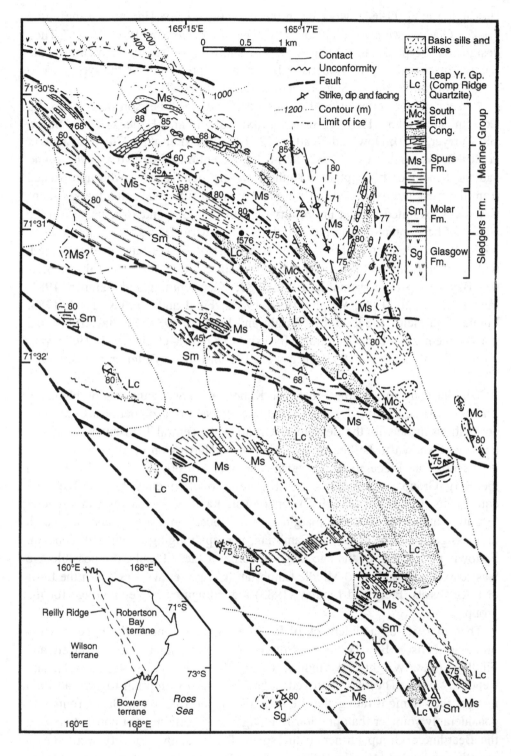

**Figure 2.19.** Geological map of Reilly Ridge. After Cooper et al. (1990). Used by permission.

plutons, the Admiralty Intrusives, and the odd middle Paleozoic volcanic center. The first systematic descriptions of the Robertson Bay Group ("slate–greywacke formation of Robertson Bay") were made by Rastall and Priestley (1921) following fieldwork from Cape Adare in 1911–12 by the Northern Party of Scott's Second (*Terra Nova*) Expedition. Klimov and Soloviev (1960) touched on the northern coastline and suggested the name "Robertson series" for these rocks.

Harrington et al. (1964), working in the vicinity of Cape Hallett and the lower Tucker Glacier, were the first to use the name "Robertson Bay Group." They designated a type locality on the western shore of Edisto Inlet, owing to the relative inaccessibility of Robertson Bay and the proximity of the type area to the then-operative Hallett Station (Harrington et al., 1967).

Subsequently, a series of field parties examined the Robertson Bay Group (Le Couteur and Leitch, 1964; Crowder, 1968; Riddols and Hancox, 1968; Sturm and Carryer, 1970; Dow and Neall, 1972, 1974; Laird et al., 1974; Bradshaw et al., 1982; Wodzicki et al., 1982). With the exception of Crowder (1968), who had the benefit of helicopters, all others were ground parties who approached from the west and examined fairly limited areas. It was found that the sedimentary rocks exhibited a great uniformity of both lithology and structure. With the advent of helicopter-supported expeditions in northern Victoria Land in the early 1980s, broad areas across the whole of the Robertson Bay terrane were examined, with specific focus on the sedimentological and structural aspects of the Robertson Bay Group (Kleinschmidt, 1981, 1983; Kleinschmidt and Skinner, 1981; Tessensohn et al., 1981; Wright, 1981, 1985a; Field and Findlay, 1982, 1983; Findlay and Field, 1982b, 1983; Wright and Findlay, 1984; Findlay, 1986) and the geochemistry of the Admiralty Intrusives (Kreuzer et al., 1981, 1987; Wyborn, 1981; Vetter et al., 1983; Borg et al., 1986b, 1987b; Vetter and Tessensohn, 1987).

Without body fossils, the age of the Robertson Bay Group was problematic. The age was generally suggested to be late Precambrian or perhaps Cambrian; correlations made with the Beardmore Group of the central Transantarctic Mountains were known to be late Precambrian (e.g., Harrington et al., 1967; Gair et al., 1969). Late Precambrian acritarchs (Soloviev, 1960; Iltchenko, 1972) from the Berg Group on the Oates Coast were also often cited as possible time equivalents. Cooper et al. (1982) reported acritarch assemblages from several localities along the western margin of the Robertson Bay terrane, to which they assigned an age of Lower Cambrian, perhaps ranging down into Vendian. Following a revision of the age range of some of these fossils, the possible age was relaxed to Vendian to Middle Cambrian (Cooper et al., 1983). On the basis of trace fossils, Field and Findlay (1983) argued against a Vendian age for the group.

Then in 1982–83 a slideblock of fossiliferous limestone in the upper portion of the Robertson Bay Group was discovered on Handler Ridge (Burrett and Findlay, 1984; Wright and Findlay, 1984; Wright et al., 1984; Buggisch and Repetski, 1987). The fossils proved to be uppermost Cambrian to Tremadocian in age, thus demonstrating that at least the upper Robertson Bay Group was considerably younger than previously thought, and certainly not correlative with the Beardmore Group farther to the south. Rocks at this locality were subsequently subdivided from the Robertson Bay Group and given the name "Handler Formation" (Wright and Brodie, 1987).

## Robertson Bay Group

The Robertson Bay Group is a sequence of rhythmically alternating graywackes and phyllites or slates. Except for the Handler Formation, which is not known beyond Handler Ridge, the group has not been subdivided, nor does there appear to be sufficient variation to warrant such. The base of the group is nowhere seen. Recent estimates of the minimum thickness are 2,000 m (Wright, 1981) and 3,000 m (Field and Findlay, 1983). The greatest thickness was suggested by Sturm and Carryer (1970) at 6,000 m.

The group has been subjected to very low-grade regional metamorphism, which has produced chlorite and sericite (Kleinschmidt, 1981, 1983). Axial-plane cleavage, particularly in the more pelitic rocks, is quite common. Local contact effects adjacent to Admiralty plutons have produced cordierite, hornblende, and biotite hornfelses, with cordierite porphyroblasts up to 2.5 cm in diameter (Harrington et al., 1967).

The Robertson Bay Group is typically various shades of gray or green, depending on the chlorite content, but a few beds of red argillite have been reported by various authors (Harrington et al., 1967; Sturm and Carryer, 1970; Laird et al., 1974, Field and Findlay, 1983). Volcanic rocks reported to be part of the Robertson Bay Group in the vicinity of Mt. McCarthy (Le Couteur and Leitch, 1964; Riddols and Hancox, 1968) were later shown to be Glasgow Formation in the Bowers terrane (Laird and Bradshaw, 1983). Mention of conglomerate and limestone beds north of Mt. Craven by Sturm and Carryer (1970) have not been substantiated.

In general the graywackes are fine- to medium-grained, with maximum grain size about 5 mm (Wright, 1981; Field and Findlay, 1983). Quartz is the predominant clast type, with minor feldspar and rock fragments. The rock fragments include quartz siltstone, pelite, and minor bits of chert, basic igneous rock, volcanics, and calcite. A metamorphic continental source for the sediments has been suggested (Field and Findlay, 1983; Wright, 1985a).

Bedding thickness in the Robertson Bay Group varies from a few centimeters to a few meters, with the average graywacke units about 30 cm thick. Rhythmic alternation of graywacke and pelite is the norm, but some amalgamated beds in which pelites have been eroded away are present. The upper parts of some pelites contain calcareous laminations or concretions (Harrington et al., 1967; Wright, 1981). The coarse-grained units are generally graded with sharp lower contacts. Basal sedimentary structures include load casts, flute casts, and groove marks. Other sedimentary structures are horizontal, wavy, cross, climbing, and convolute laminations and climbing ripples. Complete and incomplete Bouma sequences are seen.

Most geologists who have studied the Robertson Bay Group have agreed that it was deposited by turbidites. Field and Findlay (1983) also include deposition by fluidized sediment flow. Wright (1981, 1985a) interprets rocks in the more northern portion of the Robertson Bay terrane to have been deposited in a middle to distal deep-sea fan or deep-sea plain facies. Field and Findlay (1983), working farther to the south, suggest that some of the beds there are of a more proximal deep-sea fan facies. Wright (1981, 1985a) and Field and Findlay (1983) have measured transport directions toward the north to northwest, and both interpret the basin of deposition as deepening in that direction.

## Handler Formation

Due to distinctive sedimentary aspects, Wright and Brodie (1987) proposed that the Handler Formation be subdivided from the Robertson Bay Group, with its type locality designated as Handler Ridge. Initial observations of the sedimentology of these rocks were made by Wright and Findlay (1984). The Handler Formation is thought to be the uppermost portion of the Robertson Bay Group. The Handler Ridge locality contains a series of shallowly plunging, northwest-trending folds in which successively higher stratigraphic levels are exposed toward the west (Fig. 2.20). The western boundary of the Handler Formation is a thrust fault with Bowers Supergroup.

The eastern end of Handler Ridge is underlain by alternating beds of graded gray graywacke and slate typical of Robertson Bay Group. Rare red slate beds occur within this part of the section. Due to incomplete exposure caused by snow patches and scree, the thicknesses on Handler Ridge are estimates. The typical Robertson Bay Group is estimated to be 1,000–1,500 m with the bottom not exposed.

This is overlain gradationally by the lower unit of the Handler Formation, estimated to be 500–1,000 m thick (Fig. 2.21). The units continue to contain alternating graywacke and dark gray slate, but red-colored slates and siltstones become more prominent, and individual beds become thicker. The red color in these units may grade laterally to greens. Some soft-sediment slump folds also occur in these units. A second characteristic lithology is beds with well-rounded, pea-sized pebbles of quartz and lesser quartz-rich rock fragments. These conglomerates are typically matrix-supported by fine- to medium-grained sand, but some rare, clast-supported beds do occur. The beds are up to several meters thick, but the conglomerates also occur as discontinuous streaks within the graywacke.

The upper unit of the Handler Formation is estimated to be more than 500 m

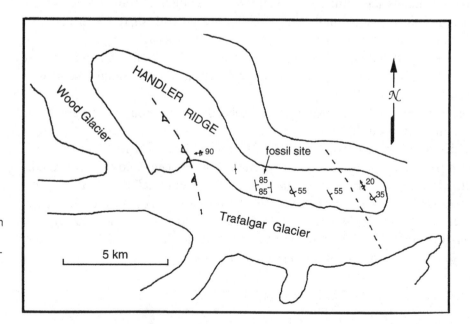

**Figure 2.20.** Index map of Handler Ridge showing thrust fault separating Bowers Supergroup to west from Handler Formation to east. Dashed line indicates gradational contact between Handler Formation and undivided Robertson Bay Group rocks. After Wright and Brodie (1987). Used by permission.

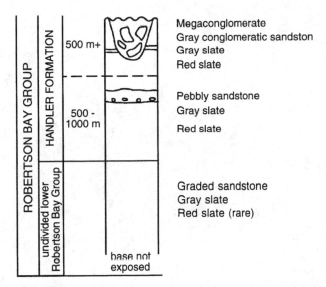

thick, with the top not exposed. Again the unit is predominantly gray graywacke and dark gray slate, with some red slate, but bedding is less clear than downsection. The distinguishing feature is the presence of chaotic blocks of limestone, quartz–pebble conglomerate, and quartz sandstone, with blocks ranging from a few centimeters to several meters. The blocks are poorly sorted, although the largest blocks occur together at distinct horizons. Some of the limestone blocks contain uppermost Cambrian to lowermost Ordovician fossils, as already discussed.

The coarse upper unit of the Handler Formation has been interpreted as a debris flow, probably in a marine channel (Wright and Findlay, 1984; Wright and Brodie, 1987). The limestone indicates shallow shelf conditions in the source area, lacking input of first-cycle detritus. A solution cavity in a limestone block filled with quartz–pebble conglomerate, as well as the red slates, suggests karsting and terra rosa development of a portion of this shelf.

The extent of the Handler Formation is at present not known. Wright and Brodie (1987) state that it occurs in spurs of the Victory Mountains from Handler Ridge at least as far north as Turret Ridge and suggest that areas in the Admiralty and Mirabito Ranges where others (Harrington et al., 1967; Sturm and Carryer, 1970; Field and Findlay, 1983) have reported red slates may also be occurrences of the formation.

## Deformation

Throughout most of its extent, the Robertson Bay Group has been deformed during a single episode of upright folding. Only along the western margin of the Robertson Bay terrane have multiple deformations and slightly higher metamorphism affected the Robertson Bay Group. Findlay and Field (1983) introduced the informal name "Millen Range schists" for these rocks, emphasizing their tectonic significance and differences from typical Robertson Bay Group. These rocks will be described separately later. Portions of the rocks described as Robertson Bay Group by Le Couteur and Leitch (1964), Crowder (1968), Riddols

**Figure 2.22.** Folded Robertson Bay Group, Buskirk Cliffs.

and Hancox (1968), Dow and Neall (1974), Bradshaw et al. (1982), and Wodzicki et al. (1982) occur within the belt of Millen Schist.

While all of the geologists who have worked in the Robertson Bay Group have reported on the structure, the studies of Kleinschmidt and Skinner (1981), Findlay and Field (1983), and Findlay (1986) are the most in-depth treatments, and each illustrates good representative examples of the folding. The most common fold style in the Robertson Bay Group is chevron, with straight limbs and sharply buckled hinges, but multihinged folds and folds with curved limbs are also found (Fig. 2.22). The wavelength of the folds is generally said to be from around 100 m to 1–2 km, with the widest range reported at a few meters to 5 km by Harrington et al. (1967).

The folds are open to tight, and true isoclinal folds are rare (Le Couteur and Leitch, 1964; Crowder, 1968). Crowder (1968) notes interlimb angles between 20° and 50°, although the smaller angles may be from the Millen Schist. Findlay (1986) records a range of interlimb angles from 25° to 150°, but notes that there is a trend toward tighter angles across the Robertson Bay terrane from northeast to southwest.

The axial planes of the folds trend consistently in a northwest–southeast direction, dipping steeply to vertically. The folds in general plunge less than 20° to either the northwest or the southeast, but plunges up to 60° have been recorded.

Figure 2.22 *(continued)*.

Kleinschmidt and Skinner (1981) explain the steep and doubly plunging aspect of folds in the Robertson Bay area as being due to a second generation of cross-folding with a northeast–southwest orientation. Findlay and Field (1983) and Findlay (1986), however, suggest that these features are due to inhomogeneous shortening during progressive deformation, citing the lack of intersecting cleavage and fold interference and the coincidence of northeast–southwest lineaments noted on *Landsat* images and topographic maps by Kleinschmidt and Skinner (1981) with a–c joints in the northwest–southeast-trending folds of the area. Findlay (1986) interprets variably and doubly plunging folds in the Mirabito Range south of Thomson Peak as sheath folds resulting from thrusting at the terrane boundary to the west.

Slatey axial-plane cleavage is well developed in most of the pelitic beds of the Robertson Bay Group, less so in the coarser-grained beds. As is normal in such alternating sequences, the cleavage is approximately parallel in the pelites and fans in the psammites. Curved cleavage has been noted as an indicator of graded bedding (Le Couteur and Leitch, 1964).

Around 40 K/Ar, whole-rock ages on slates and phyllites of the Robertson Bay Group (including Millen Schist) from throughout the Robertson Bay terrane (Fig. 2.15) have been produced by Adams et al. (1982b) and Adams and Kreuzer (1984). With the exception of one sample in the Mirabito Range, which shows effects of a pluton of Admiralty Intrusives, all of the samples are within the age range 505–455 Ma. Adams and Kreuzer (1984) note that the oldest samples are along the western boundary of the terrane and that ages are younger in the eastern portion of the terrane, leading to the conclusion that metamorphism began in the west and proceeded eastward. The entire range of ages, however, is represented in the western region, leading the authors to conclude that younger tectonic events represented in the Millen Schist may be responsible for resetting of some samples.

Using petrography and scanning electron microscopy, Wright and Dallmeyer (1991) observed that the cleavage in the Robertson Bay Group is due to growth of white mica and chlorite. Along the western boundary of the Robertson Bay terrane an overprinting crenulation cleavage is developed within the Millen Schist. Ar/Ar dating of whole-rock slates from that area showed that the axial-plane cleavage formed at 505–500 Ma and that apparently no new growth of mica accompanied the development of the crenulation.

In a more comprehensive Ar/Ar study from throughout the Robertson Bay Group, Dallmeyer and Wright (1992) affirmed the pattern seen in the K/Ar data of Adams and Kreuzer (1984) and concluded that the formation of axial-plane cleavage was diachronous from about 500 Ma in the western exposures to about 460 Ma in the easternmost areas.

Buggish and Kleinschmidt (1991) and Dallmeyer and Wright (1992) have determined illite crystallinity values for samples of Robertson Bay Group, and have found consistent values throughout their extent, indicating a uniform depth of burial across more than 100 km of outcrop perpendicular to the structural trend. Both pairs of authors suggest that one interpretation is that of flat-plane subduction with a uniform thickness of Bowers terrane overriding Robertson Bay terrane throughout its present-day exposure. The authors also offer, as a preferred alternative, tectonic thickening due to thin-skinned deformation during subduction to account for the uniformity of values. Buggish and Kleinschmidt (1991) suggest multiple ramp faults, whereas Dallmeyer and Wright (1992) suggest folding due to offscraping. A combination of these mechanisms may also be envisioned.

## Millen Schist

Along the western boundary of the Robertson Bay terrane, in contrast to the majority of the Robertson Bay Group, a zone of rocks crops out that has been subjected to multiple episodes of deformation and lower greenschist metamorphism. Although it was originally thought that the affinity of these rocks was entirely with Robertson Bay Group, some volcanics similar to Glasgow Formation have been recognized in this zone, as have metasediments indistinguishable as either Molar Formation or Robertson Bay Group (Jordan et al., 1984; Wright and Findlay, 1984). The zone apparently displays tectonic intermixing of Robertson Bay Group and Bowers Supergroup, but since the majority of the rocks were derived from the former, a discussion of their lithology and structure is presented in this section. Aspects of the deformation will be revisited in the section on terrane boundaries.

Findlay and Field (1983) introduced the name "Millen Range schists" for the suite. This was shortened to "Millen Schists" by Jordan et al. (1984) and Tessen-sohn (1984) and to "Millen Schist" by the GANOVEX Team (1987). Bradshaw et al. (1985b) and Findlay (1986) talked of the "Millen terrane," implying an allochthonous relationship to both the Bowers and Robertson Bay terranes, but Bradshaw (1987, 1989) shifted to the term "Millen Shear Zone," recognizing that the Bowers and Robertson Bay terranes had interacted throughout a broad zone, not simply along a narrow boundary fault.

The Millen Schist is well exposed throughout the Millen Range, where it reaches a width of 15 km. It extends southeastward as far as Mt. Hancox, before disappearing beneath McMurdo Volcanics on Malta Plateau. It also has been traced from west of Mt. Verhage in a northeasterly direction to the northernmost part of the Bowers Mountains around Mt. Bruce and Mt. Sheila. On the ridge east of Mt. Sheila the degree of schistosity drops off (Jordan et al., 1984). At Mt. Hager the schist zone is at least 6 km wide. The Millen Schist is apparently missing in the stretch from the King Range to Mt. Stirling.

Millen Schist consists of quartzofeldspathic schist and phyllite derivative of Robertson Bay or Molar type of rocks, with minor metabasite of volcanic origin. The schist characteristically contains segregation banding (Wodzicki et al., 1982; Wright, 1982; Jordan et al., 1984). The typical metamorphic assemblage for schistose metagraywackes is quartz + albite + white mica + epidote + opaques ± tourmaline ± carbonate, and for the metabasites a paragenesis including chlorite, epidote, and/or actinolite (Findlay, 1986).

Metavolcanic rocks are known to occur in several localities, including the eastern end of Shardik Ridge, where they contain relict clinopyroxene pheno-crysts and some agglomeratic structures (Jordan et al., 1984); the north side of Toboggan Gap, where pillow structures are displayed (Bradshaw et al., 1985a); above the thrust plane at Mt. Aorangi and Crosscut Peak, where the rocks are fine-grained greenschists (Findlay and Field, 1983; Findlay, 1986); and at the western end of Handler Ridge faulted against Handler Formation (Wright and Brodie, 1987).

Bradshaw et al. (1982) were the first to recognize more than one episode of deformation in these rocks in the area west of Mt. Verhage. There a prominent fracture–slaty cleavage, axial-planar to a northwest-trending antiform, was con-sidered second generation due to inverted beds in the antiform and limited remnant schistosity of an earlier generation.

In both the Millen Range and the northern Bowers Mountains the prominent structure is a schistosity axial-planar to the first recognized folding (Findlay and Field, 1983; Jordan et al., 1984; Findlay, 1986). This folding is generally at a mesoscopic scale, tight to isoclinal, and not well seen. In the Millen Range axes plunge shallowly northwest or southeast, whereas in the northern Bowers Mountains the fold plunge varies from horizontal to vertical. In both areas a second-generation crenulation cleavage overprints the earlier schistosity. In the northern Bowers Mountains it is associated with minor folds, whereas in the Millen Range second-generation folds are open to tight and range from a few centimeters to more than a kilometer in wavelength.

The relationship of the two cleavages in the Millen Range and the northern

Bowers Mountains to those near Mt. Verhage described by Bradshaw et al. (1982) and Wodzicki et al. (1982) is uncertain. Findlay (1986) favors correlating the two generations between the areas. However, the predominant schistosity of the Mt. Verhage area is second generation and no crenulation cleavage has been recorded, whereas in the areas to the northwest and southeast, the predominant schistosity is first generation, followed by a crenulation.

A third-generation deformation is recognized in the northern Bowers Mountains by conjugate kink folding with spaced axial-planar zones of chlorite (Jordan et al., 1984).

## Admiralty Intrusives

Stemming from the work in the Tucker Glacier area, Harrington (1958) introduced the name "Admiralty Intrusives" for all plutonic rocks in the basement of the Transantarctic Mountains. The plutons that they examined were described by Harrington et al. (1964, 1967). Following dating by Starik et al. (1959), which demonstrated that plutonic rocks in the eastern portion of northern Victioria Land were slightly older than 300 Ma, and dating by several authors (Goldich et al., 1958; Angino et al., 1962; Pearn et al., 1963) in southern Victoria Land, which gave ages around 500 Ma, Grindley and Warren (1964) recognized the distinction between the rocks in this part of northern Victoria Land and those farther to the south, and restricted the term "Admiralty Intrusives" to those plutonics of Devonian age while introducing the name "Granite Harbour Intrusives" for the Cambro-Ordovician suite. Subsequent mapping throughout the Transantarctic Mountains has demonstrated that the Admiralty Intrusives are restricted to northern Victoria Land, while the Granite Harbour Intrusives occur throughout the range.

The Admiralty Intrusives crop out as discrete plutons throughout the Bowers and Robertson Bay terranes. The pluton at Mt. Supernal possibly crosscuts the boundary between Bowers and Wilson terranes, as discussed later, but this remains controversial. The Salamander Granite complex exposed at the southern end of the Salamander Range, previously included in the Admiralty Intrusives, has recently been given separate status and will be discussed in a separate section.

Unlike the Granite Harbour Intrusives in northern Victoria Land, or elsewhere in the Transantarctic Mountains for that matter, the Admiralty Intrusives share a number of characteristics, lending themselves to generalization. As can be seen on a geological map, the Admiralty plutons are distributed in three belts trending roughly eastnortheast in the northern, central, and southern portions of the region. Throughout their occurrence the plutons are epizonal, crosscuting the country rocks discordantly, with contact aureoles up to a few tens of meters. They are also uniformly lacking in deformation effects, indicating posttectonic emplacement of the suite, although flow foliation close to contacts is present at places (Le Couteur and Leitch, 1964).

Compositionally, the Admiralty Intrusives range from monzogranite to tonalite and diorite, but they are predominantly granodiorites (Borg et al., 1986b). In most regards they can be described as I-types (Wyborn, 1981; Borg et al., 1986b;

Vetter and Tessensohn, 1987) by the classification of Chappell and White (1974). Textures are typically medium- to coarse-grained, hypidiomorphic granular. Plagioclase is euhedral to subhedral surrounded by interstitial quartz and K-feldspar. In general the rocks are equigranular, but minor occurrences are porphyritic with sparse phenocrysts of orthoclase microperthite up to 1 cm in length. Biotite is ever present and hornblende is a common constituent. Sphene and allanite are common, but not universal, accessory phases. Zircon, apatite, and opaques are ubiquitous.

From the standpoint of geochemistry, all of the analyzed samples of Admiralty Intrusives have $SiO_2 > 61\%$, and border on being calc–alkaline to calcic via the Peacock index (Borg et al., 1986b). Samples are metaluminous to mildly peraluminous (all $Al_2O_3:K_2O + Na_2O + CaO \leqq 1.1$). Xenoliths occur at most localities but are generally a minor component ($<2\%$). Most of these are medium-grained, equigranular hornblende + plagioclase, and only rarely are xenoliths of metasedimentary aspect observed.

Although the Admiralty Intrusives are characterized by a uniformity of petrology and emplacement features, Borg et al. (1986b, 1987b; Borg and DePaolo, 1991) have noted regional variations in geochemical and isotopic trends in the study described in the section on Granite Harbour Intrusives (Fig. 2.11). For the Admiralty Instrusives, compositional polarity indicates an increasing involvement of crustal material in their generation in a northnortheasterly direction, decidedly different from that of the Granite Harbour Intrusives and Wilson terrane.

The period of generation of Admiralty Intrusives was fairly short and synchronous, as indicated by numerous K/Ar mineral dates (hornblende, muscovite, biotite) that fall mainly in the interval 370–350 Ma (Kreuzer et al., 1981, 1987; Vetter et al., 1983; Stump et al., in press). Older dates outside this range include a Rb/Sr mineral isochron (WR–biot–plag–Ksp) from the Mt. Adam pluton with an age of $379 \pm 7$ Ma ($IR = 0.7093$, $MSWD = 0.16$) (Stump et al., in press), a Rb/Sr whole-rock–biotite date of $385 \pm 15$ Ma from the Mt. Burrill pluton (Stump et al., in press), and a three-point Rb/Sr whole-rock isochron from the Gregory Bluffs pluton (Yule Bay) with an age of $388 \pm 13$ Ma ($IR = 0.7141$, $MSWD = 1.7$). Each of these samples may indicate slightly older generation of magma than the bulk of the Admiralty Intrusives.

On the younger side of the age range, a K/Ar date on hornblende of $327 \pm 10$ Ma is reported by Nathan (1971a). This is discordant with a biotite date of $354 \pm 10$ Ma from the same locality. The author offers no expanation for this anomalous discordance, so the younger date must be viewed with skepticism. A Rb/Sr whole-rock–biotite date of $364 \pm 20$ Ma has been obtained from Mt. Supernal by Stump et al. (in press). In light of this and the biotite age, it is reasonable to consider the Mt. Supernal pluton to be of the same vintage as most of the rest of the Admiralty Intrusives.

## Northern Zone

**Northern Coastal Area.** In the same field season (1957–58) that Harrington et al. (1967) were collecting and naming the Admiralty Intrusives in the Tucker Glacier area, the Russians were working from shipboard along the Oates

Coast, where they collected Admiralty Intrusives from landings at Znamenskii and Sputnik Islands (Ravich et al., 1965). The rocks were found to be biotite granites, and two chemical analyses were published. As mentioned earlier, dating of these two occurrences gave the first indications that the plutonic rocks in the eastern portion of northern Victoria Land are of Devonian age, distinctly younger than the Granite Harbour Intrusives in the rest of the Transantarctic Mountains (Starik et al., 1959, 1961; Ravich and Krylov, 1964).

Several years later Australian geologists put in at Cape North and collected massive, medium-grained, biotite granodiorites from Nella Island, Thala Island, Gregory Bluffs, and Arthurson Bluff (McLeod, 1964; McLeod and Gregory, 1967). K/Ar dating of biotite from the occurrences at Nella and Thala Islands (351 and 363 Ma) affirmed the Devonian ages of these intrusives (Webb et al., 1964).

At the time of publication of the Antarctic Map Folio Series, the area between Zykov and Dennistoun Glaciers remained blank except for the landfall at Cape North. Mapping by GANOVEX I revealed that the greater area around Yule Bay is composed of a batholith of Admiralty Intrusives. Wyborn (1981) named it the Yule Bay batholith and tentatively identified four plutons: Gregory Bluff, Missen Ridge, Ackroyd Point, and Tapsell. The first three are massive gray granites, the Gregory Bluffs and Ackroyd Point contain hornblende and biotite, and the Missen Ridge, biotite alone. The Ackroyd Point biotite is red-brown, while biotite in the other plutons is dark brown. The Missen Ridge pluton is somewhat more heterogeneous and contains appreciable leucocratic dikes in its central portion. The Tapsell pluton is coarser-grained than the others and is also porphyritic, with K-feldspar phenocrysts up to 5 cm in length. Sharp, nearly vertical contacts of the plutonic rocks with Robertson Bay Group are exposed on both the lower and upper O'Hara Glacier.

Sheared in zones and foliated throughout, a biotite granodiorite crops out at Surgeon Island at the mouth of Yule Bay. The rock contains an appreciable amount of muscovite (to 10%), which Wyborn (1981) considered to be secondary in origin. Vetter et al. (1983, 1984), however, identified the rock as S-type, citing 2.5% normative corundum. There is some uncertainty about the age of this rock, for of the six samples that Vetter et al. (1984) analyzed for Rb/Sr only four fall on a straight line, whose slope indicates an age of $599 \pm 21$ Ma (MSWD = 0.5) and IR of around 0.760. Nevertheless, this age is distinctly older than that of any Admiralty Intrusives, or for that matter Granite Harbour Intrusives in northern Victoria Land. Kleinschmidt and Tessensohn (1987, p. 98) speculate that this rock is a "continental sliver caught up in the subduction process." Borg and DePaolo (1991) suggest that the Surgeon Island granodiorite is distinct and that it may belong to a separate terrane on the northeastern margin of northern Victoria Land, their Surgeon Island terrane. No contact is seen with other rocks.

The GANOVEX Team (1987) mapped a scattering of tiny plutons to the west of the Yule batholith, at Mt. Knaak, Acapulco Cliff, and Cooper Spur. The Cooper Spur pluton is a massive, hornblende-bearing, I-type granite, similar in most ways to the Admiralty Intrusives. However, a four-point Rb/Sr whole-rock isochron yielded an age of $525 \pm 15$ Ma (MSWD = 0.7) (Vetter et al., 1984). A contact of folded Robertson Bay Group crosscut by plutonic rocks was observed on a cliff face, separated by cover from outcrops where sampling was done, but it seems likely that the plutonic rocks are continuous.

The situation is enigmatic. If the date is even approximately correct, as it appears to be, then this pluton is distinctly older than Admiralty Intrusives elsewhere. Furthermore, the contact relations require that the sedimentation and deformation of the Robertson Bay Group be older than approximately 525 Ma. This is at odds with the age of the Handler Formation at Handler Ridge as dated by fossils (Burrett and Findlay, 1984; Wright et al., 1984), although it accords more with the acritarch data of Cooper et al. (1983), indicating that the Robertson Bay Group may span most or all of the Cambrian. However, if the observation that the Cooper Spur granite crosscuts *folded* Robertson Bay Group is correct, then deformation of older portions of the Robertson Bay Group must have begun before deposition of younger portions. Alternatively, the sedimentary rocks at Cooper Spur may be an older sequence than the Robertson Bay Group and part of the Surgeon Island terrane, as suggested by Borg and DePaolo (1991).

**Lillie Batholith.** A sizable body of plutonic rock (480 km$^2$), named the Lillie batholith by Sturm and Carryer (1970), crops out at the northwestern end of the Everett Range between the Lillie and Ebbe Glaciers. A probable extension of this batholith occurs to the southwest around Champness Glacier (the Champness Granodiorite of Dow and Neall, 1974). The body is massive and gray, and apparently composite, with compositions ranging from monzogranite through granodiorite to tonalite (Vetter et al., 1983; Borg et al., 1986b). Hornblende is present in most of the phases. Veins and dikes are apparently more numerous in this batholith than most other bodies of Admiralty Intrusives (Wyborn, 1981). Of note in the area to the west of Mt. Dockery are a number of irregular, steeply dipping quartz veins containing molybdenite crystals up to 2 cm in length (Wyborn, 1981). A sharp, subhorizontal contact of the Champness Granodiorite with overlying Robertson Bay Group is well exposed on Griffith Ridge (Vetter et al., 1983).

## Central Zone

**Pearl and Inferno Peak Plutons.** A pair of hornblende-bearing granodiorite plutons crops out around the head of Pearl Harbour Glacier. Le Couteur and Leitch (1964) named these the Pearl pluton for the body centering on Mt. Holdsworth, and the Inferno Peak pluton for the one centering on Inferno Peak. A phase of pink granite with phenocrysts of orthoclase occurs along the western border of the Inferno Peak pluton. Leucocratic dikes are fairly common, particularly toward the margins of the plutons.

**Mt. Midnight Pluton.** Harrington et al. (1967) mapped several plutons in the area surrounding Tucker Glacier and Moubray Bay, all of which they interpreted as exposures of a continuous stratiform sheet belonging to their Tucker Granodiorite. Le Ceuter and Leitch (1964) and Crowder (1968) disputed this interpretation, saying that the plutons that they examined were discrete bodies. The pluton at the confluence of Tucker and Leander Glaciers was named the Mt.

Midnight pluton by Le Ceuter and Leitch (1964). Although Le Couteur and Leitch (1964) and Harrington et al. (1967) describe it as gray coarse-grained granodiorite, Borg et al. (1986b) have collected medium-grained monzogranite from its northwestern side.

**Mt. Adam Pluton.** Admiralty Intrusives that may be an extension of the Mt. Midnight pluton, or a separate body, crop out in the vicinity of Mt. Adam, Mt. Royalist, and Mt. Ajax, the highest peaks in the interior of northern Victoria Land, rising conspicuously above the accordant peaks and ridges of most of the Robertson Bay terrane. These rocks were observed in passing by Harrington et al. (1967) and Crowder (1968). They were collected by Borg et al. (1986b) and mapped by Findlay and Jordan (1984). The rock type is uniformly described as gray hornblende-bearing granodiorite. A tiny pluton of Admiralty Intrusives was mapped at Mt. Cherry Garrard by the GANOVEX Team (1987).

## Southern Zone

A third tier of Admiralty plutons crops out in the southern portion of northern Victoria Land, where they intrude not only the Robertson Bay terrane, but also the Bowers terrane and possibly the Wilson terrane.

**Honeycomb Ridge.** The easternmost body occurs at Honeycomb Ridge adjacent to Moubray Bay. Harrington et al. (1967) describe this body as somewhat different from other occurrences of Admiralty Intrusives, owing to numerous inclusions, as well as both melanocratic and leucocratic dikes. They provide several sketches showing field relations of the various generations of dikes and veins. Borg et al. (1986b) have analyzed tonalite from the southern end of the body.

**Tucker Pluton.** The main mass of Harrington et al.'s (1964, 1967) Tucker Granodiorite crops out in a pluton around Mt. Northampton and Oread Spur. Flow banding is prevalent in portions of the body. It includes both hornblende-bearing granodiorite and biotite-bearing monzogranite (Borg et al., 1986b).

The outlines of three plutons were mapped through aerial reconnaissance by Crowder (1968), one on the ridge dividing Rudolph and Stafford Glaciers, one at the head of Borschgrevink Glacier centering on Mt. Lepanto, and one at the head of Hand Glacier centering on Mt. Burrill. The GANOVEX Team (1987) refined these boundaries. Borg et al. (1986b) described granodiorite from the Mt. Lepanto pluton, and both granodiorite and tonalite from the Mt. Burrill pluton, all of which contain hornblende and biotite.

Cropping out around Cape Crossfire and intruding Bowers terrane is a body of Admiralty Intrusives that may connect beneath the Cenozoic volcanics of the Malta Plateau with the Mt. Burrill pluton (GANOVEX Team, 1987). A small pluton of medium-grained granite intrudes Mariner and Leap Year Groups on the cliffs above Mariner Glacier on the west side of Lawrence Peaks (GANOVEX Team, 1987).

**Supernal Granite.** The Supernal Granite was mapped and named by Riddols and Hancox (1968) for the pluton that comprises the Mt. Supernal massif and extends in an irregular fashion eastward to the area around Mt. Montreuil. Borg et al. (1986b) describe hornblende–biotite granodiorites from the former area and a tonalite from the latter.

Borg and Stump (1987) examined contacts at the southern end of the Supernal massif and at the end of the ridge trending eastnortheast from the main summit. At both places the Supernal Granite intrudes greenschist-grade mafic volcanic and volcaniclastic rocks. The GANOVEX Team (1987) did not visit either of these localities, but mapped the end of the spur immediately to the north of the latter as Dissent Formation. This is presumably the locality cited by Riddols and Hancox (1968) in which the Supernal Granite intrudes "Robertson Bay Group" (later identified as Sledgers Group throughout the upper Mariner Glacier area).

## Salamander Granite Complex

A composite pluton of granite crops out in the southern portion of the Salamander Range. The rocks were first sighted by Le Couteur and Leitch (1964). They were visited briefly by Crowder (1968), who described a biotite adamellite. In their synthesis of northern Victoria Land, Gair et al. (1969) mapped the pluton as belonging to the Admiralty Intrusives. Dow and Neall (1972) depicted the rocks tentatively as Granite Harbour Intrusives, but in Dow and Neall (1974) their map shows the area as "not mapped." The second field party to visit these intrusive rocks was Laird et al. (1974), who analyzed returned samples. Although the authors' chemical and modal data indicated monzogranites, they named the pluton the Salamander Granodiorite. A K/Ar biotite date of 337 ± 7 Ma warranted assigning the rocks to the Admiralty Intrusives.

Borg et al. (1986b) mapped and collected the pluton in detail, suggested that "Salamander Granite Complex" is a more apt name, and distinguished it from the Admiralty Intrusives on the basis of petrography, geochemistry, and geochronology. In the field the Salamander Granite Complex contains two distinct phases, whose contact is horizontal or slightly domal, and is marked by a zone, 2–3 m thick, of fine-grained aplite containing abundant mirolitic cavities. No internal contacts were found within either the upper or lower phase, so there is no evidence for more than two plutons within the complex. The contacts between the aplite and the upper and lower phases are vague and may at places be gradational; consequently the relative ages of the upper and lower phases are uncertain from field relations.

No contact with the country rock surrounding the Salamander Granite Complex is exposed in the field. Lanterman metamorphics crop out across a snowfield to the northeast of the granites and throughout the rest of the Salamander Range. The GANOVEX Team (1987) drew an extension of the Lanterman fault between the granites and the metamorphics on their map, implying that the Salamander Granite Complex is in the Bowers terrane, but whether it is there, intrudes the Wilson terrane, or straddles the two is an unresolved question. A sill of Ferrar dolerite cuts between the phases in the vicinity of Mt. Apolotok.

The lower phase of the Salamander Granite Complex is composed of subequigranular to porphyritic hornblende–biotite monzogranite. Although modes somewhat overlap those of the Admiralty Intrusives, the lower phase is in general more abundant in alkali feldspar. Quartz grains are large, rounded, and occasionally embayed, unlike quartz found in the Admiralty Intrusives. The groundmass of this phase is generally medium-grained, but has an unusual feature, also distinct from that seen in the Admiralty Intrusives, in that it contains pea-sized clots of fine-grained plagioclase, biotite, and rare hornblende. Another distinctive feature is scattered pockets or bubbles of pegmatites ranging from a few tens of centimeters to several meters in diameter (Fig. 2.23). These pegmatites contain euhedral quartz + feldspar ± muscovite growing into open space, implying a high level of emplacement.

The upper phase of the Salamander Granite Complex is composed of biotite monzogranite to syenogranite. The upper phase is more equigranular and coarser-grained than the lower phase, and is decidedly more felsic, with pink feldspar and smokey quartz comprising 98% of the rock. Borg et al. (1986b) have modeled the upper phase as resulting from fractional crystallization of the lower phase. Although the model is an oversimplification, the implication is that the two phases are genetically related.

Stump et al. (in press) presented two Rb/Sr dates on the Salamander Granite Complex, which were interpreted as the time of final cooling. A Rb/Sr whole-rock–mineral isochron (WR–biot–plag–Ksp) on the lower phase yielded an age of $319 \pm 5$ Ma (MSWD = 0.49), and a Rb/Sr whole-rock–biotite age of $319 \pm 26$ Ma was obtained on the upper phase. These ages, as well as the 337-Ma K/Ar date of Laird et al. (1974), are considerably younger than any of those on the Admiralty Intrusives, and together with the petrographic and geochemical data indicate that the Salamander Granite Complex is distinct from the widespread Admiralty Intrusives.

## Mid-Paleozoic Volcanism

Calc–alkaline volcanism occurred at local sites in each of the three terranes in northern Victoria Land during the mid-Paleozoic, at Gallipoli Heights in the Wilson terrane, Lawrence Peaks in the Bowers terrane, and Mt. Black Prince in the Robertson Bay terrane. The rocks at Gallipoli Heights have been variously called Gallipoli porphyries (Sturm and Carryer, 1970), Gallipoli rhyolites (Dow and Neall, 1972, 1974), and Gallipoli volcanics (Grindley and Oliver, 1983). The last term is more generic and better encompasses the variety of rock types found in the complex. The rocks unconformably overlie eroded Granite Harbour Intrusives at one locality, where the contact is vertical, and show evidence of some syn- or postvolcanic movement (Stump et al., 1983a).

In their northern exposures where the unconformity is exposed, the Gallipoli volcanics are about 2 km thick and steeply tilted, with dark andesitic flows and agglomerates in the basal portion interbedded with breccias and conglomerates containing granitic clasts (Grindley and Oliver, 1983). The andesites are followed by dacite flows and ignimbrites interbedded in the upper portion with volca-

**Figure 2.23.** Pocket pegmatite in Salamander Granite, east of Mt. Apolotok. Pocket knife for scale.

niclastic sandstones and carbonaceous siltstones. The tilting is probably synchronous with and related to volcanic processes in the area.

In the southern area of the Gallipoli Heights, the volcanics are a flat-lying rhyolitic ignimbrite perhaps 250 m thick, bright reddish orange, and highly welded. Placioclase and embayed quartz phenocrysts are set in a devitrified matrix containing flattened pumice and remnant glass shards. Two vent areas characterized by large blocks of sandstone and shale in addition to reddish ignimbrite and dark-colored andesite have been identified (Grindley and Oliver, 1983). Dikes of various compositions cut much of the complex.

Faure and Gair (1970) analyzed two samples of rhyolite from the Gallipoli volcanics for rubidium and strontium, obtaining a two-point whole-rock isochron date of $367 \pm 39$ Ma, with IR of 0.7057. [The original date of $375 \pm 40$ Ma by Faure and Gair (1970), using $\lambda = 1.39 \times 10^{-11}$/yr, was recalculated in error by Grindley and Oliver (1983) as 382 Ma.]

Volcanic rocks at Lawrence Peaks and Mt. Anakiwa were discovered by Laird and Bradshaw (1982). Like the Gallipoli volcanics, the Lawrence Peaks volcanics contain rhyolites, dacites, and andesites, with the silicic rocks being dominant. Chemical analyses indicate that the rocks belong to the high-potassium calc–alkaline series (Weaver et al., 1984b). The rocks include flows and pyroclastics, as well as volcanic breccias. No contact with underlying Bowers Supergroup has been seen, but an unconformity is inferred on the basis of the topographic relationship of the two units. Portions of the Lawrence Peaks volcanics dip at shallow to moderate angles. Weaver et al. (1984b) ascribe this to folding, whereas Findlay and Jordan (1984) suggest that it represents an original emplacement feature.

The Mt. Black Prince volcanics were first described by Findlay and Field (1982a) and later by Findlay and Jordan (1984). The rocks are predominantly altered andesite and basaltic andesite, with lesser dacite and rhyodacite, and interbedded volcaniclastic units. Portions of the andesites are vesiculated. The geochemistry of the rocks shows them, like the Lawrence Peaks volcanics, to have high-potassium calc–alkaline affinities.

Thin, interbedded sediments from the north side of Mt. Black Prince contain plant fossils identified as Givetian–Lower Carboniferous in age (Findlay and Jordan, 1984). Two whole-rock K/Ar dates of $323 \pm 6$ and $375 \pm 6$ Ma have been reported by Adams et al. (1986).

An as yet unvisited contact between Black Prince volcanics and the Mt. Adam pluton (Admiralty Intrusives) occurs across 1,800 m of relief on Mt. Ajax. Findlay and Field (1982a) allowed that it might be either intrusive or unconformable; however, Findlay and Jordan (1984) suggested on the basis of some granite pebbles present in the basal portion of the volcanics that the volcanics were erupted onto the granite. The date of $379 \pm 7$ Ma from the Mt. Adam pluton (Stump et al., in press) is not helpful in determining the relative age because of the disparity in the two K/Ar dates from the volcanics.

## Terrane Boundaries

### Robertson Bay–Bowers Terrane Boundary

The boundary between much of the Robertson Bay and Bowers terranes is a remarkably linear feature expressed by narrow northnorthwest-trending glaciers of topographic contrast throughout most of its more than 350-km length. Two parallel faults were inferred by Le Couteur and Leitch (1964) beneath the ice southwest of the Millen Range, merging to a single fault passing along the northeast side of the East Quartzite Range. This inferred fault was extended northward by Gair et al. (1969) and Sturm and Carryer (1970) to the northern end of Edlin Névé. Dow and Neall (1974) named the feature the Leap Year Fault. By that time the fault remained unobserved due to cover by ice even though Robertson Bay Group and Bowers Supergroup had been traced to within a few hundred meters of each other at several localities (Dow and Neall, 1974).

The first observation of the Leap Year Fault in outcrop was made in 1981–82 along the ridge connecting Mt. McCarthy and Mt. Burton (Laird and Bradshaw, 1982; Wright, 1982; Stump et al., 1983b). There the occurrence consists of a zone approximately 500 m in width separating Molar Formation on the southwest from Robertson Bay Group (Millen Schist) on the northeast. The rocks in the fault zone are limonite-stained and intensely sheared, with a nearly vertical orientation. Subsequently, two other outcrops of the fault contact with similar nearly vertical shearing were observed in the northern Bowers Mountains near Mt. Hager and Mt. Bruce (Jordan et al., 1984).

Thrust faulting of a different character is recognized in the vicinity of the Millen Range. The first observation of low-angle faulting, probably from helicopter, was made by Crowder (1968, p. D101), who photographed the arching thrust slice on the northwest face of Crosscut Peak (erroneously identified as Mt. Aorangi). This fault and its extension, which does crop out on Mt. Aorangi, were examined and much photographed by Findlay and Field (1982b, p. 18, 1983, pp. 108, 110), Wright and Findlay (1984, p. 112), Bradshaw et al. (1985a, p. 2), and Findlay (1986, p. 107) (Fig. 2.24).

The rocks in the upper plate are metabasites, probably equivalent to the Glasgow Volcanics (Bradshaw, 1989). Rocks of the lower plate are Millen Schist. Southwest of the summit of Mt. Aorangi the fault dips steeply to the southwest. Its eroded trace presumably arches to the northeast, for the upper plate crops out with shallow northeasterly dip on the ridge system northeast of Mt. Aorangi. About 5 km to the northwest on Crosscut Peak, the fault arches from shallow southwest through shallow northeast dips. The fault appears to truncate folded beds within the Millen Schist, but the arching of the fault trace is almost certainly due to later-stage upright folding, making the fault a relatively early structure.

On Handler Ridge, about 10 km southeast of the above-mentioned localities, another fault juxtaposes Sledgers Group over Handler Formation of the Robertson Bay Group (Bradshaw et al., 1985a; Wright and Brodie, 1987). This is the northeasternmost incursion of Bowers Supergroup onto Robertson Bay Group in northern Victoria Land. In the 2–3 km of outcrop of Handler Formation approaching the fault, development of cleavage intensifies and minor folds change from plunges around 20° to nearly vertical. A zone of intense shearing approximately 100 m thick marks the fault, with stretching lineations and sense-of-shear indicators demonstrating reverse dip–slip movement in a west-over-east sense.

Findlay (1992) connects a fault exposed at Turret Ridge with the fault on Handler Ridge, interpreting it as a separate terrane boundary between Millen Schist and Robertson Bay Group. This suggestion is rejected by Wright and Dallmeyer (1992), who argue that the fault on Handler Ridge is a splay of the thrust system cropping out on Mt. Aorangi and Crosscut Peak and that the transition from Millen Schist to Robertson Bay Group is gradational.

The original mapping by Le Couteur and Leitch (1964), followed by Gair et al. (1969), showed all the rocks from Pyramid Peak southeastward through the

**Figure 2.24.** Thrust fault juxtaposing metabasites over quartzofeldspathic Miller Schist. View is to the southeast. Joice Icefall shows to lower right, and the high point in the background is Mt. Aorangi. Photo by Tom Wright.

Barker Range as Robertson Bay Group. Later mapping by Laird et al. (1974) was in agreement with the interpretation that Robertson Bay Group occurs in the vicinity of Pyramid Peak. Although subsequent mapping in the Barker Range indicated that the rocks south of Mt. Roy may be part of Sledgers Group, it was uncertain whether the rocks around Mt. Watt should remain mapped as Robertson Bay Group or as Molar Formation (Field and Findlay, 1982; Laird and Bradshaw, 1982, 1983). Accepting that some or all of the rocks between Pyramid Peak and Mt. Watt are Robertson Bay Group, the zone of intermingling of Bowers and Robertson Bay rocks is up to 25 km wide in the vicinity of eastern Evans Névé. This alternation of units may be due to either a single undulating thrust plane or a series of thrust slices (Bradshaw et al. 1985a).

Throughout much of its trace, the boundary fault between Bowers and Robertson Bay terranes is linear and juxtaposes Robertson Bay Group and Camp Ridge Quartzite, the highest formation of the Bowers Supergroup. This requires 3–5 km of strata within the Bowers Supergroup to have been omitted along the contact, in contrast to the low-angle thrust faults in which Sledgers Group contacts Robertson Bay Group or Millen Schist (Bradshaw, 1987). The interpretation is that the low-angle thrust faulting is an earlier feature associated with the amalgamation of the terranes, while steep faulting occurred later (Gibson and Wright, 1985; Wright, 1985b), possibly with a component of strike–slip motion. The name "Leap Year Fault" is retained for the later, upright faulting. Bradshaw (1989) introduced the term "Millen Shear Zone" to describe the area of the terrane boundary, but this apparently includes the low-angle faults as well as the Millen Schist.

## Wilson–Bowers Terrane Boundary

The boundary between the Wilson and Bowers terranes is a complex zone of faulting that emerges from ice cover in the Lanterman and Mountaineer Ranges. The first published suggestion of the fault was that of Crowder (1968), who referred to the in press work of Sturm and Carryer (1970). These authors inferred a fault to the west of a zone of flattened conglomerates along the eastern portion of the Lanterman Range. They also suggested that a concealed fault lay beneath the eastern margin of the lower Rennick Glacier based on shear zones and slickensides at a number of localities in the western Bower Mountains. In their mapping of approximately the same area, Dow and Neall (1974) named the Lanterman fault for the segment passing through the eastern Lanterman Range and the Bowers fault for the segment along Rennick Glacier. Bradshaw et al. (1982) applied the more apt name "Lanterman fault zone." Mapping by Bradshaw et al. (1982), Gibson (1984, 1987), and Cooper et al. (1990) has shown that the zone adjacent to the Lanterman fault, particularly in the Bowers terrane, is characterized by numerous fault slivers juxtaposing a variety of units, some with opposite facing directions (Fig. 2.19).

In their synthesis of northern Victoria Land, Gair et al. (1969) carried the Lanterman fault to the east of the entire Salamander Range, whereas Dow and Neall (1974) mapped it cutting through the southern end of the range to the west of Mt. Apolotok. The latter positioning was favored by the GANOVEX Team (1987). Gair et al. (1969) also extended the fault to the head of Mariner Glacier, where they showed Admiralty Intrusives faulted against Robertson Bay Group. This southern locality has since been mapped as an intrusive contact of Admiralty Intrusives into Bowers Supergroup (GANOVEX Team, 1987). Two small nunataks on the eastern side of Evans Névé (spot height $2,480 \pm$ m) were mapped as Bowers Supergroup by Laird et al. (1974), implying that the fault traces to the west of these outcrops.

The discovery of Bowers Supergroup on Spatulate Ridge (Stump et al., 1983b) extended the boundary between the Wilson and Bowers terranes to the Ross Sea. The details of its location were mapped by GANOVEX III through the Mountaineer Range (Fig. 2.10), where several splays were recognized, although actual outcrops of the main boundary fault appear to be exposed only on Whitcomb Ridge south of Mt. Supernal, where gneissic Murchison Formation contacts greenschists assigned to the Bowers Supergroup (Gibson et al., 1984, Kleinschmidt et al., 1984; GANOVEX Team, 1987). At several localities, shallow to moderate southwesterly dips in Mountaineer metamorphics steepen as the boundary fault is approached to the east, becoming nearly vertical at places.

An important issue with regard to amalgamation of the terranes in northern Victoria Land is whether the Devonian Admiralty Intrusives, which are unquestionably in the Robertson Bay and Bowers terranes, are also found in the Wilson terrane. If the latter is the case, then amalgamation of Bowers and Wilson terranes must have occurred by the time of intrusion. If not, then the timing of amalgamation is not constrained by this situation. The Mt. Supernal pluton is the only intrusion that may equivocally intrude the Lanterman fault and be in part within the Wilson terrane. The observed contacts of the Supernal granodiorite with greenschist-grade metavolcanics suggest that the pluton itself is intruding Bowers Supergroup (Borg et al., 1987b; Borg and Stump, 1987). Further, the high $\varepsilon_{Nd}$ of the Supernal Granodiorite suggests no contamination by older continental rocks such as those of the Wilson terrane. Also, the straight, steep boundary between the southwestern flank of the Supernal massif and the icefall feeding the upper Meander Glacier is suggestive of a fault. Nevertheless, the fault contact on Whitcomb Ridge, already cited, would appear to project directly into the Supernal pluton and has been shown doing so on interpretative maps by Tessensohn (1984) and others of the GANOVEX Team (1987).

Thrusting has been postulated as the primary type of faulting along the Lanterman fault (Gibson, 1984, 1985, 1987; Gibson et al., 1984; Gibson and Wright, 1985). Evidence cited includes stretching lineations and conglomeratic clasts elongate in a downdip direction, along with variably plunging mesofolds whose axes all lie within the plane of foliation. Although none of these features demonstrates sense of shear, and in the few places where the fault is exposed it is nearly vertical, it is generally thought that the amphibolite-grade metamorphics of the Wilson terrane have been brought up over the Bowers terrane, with the greenschist grade along the western margin of the latter due to juxtapositioning of hotter rocks over them. Bradshaw (1989), however, raises the possibility that the Lanterman fault is due to extension.

Elsewhere Bradshaw (1987) argues that although thrusting is well documented and would have been the final sense of movement on the Lanterman fault, strike–slip faulting along the boundary may also have been important earlier in its history.

## Assembly of Terranes

Since the initial recognition of terranes in northern Victoria Land (Weaver et al., 1984a), various geologists have attempted to model the history of events that led to their present configuration. Although Weaver et al. (1984a) and Bradshaw et al. (1985b) acknowledged the roll of subduction in producing the magmatic rocks of the Wilson and Bowers terranes, they preferred a model emphasizing strike–slip movement to explain juxtapositioning of the disparate geological features across the boundary faults. Recognition of structural evidence indicating convergent movement on both faults led to a spate of models with collision as the primary mechanism bringing about the amalgamation (Gibson, 1985, 1987; Gibson and Wright, 1985; Kleinschmidt and Tessensohn, 1987; Dallmeyer and Wright, 1992; Kleinschmidt et al., 1992a). Wodzicki and Robert (1986) proposed that convergence followed by strike–slip movement occurred along the Lanterman fault, and Bradshaw (1985, 1987, 1989) modified his interpretations to include thrusting at the boundaries, but did not abandon arguments for the need to include some strike–slip movement.

It is generally agreed that west-directed subduction beneath the Wilson terrane was responsible for the tectonic development of that area. Specifically the low-pressure/high-temperature metamorphism of the western ranges paired with the intermediate-level metamorphism of the eastern ones (Grew et al., 1984) and the increasing crustal involvement in the geochemical signatures of the Granite Harbour Intrusives in a westerly direction (Borg et al., 1987b). However, the lack of a high-pressure/low-temperature belt and the occurrence of granitic plutonics to within a few kilometers of the eastern boundary of the Wilson terrane argue for the removal of a portion of the terrane.

It is also generally agreed that the volcanics of the Sledgers Group were formed through arc magmatism above a subduction zone in an oceanic setting (Weaver et al., 1984a); however, different authors set the polarity of the subduction zone in opposite directions. At first the majority of models had it westerly-directed (e.g., Weaver et al., 1984; Bradshaw et al., 1985b; Gibson and Wright, 1985; Kleinschmidt and Tessensohn, 1987). Wodzicki and Robert (1986) were the first to suggest easterly subduction beneath the Glasgow arc based on their interpretation of the Sledgers basin sloping to the west or southwest and a continental source for the Molar Formation to the east or northeast. Gibson (1985, 1987) and Bradshaw (1987) have both swung in that direction – Gibson in explaining the source of Husky and Lanterman Conglomerates as an obducted slab of Sledgers Group over Lanterman metamorphics to the west, and Bradshaw in explaining an apparent subtle polarity in the Glasgow volcanics as more boninitic in the west and more backarc-related in the east.

Although the Glasgow volcanics are intraoceanic in character, a fraction of the associated Molar Formation includes clasts of continental derivation, indicating the influence of such a source. Regardless of the polarity of subduction beneath the Glasgow arc, it had apparently extinguished by the onset of Mariner Group deposition (late Middle Cambrian), at which time a continental source of sedimentary, volcanic, and plutonic rocks to the southwest became more influential (Laird, 1989). Limestone clasts in channelized deposits of the Sledgers and Mariner Groups indicate the proximity of a carbonate shelf that is not preserved in the outcrop record. The Leap Year Group bears evidence of sources to the southwest and southeast. The Camp Ridge Quartzite is quartzose, while the Carryer Conglomerate is polymict, but both lack metamorphic clasts typical of the rocks in the Lanterman Range, so some other source is suggested. Due to both the sedimentary associations already mentioned and the apparent incompleteness of the Glasgow arc, it is suggested that a portion of the Bowers terrane was tectonically truncated (Laird, 1989).

As discussed in the preceding section, the amalgamation of the Bowers and Robertson Bay terranes appears to have resulted from low-angle thrusting, producing the zone of Millen schists at the boundary, primarily in Robertson Bay Group, but including some interleaved Bowers Supergroup, with Bowers terrane overriding Robertson Bay terrane. The high-angle Leap Year fault occurred subsequently, possibly as a result of strike–slip movement (Wright, 1985). The uppermost Cambrian–Tremadocian fossils of the Handler Formation coupled with the approximately 500-Ma onset of cleavage development in the western portion of the Robertson Bay Group (Wright and Dallmeyer, 1991) indicates that the terminal sedimentation in the Robertson Bay Group was occurring at about the same time as deformation commenced. The diachronous development of cleavage from about 500 to 460 Ma across the Robertson Bay terrane suggests incremental convergence of the terrane from the east.

When the Bowers and Wilson terranes were juxtaposed is less certain. This must have followed deposition of the Leap Year Group, perhaps as early as Early Ordovician. The K/Ar dates from the eastern portion of the Wilson terrane are similar to those throughout the Ross orogen, and may signal the arrival of the Bowers terrane. If docking was later, the event is not indicated by younger cooling ages, and thrusting must not have caused significant crustal thickening and subsequent unroofing. On the other hand, the K/Ar dates from the Bowers terrane speard evenly from around 500 to less than 300 Ma (Adams and Kreuzer, 1984). It seems unlikely that these dates resulted from slow removal of overlying Wilson terrane rocks; otherwise some of the Lanterman metamorphics adjacent to the terrane boundary would also record similar cooling ages. The younger cooling ages in the Bowers terrane may be the result of Cretaceous thermal activity in northern Victoria Land, as indicated by paleomagnetic rejuvenation (Delisle and Fromm, 1984; Schmerer and Burmester, 1986), anomalous K/Ar dates on Jurassic Kirkpatrick Basalt (Elliot and Foland, 1986), and apatite fission-track dates (Fitzgerald and Gleadow, 1988). That thermal rejuvenation focused along the Bowers terrane may be due to some fundamental difference between its underpinning and that of the adjacent terranes. If the juxtaposition of the Wilson and Bowers terranes is indeed within an Ordovician time frame, then whether or not the Supernal pluton intrudes Wilson terrane is of local interest only.

# 3 Southern Victoria Land

## Geological Summary

Although southern Victoria Land is in close proximity to the bases on Ross Island and has been studied by geologists since the heroic era, the basement geology remains perhaps the most poorly understood of that in any major area of the Transantarctic Mountains. This is owing in large part to the complexity of the rocks. Another factor, however, is the nature of the terrain itself. Paradoxically, the area with the most complete exposure in the Transantarctic Mountains is the most inaccessible to geologists. Throughout the rest of the range, the glaciers that cover much of the exposure offer highways for snowmobiles (in the old days, dog teams) into every recess of the mountains. In the Dry Valleys, however, local camps must be established by helicopter and all ground travel must be by foot, severely limiting the kind of areal coverage that characterized the initial reconnaissance mapping throughout the rest of the Transantarctic Mountains, wherein single investigators observed broad areas and were able to develop a comprehensive understanding of whole regions.

The oldest rocks in southern Victoria Land comprise a suite of metamorphics, multiply deformed and primarily of amphibolite grade, which have defied more than the broadest stratigraphic correlation. Throughout the Dry Valleys area, metamorphic rocks belong to the Koettlitz Group, whereas in the Skelton Glacier area, where the grade is mainly greenschist, the metamorphics are called Skelton Group. Due to the lack of fossils, the age and correlation of these metamorphic rocks have been speculative. Recently a date of 800–700 Ma has been published for pillow lavas in the Skelton Group (Rowell et al., 1993b), giving an indication that those rocks are Neoproterozoic in age. Whether the Koettlitz and Skelton Groups correlate is also uncertain, although the arguments presented later suggest that they do.

The metamorphic rocks are interlayered with and crosscut by a succession of pre(?)-, syn-, and posttectonic plutonics, collectively called the Granite Harbour Intrusive Complex, the same nomenclature applied to plutonics of the Ross orogen throughout the Transantarctic Mountains. These rocks share a common cooling history, with K/Ar systems locking in around 500 Ma. The time of initial generation of the magma is less certain, but several U/Pb dates fall in the range of approximately 590–550 Ma. An undeformed pluton dated at 550 Ma crosscuts twice-folded Skelton Group (Rowell et al., 1993b). Although the plutonic episode appears to span the Cambrian, it is uncertain whether any compressive deformation occurred in southern Victoria Land during that interval, the latter portion of which is the traditional time of the Ross orogeny.

**Figure 3.1.** Upper Wright Valley viewed from Bull Pass, showing Lake Vanda and, behind to the left, the Dais. The area is underlain by various phases of Granite Harbour Intrusives. A mafic dike swarm cuts across the valley in the vicinity of the lake. Beacon Supergroup and sills of Ferrar dolerite top the basement in the high wall of the western Asgard Range.

## Chronology of Exploration

The contributions of the *Southern Cross* expedition had been the collection and description of a variety of rocks from moraines and in situ specimens of Robertson Bay Group and McMurdo volcanics from several islands, but the fundamental geological relationships in the Transantarctic Mountains were discerned by the subsequent expeditions led by Scott and Shackleton between 1901 and 1912. Although the goal of these groups, either implicit or explicit, was to push into the interior and reach the South Pole, scientific discovery was the fundamental purpose of the undertakings. Geologists, physicists, biologists, and medical doctors were members of the expeditions.

Scott's first expedition, 1901–4, sailed on the *Discovery*. Following the route of Ross and Borchgrevink, it reached the continent at Cape Adare and sailed southward as close to the coast as possible. But where previous ships had followed a more easterly course south of Cape Washington that carried them away from the mountains, the *Discovery* found open water and was able to hug the coastline down to the Drygalski Ice Tongue. The tabular nature of the Prince Albert Mountains was noted at a distance, as was the fact that the uplands were built of flat-lying sedimentary rocks. Farther south a landing was made at Granite Harbour, where the first in situ granite was recorded and collected. Although the expedition was tempted to make this secure harbor its base, Scott wanted to

**Figure 3.2.** Victoria Valley and Lake Vida viewed eastward from Sponsers Peak. Granite Harbour Intrusives underlie most of the area. Dikes cross-cut the foreground and the end of the St. Johns Range to the left. The "basement sill" of Ferrar dolerite cuts through granitic rocks of the Olympus Range to the right, more than 100 m below the Kukri peneplain.

continue as far south as possible to minimize the distance to the Pole. Having retraced Ross's course along the front of the barrier, and having gone up in the first balloon ascent on the continent, Scott returned to Ross Island. The sea ice had broken out as far around the island as the present-day McMurdo Station, and the *Discovery* anchored in the lee of Hut Point in the protected cove that is used today for docking tanker and supply ships for the U.S. base.

Some local sledging preceded the winter-over in 1902. Then in the spring parties set out in two directions. Those to the south laid depots in support of the polar party, Scott, Shackleton, and Wilson, who made it as far as Nimrod Glacier on 30 December 1902 before running short of rations. The others crossed McMurdo Sound with the goal of finding a route through the mountains to the interior. It was on these western forays that geological observations were conducted. H. T. Ferrar was the geologist for the expedition, and he accompanied several of the trips; wherever any party touched rock, collections were made.

The first crossing of McMurdo Sound in the spring of 1902 was led by Lieut. Armitage and included Ferrar and four seamen. They managed to make it partway up Ferrar Glacier before turning back due to rough ice conditions. Then between November and January Armitage led another party (not including Ferrar) partway up Blue Glacier, crossed a steep divide into the upper Ferrar Glacier (Descent Pass), and continued westward onto the polar plateau.

The following summer Scott led a party of nine up Ferrar Glacier and onto the

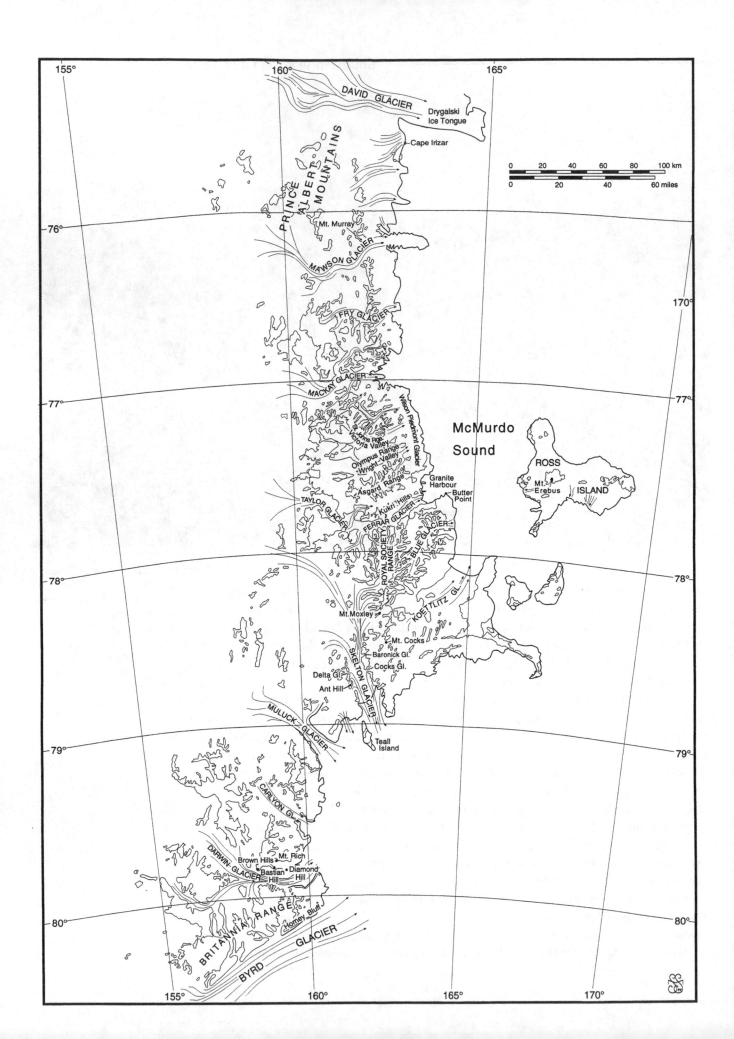

DAVID GLACIER

Drygalski
Ice Tongue

Cape Irizar

0    20    40    60    80    100 km

0         20        40        60 miles

PRINCE ALBERT MOUNTAINS

Mt. Murray

MAWSON GLACIER

FRY GLACIER

MACKAY GLACIER

Wilson Piedmont Glacier

McMurdo
Sound

ROSS

Mt.
Erebus

ISLAND

Jones Rge
Victoria Valley

Olympus Range
Wright Valley

Asgard Range

Granite
Harbour

Butter
Point

TAYLOR GLACIER

Kukri Hills

FERRAR GLACIER

BLUE GLACIER

ROYAL SOCIETY RANGE

KOETTLITZ GL.

Mt.Moxley

Mt. Cocks

Baronick Gl.

Cocks Gl.

SKELTON GLACIER

Delta Gl.

Ant Hill

MULUCK GLACIER

Teall
Island

CARLYON GL.

DARWIN GLACIER

Brown Hills   Mt. Rich

Bastian
Hill

Diamond
Hill

BRITANNIA RANGE

Horney Bluff

BYRD        GLACIER

plateau in an attempt to penetrate into the interior as far as possible. A three-man party led by Ferrar split off from the group at the head of the glacier and undertook a geological reconnaissance back through the mountains, including a side trip into the upper Taylor Valley along Taylor Glacier.

The field observations from the *Discovery* expedition were recorded by Ferrar (1907) and descriptions of the returned specimens by Prior (1907). All of the major rock groups were described, including the alkaline volcanics of the McMurdo Volcanic Group, the dolerites of the Ferrar Group, the sedimentary sequence of the Beacon Supergroup, and the metamorphic and plutonic rocks of the basement. The name "Beacon Sandstone" was applied at this time to the sedimentary rocks, and plant fossils were discovered therein, although they were too scrappy for positive identification. The marked unconformity between the Beacon rocks and the basement was also recorded. A peculiar sill of dolerite within the crystalline basement was observed running parallel to the unconformity several hundred meters beneath it. Subsequent investigations have shown that this "basement sill" occurs throughout much of the Dry Valleys area.

The basement rocks recorded by Ferrar (1907) and Prior (1907) included a metamorphic suite of conspicuous, white, "crystalline limestone" (i.e., marble), banded gneiss, and mica schist. Augen gneiss, thought to have been of igneous origin, was also described. Plutonic rocks were primarily granites, with biotite or biotite and hornblende. The rocks at Granite Harbour were described as gray biotite granite intruded by pink granitic veins.

Shackleton's expedition of 1907–9 sailed aboard the *Nimrod*. The appointed geologist was Raymond Priestley; however, in Australia the expedition was joined by Professor Edgeworth David and a young physics student, Douglas Mawson. Professor David was to return to Australia aboard the *Nimrod* once the expedition had been landed at Ross Island, but he was so well liked by the group that they prevailed upon him to remain in Antarctica with them. Shackleton had stated three goals for the expedition: to reach the geographic South Pole, to reach the South Magnetic Pole, and to explore Edward VII Land to the east of the previous expeditions' end points along the Ross Ice Shelf.

Before the first winter-over, a party of six including David and Mawson made the first ascent of Mt. Erebus, peered into its steaming crater, and collected some of the abundant anorthoclase crystals that form a skree at the summit.

The following summer Shackleton and his party of four trekked through the Gateway, up the Beardmore Glacier, and onto the polar plateau to within 180 km of the South Pole, collecting rocks along the Beardmore upon their return. At the same time a party of three, including Priestley, retraced the route of the previous expedition up the Ferrar Glacier, recording geological information and collecting samples. The greatest success of the *Nimrod* expedition, however, was that of the Northern Party, composed of David, Mawson, and A. F. Mackay, who reached the magnetic pole. They crossed McMurdo Sound to Butter Point and made geological observations northward along the coast, crossing the Drygalski Ice Tongue and ascending to the plateau via a small glacier named Backstairs Passage between Mt. Gerlache and Mt. Crummer.

The geological results of the *Nimrod* expedition are copiously recorded by David and Priestley (1914), with petrographic descriptions from southern Victoria Land by Mawson (1916) and shorter notes by Priestley and David (1912).

**Figure 3.3** (*facing page*). Location map of southern Victoria Land.

GRANITE

HARBOUR

MACKAY GLACIER

Cape
Roberts

WILSON PIEDMONT GLACIER

Marble
Point

NEW

HARBOUR

BENSON GLACIER

Sperm Bluff

MILLER GLACIER

DEBENHAM GLACIER

VICTORIA UPPER GLACIER

St. Johns Range

Sponsers
Pk.

Purgatory
Pk.

Lake
Vashka

VICTORIA VALLEY

Insel Range

Lake Vida

CLARK GLACIER

Mt. Doorly

Mt. Theseus

Mt. Loke

Bull Pass

OLYMPUS RANGE

WRIGHT UPPER GLACIER

Dais

Lake Nanda

ASGARD RANGE

RESERVE GL.

CROSSED GL.

COMMONWEALTH GL.

Mt.
Falconer

CANADA GL.

TAYLOR VALLEY

CRESCENT GL.

Nussbaum
Reigel

RHONE GL.

STOCKING GL.

Lake Bonney

KUKRI HILLS

Con-Rod
Hills

BLUE GLACIER

Pearse
Valley

Friis
Hills

WRIGHT VALLEY

TAYLOR GLACIER

Solitary Rocks

Cavendish
Rocks

Cathedral Rocks

FERRAR GLACIER

ROYAL SOCIETY RANGE

HOBBS GLACIER

Hobbs
Pk.

SALMON GL.

Garwood Valley

Marshall Valley

Miers Valley

WALCOTT GLACIER

Rucker Rdg.

RADIAN GL.

Mt. Dromedary

SKELTON GLACIER

KOETTLITZ GLACIER

159°

0   5   10   15   20 km

0   5   10   15 miles

S. Selkirk '94

Scott's ill-fated *Terra Nova* expedition (1910–13) continued the British tradition of geological and other scientific acomplishment. Following the placement of the expedition at Cape Evans, the *Terra Nova* sailed a party of four, including geologists T. Griffith Taylor and F. Debenham, across McMurdo Sound, dropping them at Butter Point. This party followed the established route up Ferrar Glacier and down Taylor Glacier. From the terminus of the glacier they explored central Taylor Valley as far as the eastern end of the Kukri Hills. Then after backtracking to the mouth of Ferrar Glacier, they examined the foothills of the Royal Society Range past Blue Glacier and up the west side of the Koettlitz.

The first full summer a geological party crossed McMurdo Sound to Butter Point and followed David's tracks to Granite Harbour, where they spent six weeks south of Mackay Glacier and inland as far as Mt. Seuss (Fig. 3.6). The results of work by this expedition on the pre-Beacon rocks in southern Victoria Land are found in Smith and Debenham (1921) and Smith (1924, 1964).

Following the heroic era of Antarctic exploration, expeditions did not return to southern Victoria Land until the advent of the IGY (1957–59). Aerial photography provided a base for geological studies and topographic mapping, and revealed that Victoria Valley was only the southernmost of a major network of ice-free valleys to the west of the Wilson Piedmont Glacier. As part of the Commonwealth Trans-Antarctic Expedition that crossed the East Antarctic Ice Sheet in 1957–58, New Zealand parties worked in southern Victoria Land during the 1955–56, 1956–57, and 1957–58 seasons. Geologists B. M. Gunn and Guyon Warren mapped a considerable portion of the area between Mawson and Mulock Glaciers during these seasons, and combined their field results with those of previous and contemporary parties to produce a monograph and a map at a scale of 1:250,000, which stand as the geological benchmarks for southern Victoria Land (Gunn and Warren, 1962).

With the advent of helicopter support from McMurdo Station following the IGY, geologists from New Zealand and/or the United States have worked in the McMurdo Sound area practically on a yearly basis. The initial reconnaissance mapping of Victoria Valley was done in 1957–58 by Webb and McKelvey (1959), two New Zealand undergraduate students from Victoria University-Wellington, who began a tradition of yearly Antarctic expeditions from that university, not to mention highly successful careers of their own in Antarctic geology. In 1958–59 McKelvey and Webb (1961, 1962) mapped in Wright Valley, while a U.S. party studied Victoria Valley (Hamilton and Hayes, 1960; Hamilton, 1961). In 1959–60 New Zealanders were back in Victoria Valley (Allen and Gibson, 1962). That season and the following, another U.S. party worked in the lower Taylor Valley (Angino et al., 1962). During 1960–61 a New Zealand party mapped the area between Blue and Koettlitz Glaciers (Blank et al., 1963). In 1962–63 another New Zealand party mapped both along the north side of Darwin Glacier (Haskell et al., 1964, 1965a) and in the middle and lower Taylor Valley (Haskell et al., 1965b). In 1964–65 two detailed investigations were carried out, one by a New Zealand party on the orbicular granodiorite in central Taylor Valley (Palmer et al., 1967) and the other, a structural one, by an Australian party in the Garwood Valley and at Nussbaum Riegel in Taylor Valley (Williams et al., 1971). In 1965–66 a New Zealand party mapped and sampled in

**Figure 3.4** (*facing page*). Location map of McMurdo Sound area, southern Victoria Land.

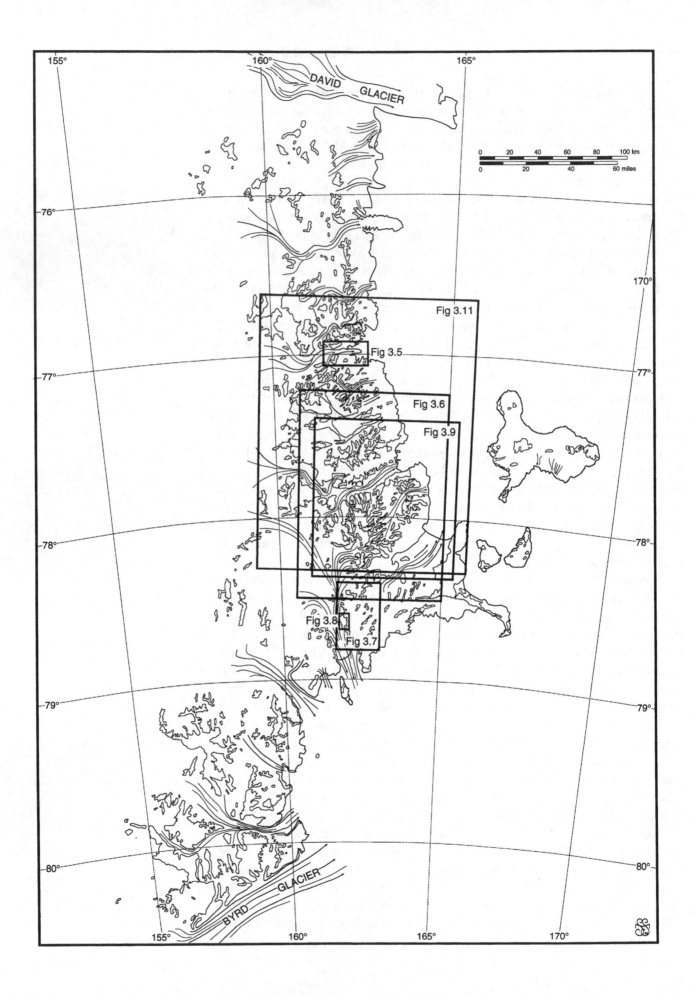

detail the Mt. Falconer pluton in the lower Taylor Valley (Ghent and Henderson, 1968; Ghent, 1970). Russian exchange scientists participating in the U.S. program undertook field studies throughout the Dry Valleys area in 1966–67 (L. V. Klimov) and 1968–69 (B. G. Lopatin), producing the first detailed geological map of the area (Lopatin, 1972) (Fig. 3.7). From 1967–68 to 1970–71 parties from the University of Wyoming undertook a series of detailed mapping projects throughout the area from Victoria Valley to Skelton Glacier (Smithson et al., 1969, 1970, 1971a, b, 1972; Murphy et al., 1970; Flory et al., 1971; Murphy, 1972). A comprehensive study of the metasedimentary rocks between Skelton Glacier and the Dry Valleys was conducted by New Zealand geologists between 1977 and 1981, resulting in a synthesis of information about the Koettlitz Group (Findlay et al., 1984). Recently New Zealand parties have conducted detailed mapping in an effort that may eventually lead to geological coverage of all of southern Victoria Land at a scale of 1:50,000 (McElroy and Rose, 1987; Woolfe et al., 1989; Allibone et al., 1991).

## Koettlitz and Skelton Groups

The oldest rocks in southern Victoria Land are multiply deformed metasediments referred to as the Koettlitz and Skelton Groups. Gunn and Warren (1962) originally designated Skelton Group for all pre-Devonian metasedimentary rocks throughout southern Victoria Land, but Grindley and Warren (1964) divided them into two groups, limiting Skelton Group to the greenschist-grade metasediments in the vicinity of Skelton Glacier and elevating the formational name "Koettlitz Marble" of Gunn and Warren (1962) to "Koettlitz Group" for the amphibolite-grade metasediments found in the region between Koettlitz and Mackay Glaciers.

One small outcropping of metasedimentary rock is known in the Prince Albert Mountains between Mackay and Mawson Glaciers in a narrow portion of the Transantarctic Mountains where the basement is otherwise all plutonic rock. This is at Mt. Murry, where gray semipelitic schists are rafted in a granodiorite (Ricker, 1964; Skinner and Ricker, 1968). These rocks, which have been metamorphosed to the pyroxene–hornfels facies, were included in the Priestley Formation cropping out 150 km farther north and first identified by the same authors.

The area between Skelton and Darwin Glaciers is thought to be underlain entirely by plutonic rocks; however, much of the Britannia Range between Darwin and Byrd Glacier is a poorly studied suite of amphibolite-grade gneisses that Borg et al. (1989) have called the Horney Formation.

## Skelton Group

Skelton Group occurs in limited outcrops at Teall Island, on both margins of Skelton Glacier, and around Mt. Cocks on the divide between Skelton and Koettlitz Glaciers (Fig. 3.8). Gunn and Warren (1962) identified two formations: Teall Greywacke and Anthill Limestone, designating a type locality for the former on east-central Teall Island and for the latter at Ant Hill on the west side of Skelton Glacier. Skinner (1982) erected the Cocks Formation for "greywacke-

**Figure 3.5** (*facing page*). Location map for figures in Chapter 3.

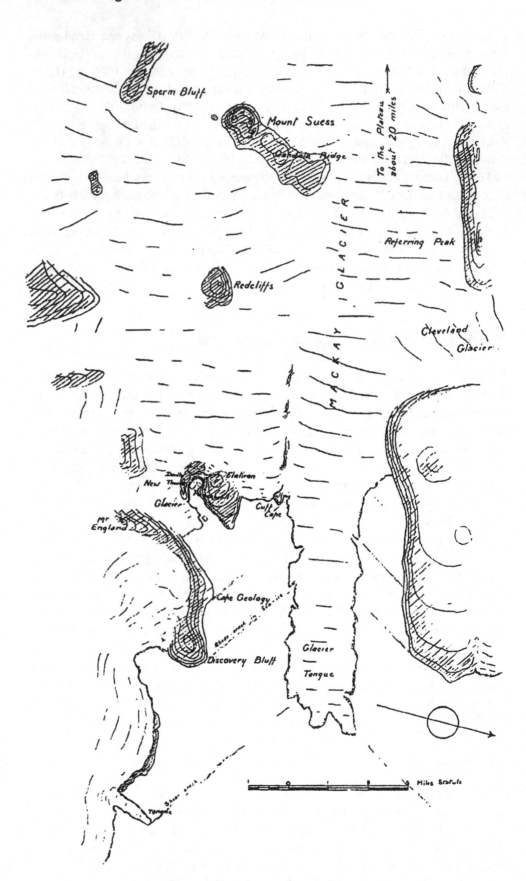

**Figure 3.6.** Sketch map of part of Granite Harbour by Smith (1924).

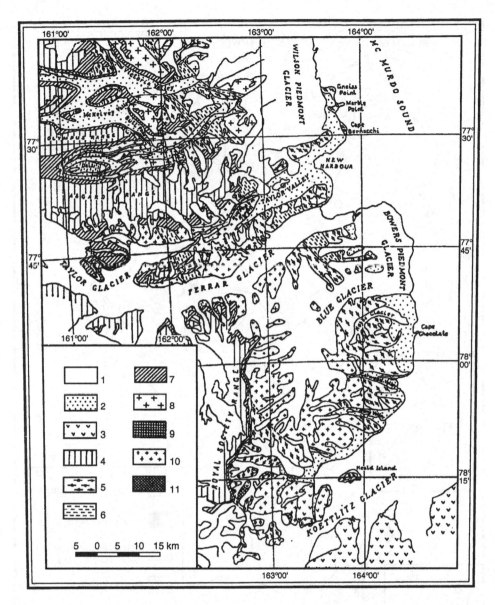

**Figure 3.7.** Geological sketch map of the area between Koettlitz Glacier and Victoria Valley. 1, Ice; 2, fluvio-glacial deposits; 3, volcanic complex; 4, Beacon Supergroup sediments; 5, marble–schist series; 6, gneiss series; 7, Ferrar dolerite; 8, posttectonic granite; 9, posttectonic diorite; 10, syntectonic quartz diorite, grano-diorite, and granite; 11, syntectonic gabbro-diorite. After Lopatin (1972). Used by permission.

like metasedimentary and metavolcanic rocks" with a type locality immediately north of the confluence of Cocks and Skelton Glaciers where it contacts Anthill Limestone. He also called for abandonment of the name "Teall Greywacke," arguing that the rocks on east-central Teall Island were of a similar calcareous content to calc–silicate rocks of Anthill Limestone and that other less calcareous lithologies at northern and southern Teall Island and Red Dog Bluff (the only other localities of Teall Greywacke mapped by Gunn and Warren) should be included in Cocks Formation. Skinner (1982) reinforced this proposal on struc-tural grounds by identifying three episodes of deformation in Anthill Limestone and the central Teall Island "Teall Greywacke," but only two episodes in Cocks Formation including the "Teall Greywacke" at northern and southern Teall Island and Red Dog Bluff.

Although Grindley and Warren (1964) originally distinguished Skelton Group

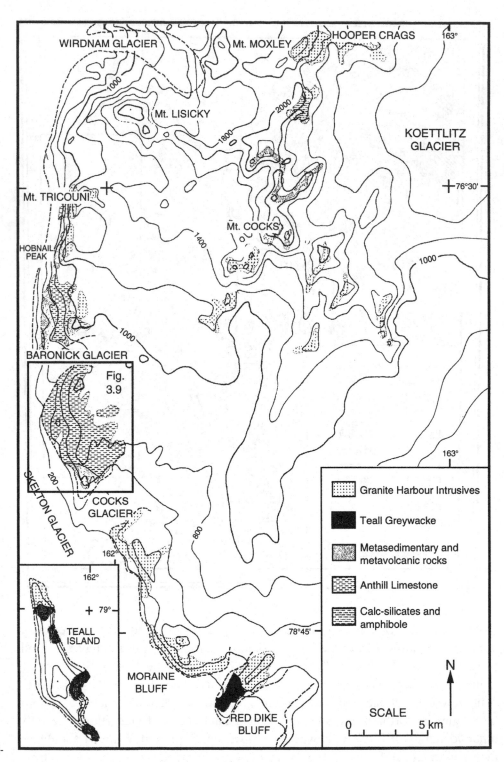

**Figure 3.8.** Generalized geological map of area between Skelton and Koettlitz Glaciers, with Teall Island inset. After Skinner (1982). Used by permission.

from Koettlitz Group on the basis of metamorphic grade (greenschist vs. amphibolite), mapping by Skinner (1982) has shown that metamorphism increases within Skelton Group in a northerly direction along the divide between Skelton and Koettlitz Glaciers, with amphibolite grade being reached near Mt. Moxley.

## Anthill Limestone

As described by Gunn and Warren (1962) the Anthill Limestone is predominantly a well-bedded white to gray limestone with minor mudstone, siltstone, and quartzite. Bedding ranges from several centimeters to about a meter in thickness. An estimate of a minimum of 3,300 m thickness was made for a section between Dilemma and Delta Glaciers. No fossils are known.

In a petrographic study of "Teall Greywacke" that included some of Gunn and Warren's original samples, Skinner (1982) found that samples from central Teall Island contain approximately 25% quartz clasts and considerable calcite (35–60%), similar to calcarenite from Anthill Limestone north of Cocks Glacier. Felsic volcanic rock fragments comprise a minor fraction of most of the samples.

On the west side of Skelton Glacier, the Anthill Limestone has been metamorphosed only slightly with recrystallization of calcite and the presence of some fine biotite. On the east side of Skelton Glacier north of the Cocks, the rocks contain plentiful diopside. Near Mt. Moxley sillimanite–muscovite calc–silicate rocks are interpreted to be amphibolite equivalents of Anthill Limestone (Skinner, 1982).

## Cocks Formation

At two localities, north of Cocks Glacier and north of Baronick Glacier, Skinner (1982) has described the contact between Anthill Limestone and Cocks Formation as an unconformity; however, Rees et al. (1989a) interpret the contact as tectonic (Fig. 3.9). Close to the contact the Cocks Formation contains lenticular conglomeratic horizons up to 3 m thick with clasts of metaandesite, granite, and lesser limestone up to 35 cm in length, set in a calcareous, tuffaceous (?), quartz–sandstone matrix. This is overlain by well-bedded volcaniclastic sandstone in which Skinner (1982) has identified graded bedding and cross-bedding, although Rees et al. (1989a) found sedimentary structures to be obscure. Thin lenses of marble are interbedded with the conglomerate and sandstone.

Porphyritic pillow lava was also noted at one horizon. Rowell et al. (1993b) analyzed Sm/Nd from this rock on three magnetically separated splits. The ratios were not sufficiently different to define an isochron; however, the authors determined a neodymium model age ($T_{DM}$) of 700–800 Ma, indicating a Neoproterozoic time frame for the Skelton Group. Rowell et al. (1993b) suggested that the basalt is of continental rift affinity, owing to its major-element composition.

On the basis of petrographic observation, Skinner (1982) argued that most of the sandstones in the Cocks Formation have insufficient matrix to be classified as true graywackes, yet in his modal analyses he shows biotite, actinolite, or diopside, all of which are presumably metamorphic minerals derived from matrix, accounting for 20–40% of the samples. Quartz is the predominant, detrital mineral. Feldspar is also present in all samples, while igneous and sedimentary rock fragments make up a minor percentage of some (Skinner, 1982). Pilotaxitic volcanic grains are found at Red Dog Bluff, but apparently not at Teall Island.

Skinner (1982) also recognized an eastern belt of Cocks Formation along the divide between Skelton and Koettlitz Glaciers from Mt. Cocks to Mt. Moxley in which the metamorphic grade is higher than along Skelton Glacier. Around Mt. Cocks the rocks are biotite schists, quartzose marbles, and actinolite amphibolites, increasing to amphibolite grade northward to Mt. Moxley where sillimanite–muscovite has been identified in calc–silicate rocks.

## Koettlitz Group

Rocks of the Koettlitz Group, a suite of polydeformed schist, calc schist, marble, and gneiss, crop out throughout the area from Mackay to Koettlitz Glacier in

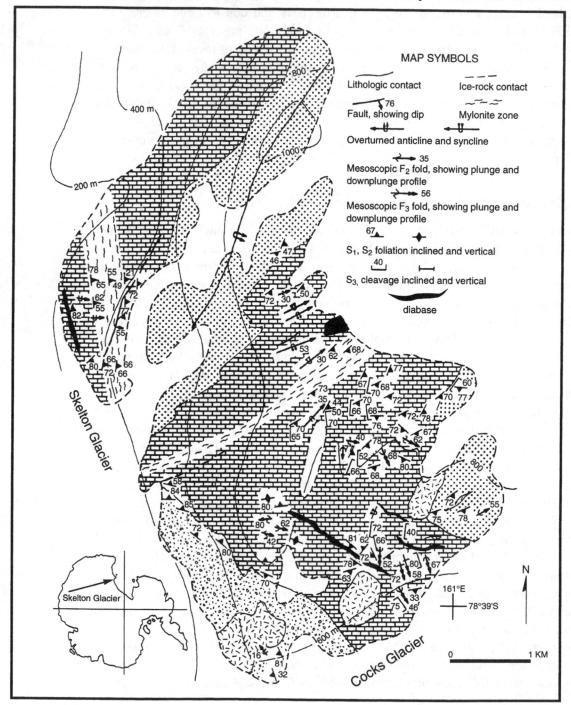

several northwest-trending belts separated by plutonics. The most comprehensive study of this complex assemblage of rocks is that of Findlay et al. (1984), who remapped large portions of the area between Skelton Glacier and Wright Valley either in detail or in reconnaissance fashion (Fig. 3.10). All Koettlitz Group rocks of Wright and Victoria Valleys had previously been assigned to the Asgard Formation (Allen and Gibson, 1962; McKelvey and Webb, 1962). In the Koettlitz–Blue Glaciers area Blank et al. (1963) described, but did not separately map, five formations: Marshall Formation, Miers Marble, Garwood Lake Formation, Salmon Marble, and Hobbs Formation. On the basis of interpretations of cut-and-fill structures and graded bedding, they proposed that this was the stratigraphic order with Hobbs Formation on top. Mortimer (1981) and Findlay et al. (1984) rejected the supposition that such sedimentary structures could be seen in these amphibolite-grade rocks and concluded that the evidence to date is not sufficient to determine the stratigraphic order of the formations.

Haskell et al. (1965b), working in the middle and lower Taylor Valley, recognized similar lithologies, but did not apply specific names. Instead, they suggested correlation of the rocks around Nussbaum Reigel with Asgard Formation to the north and Salmon Marble and basal Hobbs Formation to the south. Working between Miers and Salmon Valleys Mortimer (1981) concluded that Miers Marble and Salmon Marble were one and the same formation, retaining the name "Salmon Marble." Likewise, he found Garwood Lake Formation and Marshall Formation to be the same, and retained the name "Marshall Formation." Mortimer (1981) also introduced a new name, "Penance Pass Formation," for rocks structurally overlying Salmon Marble, but this was rejected by Findlay et al. (1984), who apportioned these rocks between Salmon Marble and Hobbs Formation. The final subdivision of Koettlitz Group by Findlay et al. (1984) recognizes three formations of uncertain stratigraphic order: Marshall Formation, Salmon Marble Formation, and Hobbs Formation, with the latter two divided into several members.

## Marshall Formation

The Marshall Formation is limited to the vicinity of Marshall and Garwood Valleys. It contains biotite schists, some biotite gneisses, thin marbles, calc–silicate rocks, and minor amphibolite and quartzite (Blank et al., 1963; Mortimer, 1981; Findlay et al., 1984).

## Salmon Marble Formation

"Salmon Marble" is the name applied to all of the white to light-colored marbles that occur throughout broad areas of the Dry Valleys. They crop out in a band from Marshall Valley, through Salmon Valley to Hobbs Peak, through Beattle Peak in the Con-Rod Hills, and across the Kukri Hills to Nussbaum Riegel in Taylor Valley. Other outcrops occur in the vicinity of Radian and Walcott Glaciers. Blank et al. (1963) reported cylindrical siliceous objects in Salmon Marble from Salmon Valley, which they interpreted to be poorly preserved archaeocyathids. It is likely that they were influenced by the discovery at that time of archaeocyathids in the Shackleton Limestone several hundred kilometers

**Figure 3.9** *(facing page)*. Generalized map of the Skelton–Cocks Glacier area. Block pattern: limestone of Anthill Limestone; dotted pattern: quartzite of Anthill Limestone; speckled pattern: argillites, sandstones, conglomerates, and pillow basalts of Cocks Formation; black areas: diabase; short-line pattern: biotite granite. After Rees et al. (1989a). Used by permission.

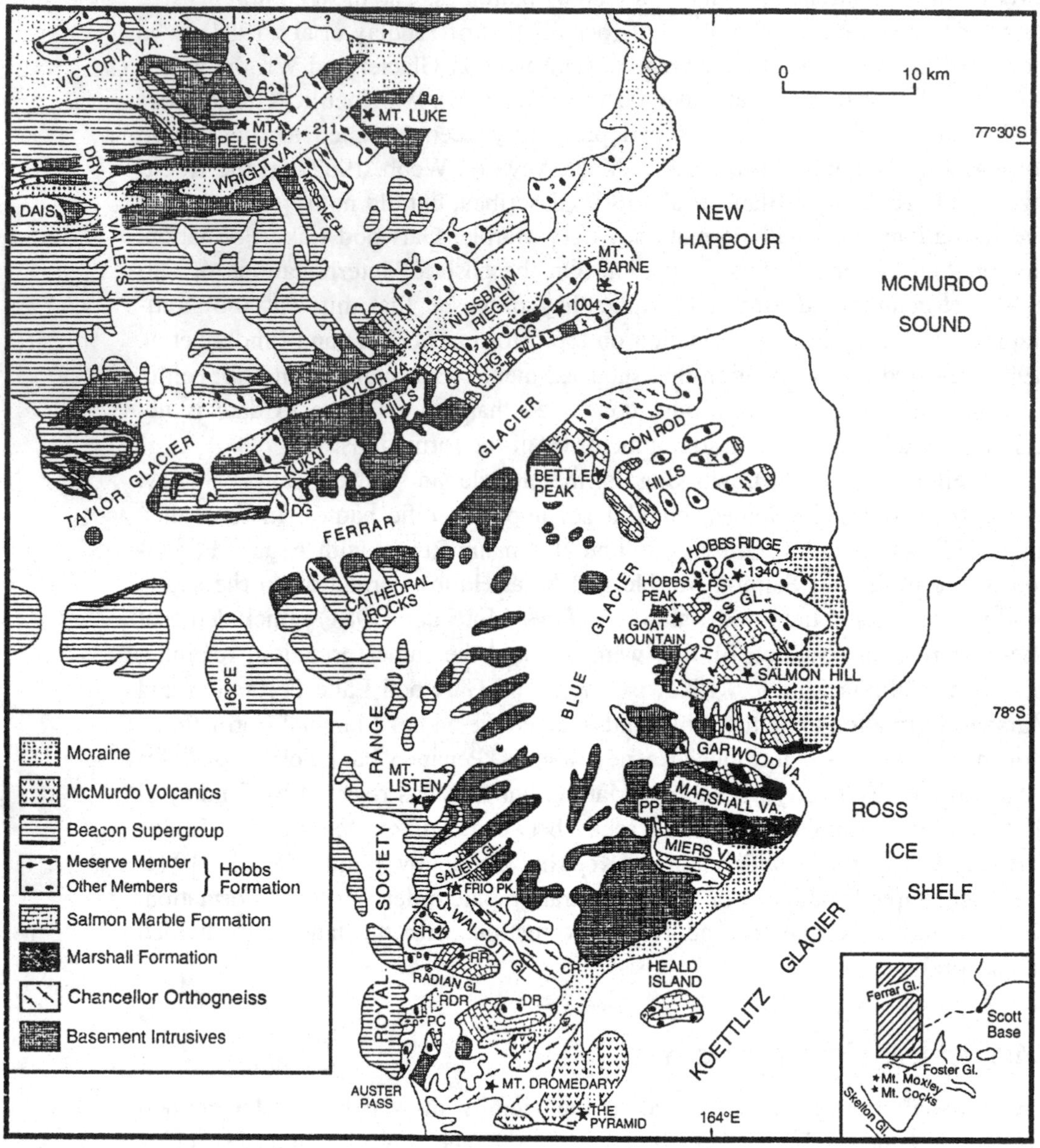

**Figure 3.10.** Geological sketch map of area between Koettlitz Glacier and Victoria Valley. Compare with Figure 3.6. The region north of Taylor Valley is from Lopatin (1972). PG, Pipe-cleaner Glacier; DR, Dismal Ridge; RDR, Radian Ridge; RR, Rucker Ridge; SR, Substitution Ridge; CR, Chancellor Ridge; PP, Penance Pass; PS, Pillow Saddle; DG, Dun Glacier; HG, Howard Glacier; CG, Crescent Glacier. After Findlay et al. (1984). Used by permission.

to the south (Laird and Waterhouse, 1962). At any rate, this suggestion of fossils in the Koettlitz Group fixed the notion for many years that the Koettlitz and Skelton Groups are at least in part of Cambrian age. After further examination of the occurrence, Findlay et al. (1984) reinterpreted the objects to be aplitic rods formed during polyphase folding of aplitic pegmatites.

Findlay et al. (1984) have designated two members of the Salmon Marble: Heald Member and Dismal Member.

**Heald Member.** The Heald Member is the more widespread. It is white, pale pink, or light gray, moderately to coarsely crystalline marble with characteristic thin layers of pyrite, white mica, and diopside. At some localities thin layers of biotite schist, calc–silicate hornfels, and/or quartzite are intercalated (Haskell et al., 1965b).

**Dismal Member.** The Dismal Member of the Salmon Marble is the transitional unit into the Hobbs Formation. It is more colorful and more layered than the Heald Member, with cream-orange, gray, and black portions containing thin interbeds of rusty-weathering fissle pelites. From Pillow Saddle on Hobbs Ridge, Findlay et al. (1984) report three bands of amphibolitic ellipsoidal objects, which they interpret as either possible lava pillows or boudins.

## Hobbs Formation

Hobbs Formation is the most widespread of the formations of the Koettlitz Group. Findlay et al. (1984) have broken out four members: Radian Schist Member, Con-Rod Hills Member, Rucker Member, and Meserve Member. The formation crops out adjacent to Salmon Marble both in the Radian–Walcott Glaciers area and in the band that extends from Garwood Valley to Taylor Valley. Findlay et al. (1984) carry the Meserve Member into Wright Valley, where it includes a portion of the Asgard Formation. In addition, Hobbs Formation crops out in the narrow belt from Cathedral Rocks through the western Kukri Hills and upper Taylor Valley. Findlay et al. (1984) have also mapped the western portion of the Dais in upper Wright Valley as belonging to the Meserve Member of Hobbs Formation.

Hobbs Formation is predominantly mica schist and lesser calc schist and hornfels, with minor marble and quartzite. Matrix-supported conglomeratic layers occur fairly commonly in Hobbs Formation south of Blue Glacier (Blank et al., 1963), but are not known in the occurrences north of there. Skinner (1992) has speculated that these diamictites may be of glaciogenic origin.

**Radian Schist Member.** The Radian Schist Member follows the Dismal Member of the Salmon Marble. It consists of Brown-weathering biotite–actinolite, biotite–garnet, and biotite–aluminosilicate schists. Biotite gneisses and hornblende amphibolites are a minor part of the sequence in the eastern Con-Rod Hills.

**Con-Rod Hills Member.** The Con-Rod Hills Member is characterized by conglomeratic portions. Findlay et al. (1984) recognize a coarse-grained and a fine-grained facies. The course-grained facies, which crops out as far north as Hobbs Ridge, contains clasts up to 10–15 cm of granite, quartz, biotite pelite, amphibolite, and rare marble, set in a matrix primarily of tremolite–actinolite. The fine-grained facies crops out in the Koettlitz–Blue Glacier area and northward to the Kukri Hills. It is a rusty-weathering quartz tremolite–actinolite schist with sparse pebbles, at places with thin biotite pelites and marbles.

**Rucker Member.** Like the Con-Rod Hills Member with which it is gradational, the Rucker Member crops out between Radian Ridge and the Kukri Hills. It is a rusty-weathering tremolite–actinolite amphibolite with diopside streaks, containing rare pebbles of quartz, granite, and pelite.

**Meserve Member.** Findlay et al. (1984) identify their Meserve Member as a portion of the Asgard Formation (McKelvey and Webb, 1962) in Wright Valley, as well as the upper (*sic*) Taylor Valley belt of Haskell et al. (1965b), which continues southward to Cathedral Rocks. [The correlation presumably was with the *middle* Taylor Valley belt of Haskell et al. (1965b), since they identified no upper Taylor Valley belt.] Allibone et al. (1991) map the middle Taylor Valley belt as Koettlitz Group undifferentiated, since they feel correlation of the Meserve Member has not been well established.

Description of the Meserve Member is presented as a paced section on the north side of Wright Valley where the rocks are primarily a variety of gray and pink gneisses, with biotite or biotite + hornblende (Findlay et al., 1984). Interbedded with these are several thin white marble beds, as well as some calc–silicate units.

## Asgard Formation

Although Findlay et al. (1984) carry their subdivision of Koettlitz Group into Wright Valley, it is a fairly tentative reach from the sequences south of Ferrar Glacier, which have been studied in greater detail. Furthermore, they have not studied the Koettlitz Group where it crops out in Victoria Valley. The tie to Wright Valley is on the south side of the valley in the area between Mt. Loke and Meserve Glacier, where at the eastern end they have correlated a 300-m section of schists with the Radian Schist Member of Hobbs Formation. West of this for 650 m are rocks they correlate with the fine-grained facies of the Con-Rod Hills Member. The Rucker Schist Member is apparently absent, and the rocks that pick up to the west they have called the Meserve Member.

Descriptions of Asgard Formation in Wright Valley follow those of McKelvey and Webb (1962), who presented a reconnaissance map of the entire valley, and Murphy (1972), who mapped a small area on the south side of Wright Valley between Goodspeed and Meserve Glaciers. Asgard Formation in Victoria Valley was mapped by Allen and Gibson (1962), and recently a preliminary report on the St. Johns Range has been presented by Cook (1991).

Asgard Formation consists of interbedded schists, gneisses, hornfelses, and marbles. The marbles are white and granular, and crumble before producing skree. They are thinner and less extensively developed than the Salmon Marble Formation exposed from Nussbaum Riegel southward to Koettlitz Glacier. Calc–silicate schists and hornfelses contain a variety of metamorphic assemblages with such characteristic minerals as diopside, tremolite, scapolite, wollastonite, hornblende, garnet, and sphene. Primary bedding is most apparent in the calc–silicates where compositional layering is accentuated. Foliation in the schists and gneisses typically parallels this, but in some cases may actually be transposed bedding, as indicated by rootless folds within the foliation (Murphy, 1972).

The most common schists and gneisses are quartzofeldspathic with biotite or biotite + hornblende. Most are various shades of gray; some are pink. Murphy (1972) has identified pelitic schists and gneisses containing quartz + biotite + orthoclase + sillimanite ± cordierite ± garnet. On the basis of these assemblages, he has argued that metamorphism of the Asgard Formation in Wright Valley occurred at approximately 4 kb and 675°–700° C.

In Victoria Valley, Asgard Formation crops out in a northwest-trending belt along the axis of the St. Johns Range, as well as at Sponsers Peak. The sequences include marble, calc schist, pelitic schist, and amphibolite. At Sponsers Peak marble is the predominant lithology, whereas in the St. Johns Range marble, calc schist, and pelitic schist are in approximately equal portions (Cook, 1991). Characteristic calc–silicate minerals include diopside, garnet, wollastonite, epidote, and idocrase. Amphibolite and associated garnet–hornblende schist occur in abundance at Sponsers Peak and to a lesser degree in the St. Johns Range. These are interpreted to be metamorphosed basic igneous rocks (Cook, 1991).

## Correlation of Koettlitz and Skelton Groups

Little has been offered regarding the sedimentary conditions under which the Koettlitz and Skelton Groups were deposited, except that they were marine (Laird, 1981a). The stratigraphic order of the Koettlitz Group is unknown (Findlay et al., 1984), and although Skinner (1982) has identified tops indicators in the Cocks Formation and Anthill Limestone, and has said that the former overlies the latter unconformably, Rees et al. (1989a) disagree on all counts. Consequently, the stratigraphic relationship of the Koettlitz and Skelton Groups remains uncertain. Koettlitz Group has been mapped as far south as Mt. Dromedary (Blank et al., 1963; Blattner, 1978) to within 6 km of Mt. Moxley, where Skinner (1982) has identified amphibolite-grade rocks that he assigns to Skelton Group. Skinner (1983b) correlated what he considered to be the lower portion of the Anthill Limestone with Koettlitz Group, but due to lack of continuity of outcrop and uncertainty of stratigraphic order in Koettlitz Group in the Koettlitz–Blue Glacier area, Findlay et al. (1984) would take the correlation in detail no farther than that. If the rocks along the Skelton–Koettlitz Glaciers divide are in fact higher-grade equivalents of the Skelton Group cropping out along Skelton Glacier, it is reasonable to assume that Skelton Group is a continuation of Koettlitz Group only 6 km to the north. The lithologic components of both groups are similar. They both contain clastic metasediments, calc–silicates, and marbles. Conglomerates are found in both, as are volcanic rocks and amphibolites. Still, the connection has not been proved conclusively, so caution must be exercised when generalizing about depositional history in this portion of the Transantarctic Mountains.

In looking for correlatives for Skelton Group south of Byrd Glacier, various authors (e.g., Grindley and Warren, 1964) have suggested matching the Teall Greywacke with Goldie Formation (Neoproterozoic) and Anthill Limestone with Shackleton Limestone (fossil-dated Early Cambrian). Skinner (1982) suggested a correlation of Anthill Limestone and Cocks Formation with the Byrd Group (Shackleton Limestone plus clastic Dick Formation and Douglas Conglomerate), but later he backed away from this, arguing that Skelton Group was Proterozoic in age (Skinner, 1983b).

A pluton (discussed later) that crosscuts Skelton Group has been U/Pb-dated at $550.5 \pm 4$ Ma, requiring that the Skelton Group be Proterozoic in age and precluding a correlation with Byrd Group (Rowell et al., 1993b). The neodymium model age ($T_{DM}$) of 700–800 Ma on the pillow basalt in Skelton Group (Rowell et al., 1993b) brackets the age of pillow basalt in Goldie Formation (Beardmore

Group) at Cotton Plateau of $762 \pm 24$ Ma (Borg et al., 1990). This strongly suggests that the Beardmore and Skelton Groups were being deposited in the same Neoproterozoic time frame.

## Horney Formation

Complexly folded, amphibolite-grade gneisses and schists, informally named the Horney Formation, crop out in the Britannia Range on the north side of Byrd Glacier (Fig. 3.11). Borg et al. (1989) state that the rocks resemble the Miller Formation of the Miller Range, but no systematic description or mapping has yet been done on these metamorphics.

## Granite Harbour Intrusives

Tremendous volumes of magma were intruded in southern Victoria Land between the Neoproterozoic and Ordovician. In fact, large segments of the region are virtually devoid of all but plutonic rocks, as in the 200 km between Skelton and upper Byrd Glaciers, and the 180 km from Granite Harbour to David Glacier, and beyond into Terra Nova Bay. In the area north of Koettlitz Glacier the major plutons are interspersed between elongate belts of Koettlitz Group that trend northwest through the Dry Valleys and become completely engulfed in magma before reaching Granite Harbour, save for some rafts and xenoliths of metasediments that remain. In this central region of southern Victoria Land the history of intrusion is complex, with early stages caught up in and perhaps contributing to the deformation, followed by voluminous posttectonic plutonism, that ends with extensive, northeast-trending dike swarms. Although S-type plutonics dominate the Granite Harbour Intrusives in northern Victoria Land north of Terra Nova Bay, from there southward through all of southern Victoria Land the main plutonic rocks are I-types.

Scott's *Discovery* expedition was the first to sail into Granite Harbour. A party was launched to shore at Discovery Bluff on the south side of the harbor, as Ferrar (1907, p. 33) described it, "a prominent headland some 500 feet high and two miles long; in form it is distinctly like a bursting cabbage." The rocks were a gray biotite granite cut by composite dikes of pink coarse-grained granite that shot up the face of the cliff at about 100-m spacings. This was the first in situ occurrence of granite recorded in the Transantarctic Mountains, and Granite Harbour was named for it. During Scott's *Terra Nova* expedition, Debenham worked around Granite Harbour and up Mackay Glacier for six weeks, collecting specimens and mapping. This resulted in a sketch map of the area inland as far as Mt. Gauss and Sperm Bluff (Fig. 3.6), a plane table map of Devil's Punchbowl, and a geological map of the Flatiron (Smith, 1924). Smith (1924) referred to the gray biotite granite found throughout southern Victoria Land as "Granite Harbour type."

Gunn and Warren (1962) later designated the Granite Harbour Intrusive Complex to encompass all of the hypabyssal and plutonic intrusive rocks of the Ross orogen (pre-Kukri peneplain). This was in opposition to Admiralty Intrusives, which Harrington (1958) had defined in a like manner four years earlier, stem-

**Figure 3.11.** Layered gneisses of the Horney Formation at Horney Bluff. Cliff approximately 150 m high.

ming from work in the Admiralty Mountains in northern Victoria Land. Grindley and Warren (1964) retained the name "Granite Harbour Intrusives" for the rocks that preceded and accompanied the Cambro-Ordovician Ross orogeny, and restricted "Admiralty Intrusives" to the plutons exposed in the eastern portion of northern Victoria Land, based on the early geochronological work of Starik et al. (1959), whose K/Ar dates of 313 and 318 Ma were the first indication that a distinctly younger suite of plutonic rocks existed in that area.

Gunn and Warren (1962) subdivided the Granite Harbour Intrusive Complex between Mawson and Mulock Glaciers into pretectonic, syntectonic, and posttectonic phases, named both formations and individual plutons within them, and spawned a nomenclature that has become ever more Byzantine as subsequent studies have added their own local names. The original work of Ferrar (1907) recognized the widespread occurrence of "grey granite" and "pink granite," with the latter "overlying" the former in the Ferrar Glacier area. The subsequent pioneers, Mawson (1916) and Smith (1924), described similar lithologies. In Gunn and Warren's (1962) proposed threefold subdivision, foliated, "syntectonic" gray granite was called Larsen Granodiorite and nonfoliated, "posttectonic" pink granite was called Irizar Granite. The names were derived from Mt. Larsen and Cape Irizar, localities in the vicinity of the Drygalski Ice Tongue described by Mawson (1916) but not visited by Gunn and Warren. The main phase of pretectonic intrusives identified by Gunn and Warren (1962) was "granodiorite gneisses." Showing uncharacteristic restraint, the authors chose not to name these rocks.

Such was not the case in Wright and Victoria Valleys, where McKelvey and Webb (1962) introduced the names "Wright Intrusives" and "Victoria Intrusives." The Wright Intrusives included two major phases, the Olympus Granite Gneiss and the Dais Granite, the latter being the gray granite of the early workers. The Wright Intrusives also included two sets of dike swarms: the Loke Microdiotrite, which intrudes Olympus Granite Gneiss in the lower Wright Valley with a consistent northwest strike and low-angle dip to the southwest, and the Theseus Granodiorite, which intrudes Dais Granite, Olympus Granite Gneiss, Koettlitz Group, and Loke Microdiorite and ramifies with no consistent trend. The main phase of the Victoria Intrusives is the Vida Granite, synonymous with the pink granite of the early workers. The youngest plutonic rocks, also included in the Victoria Intrusives and named the Vanda Lamprophyre and Porphyry by McKelvey and Webb (1962), are a suite of lamprophyres and felsic porphyries that form intense dike swarms throughout large portions of the Wright and Victoria Valleys, striking consistently northeast with steep dips. Haskell et al. (1965b), working in the lower Taylor Valley, partially adopted the nomenclature of Gunn and Warren (1962) as well as McKelvey and Webb (1962). They retained "Olympus Granite Gneiss" and correlated Larsen Granodiorite with Dais Granite, and Irizar Granite with Vida Granite, reinforcing the twofold, gray-versus-pink, lithology-based classification.

In the early 1980s New Zealand geologists began to find this simplified system unworkable. Skinner (1983b) proposed a classification of plutonic rocks in southern Victoria Land based on their relationship to deformational episodes. Findlay (1985) discussed the accumulating evidence that the simple gray-versus-pink and foliated-versus-nonfoliated classification could not be universally applied. His contribution toward clarification was to introduce a new subdivision: Kukri Hills Group, Larsen Intrusive Group, and Victoria Intrusive Group. The Kukri Hills Group consisted of dark- and light-colored gneisses associated with the Koettlitz Group. The Larsen Intrusive Group included two portions, "the Dais Granodiorite . . . constituted by the Dais Granite and Olympus Granite gneiss as defined originally by McKelvey and Webb (1962)" and the younger Briggs Hill Granodiorite (Findlay, 1985, p. 12). Several localities of pink rocks were cited that were included in the Briggs Hill Granodiorite. Likewise, although much of

Findlay's (1985) Victoria Intrusive Group is composed of pink Vida Granite, appreciable portions of gray granite also occur within this group.

The most recent attempts to bring order to the plutonic morass in southern Victoria Land are those of Allibone et al. (1991), Cox and Allibone (1991), and Smillie (1992). These authors abandoned a lithology-based classification altogether and proceeded to describe the characteristics of individual plutons and, through contacts, their temporal relationships. Then, on the basis of geochemical characteristics, they attempted to group the rocks into suites, specifically Dry Valley 1 (DV1) and Dry Valley 2 (DV2). Although this approach is promising, the work to date is limited mainly to the area from the western Kukri Hills northward across the west-central Asgard Range into a small area of Wright Valley, and a complete application of this scheme throughout southern Victoria Land awaits considerably more detailed field mapping coupled with geochemical analysis.

Consequently, for a comprehensive overview of plutonic rocks in southern Victoria Land, we must still rely on the mapping and descriptions of the earlier workers, but we will return to the work of Allibone et al. (1991), Cox and Allibone (1991), and Smillie (1992) at the end of the section.

## Orthogneisses

Granitoid gneisses crop out at a number of locations throughout the Dry Valleys region. In many places they are intimately associated with Koettlitz Group and have been deformed along with the metasediments. Whether all of these gneisses are of the same intrusive phase is uncertain, and some disagreement exists as to their genesis, with certain workers arguing that they are metasomatized metasediments (i.e., paragneisses) rather than orthogneisses (Smithson et al., 1971b, 1972).

Compositions of these rocks vary from monzodiorite to granite (Cox and Allibone, 1991). The characteristic mineral assemblage is plagioclase + quartz + orthoclase + biotite, typically including hornblende. Clinopyroxene, apatite, zircon, titanite, and allanite have been reported as accessory phases. The mafic minerals are partially altered to chlorite in many cases. Most of the occurrences contain megacrysts of K-feldspar, up to 5 cm in length, often with an augen shape. Cataclastic textures are common with granulation of quartz and distortion or kinking of feldspar (McKelvey and Webb, 1962; Haskell et al., 1965b).

Gneissic banding is steeply dipping and, throughout much of the Dry Valleys area, strikes northwest, parallel to the orientation of foliation in the Koettlitz Group. The bodies of gneiss are interlayered with Koettlitz Group on all scales from 2 to 200 m thickness; contacts are usually concordant and typically gradational.

Regarding the origin of these rocks, Gunn and Warren (1962) considered them all to have been of igneous protolith, originally intrusive into Koettlitz and Skelton Group metasediments. McKelvey and Webb (1962) were ambivalent, suggesting that their Olympus Granite Gneiss may have been derived by metasomatism of Asgard Formation, while also interpreting the contact between Dais Granite and Olympus Granite Gneiss as gradational. They also pictured (p. 152) inclusions of Asgard Formation in Olympus Granite Gneiss. Murphy (1972) and

Smithson et al. (1971b, 1972) argued that the augen gneisses found particularly in the lower Wright Valley were formed by metasomatism of calc–silicate rocks during strong deformation. Findlay (1985) felt that most of the gneisses in the region were originally intrusive, citing crosscutting relationships, but allowed that some of the concordant gneisses might have been derived from sedimentary rocks.

Cox and Allibone (1991) summarized arguments in favor of the gneisses being of igneous origin: (1) crosscutting relations at limited places with the metasediments, (2) inclusions of mafic porphyritic rocks in the gneisses unlike any lithologies in the Koettlitz Group, (3) euhedrally zoned inclusions in the K-feldspar augens, indicating growth as phenocrysts rather than porphyroblasts, and (4) systematic variations in abundance of megacrysts, suggesting flow differentation. In addition, Cox and Allibone (1991) identified three bodies of homogeneous gneiss large enough to be mapped as separate plutons (Dun pluton, Calkin pluton, plus a third unnamed body), and argued that their size precludes a sedimentary origin, since one would expect such a protolith to result in inhomogeneous gneisses on such a scale.

Skinner (1983b) subdivided and named several phases of orthogneiss in the Koettlitz–Ferrar Glaciers area, including leucocratic Portal Augen Gneiss and Chancellor Orthogneiss and melanocratic Renegar Mafic Gneisses. In their mapping of the western Kukri Hills and central Asgard Range, Allibone et al. (1991) mapped several limited occurrences of orthogneiss in association with Koettlitz Group, as well as the Calkin and Dun plutons just mentioned; however, they included considerable areas previously mapped as Olympus Granite Gneiss by McKelvey and Webb (1962) as parts of their Bonney and Valhalla plutons. This was anticipated by Skinner (1983b), who stated that the degree of gniessosity and augen development varied considerably within both the Olympus Granite Gneiss and the Dias Granite, such that the latter is more gneissic than the former in some places. Similarly, Findlay (1985) included the Olympus Granite Gneiss in his Larsen Intrusive Group along with Dais Granite, distinct from the orthogneisses that he classified as Kukri Hills Group.

The conclusion reached by Cox and Allibone (1991) is that the orthogneisses are the early phase of a petrogenetic suite of I-type intrusives that spanned the time of deformation of Koettlitz Group and perhaps were an important contributor to that deformation.

## Gray Granitoids

The most widespread family of intrusive rocks in southern Victoria Land, occurring in plutons from Priestley to Byrd Glacier, are texturally varied, weakly to strongly foliated gray granitoids composed essentially of quartz, K-feldspar, plagioclase, and biotite, with hornblende present or not and accessory phases of allanite, sphene, zircon, apatite, and iron oxide (Skinner, 1983b). K-feldspar megacrysts up to 5 cm in length are common to many of the occurrences, with alignment parallel to the foliation. Rock types are most commonly granodiorite, but range from monzogranite to tonalite and quartz monzodiorite.

Skinner and Ricker (1968) described Larsen Granodiorite as the most prevalent rock type in the area between Priestley and Mawson Glaciers, noting that it is extremely variable in composition and contains portions that grade from gray to pale pink coloration. Gunn and Warren (1962) mapped localities of this rock around the perimeter of the Dry Valleys from Mawson to Mulock Glaciers. McKelvey and Webb (1962) and Allen and Gibson (1962) map Dais Granite around the mouth of Victoria and Wright Valleys and in upper Wright Valley around the Dais beneath the basement sill of Ferrar dolerite. Between these two occurrences and surrounding a belt of Koettlitz Group is Olympus Granite Gneiss. Skinner (1983b) and Findlay (1985) have interpreted the Olympus Granite Gneiss as a more foliated border phase and the Dais Granite as a less foliated, interior phase of the same intrusion. Mapping by Allibone et al. (1991) indicates that portions of their Bonney and Valhalla plutons in Wright Valley include both Dais Granite and Olympus Granite Gneiss as previously mapped.

Haskell et al. (1965b) mapped Larsen Granodiorite in the central Taylor Valley area, which was subsequently mapped as the Bonney Pluton by Allibone et al. (1991). This belt of plutonic rocks extends southward across the Kukri Hills and into the Blue Glacier area, where it divides the two major belts of Koettlitz Group. There Blank et al. (1963) described a complex batholith with coarse- and fine-grained facies and both crosscutting and gradational contacts with the Koettlitz metasediments. Mortimer (1981) related foliated granodiorite in the Garwood–Miers Valleys area to Dais Granite, and also noted an older nonfoliated pluton crosscut by his Dais Granite, which he called Buddha Diorite. Skinner (1983b) and Findlay (1985) described the plutonic rocks of the Blue Glacier area as Larsen Granodiorite. Smillie (1992) notes that some of the rocks in the Miers Valley area are identical to those of the Bonney pluton mapped in the central Wright and Taylor Valleys area, and that if the body is in fact continuous over that area then this single pluton is approximately 100 km in length and up to 25 km in width, bounded by Koettlitz Group and intercalated orthogneiss on both its eastern and western sides.

Farther south in the Darwin Glacier area, Haskell et al. (1964, 1965a) described a well-foliated biotite–hornblende granodiorite, typically containing phenocrysts of K-feldspar to 3 cm in length, which they named the Carlyon Granodiorite. It contains a belt of darker, more gneissic rocks extending from Bastion Hill to Diamond Hill. Haskell et al. (1965a) say that the Carlyon Granodiorite most resembles descriptions of the Dais Granite of McKelvey and Webb (1962) and the more gneissic band, the Olympus Granite Gneiss. Haskell et al. (1965a) also described a nonfoliated to weakly foliated, porphyritic, hornblende-bearing granite with pink and white orthoclase crystals, which they called the Mt. Rich Granite. They stated that this rock is gradational with the Carlyon Granodiorite. These rocks are in turn intruded by dikes of leucogranite, pegmatite, lamprophyre, and meladiorite. Borg et al. (1989) mapped an occurrence of Carlyon Granodiorite at the western end of the Britannia Range where it intrudes Horney Formation.

Orbicular inclusions are a curiosity found within the gray granitoids at several localities in southern Victoria Land, for example in the Granite Harbour area at the Flatiron (Smith, 1924) and Benson Glacier (Gunn and Walcott, 1962; Stump and Maccracken, 1983), in the St. Johns Range (Cook, 1991; Waters, 1991), and in Taylor Valley (Haskell et al., 1965b; Palmer et al., 1967; Dahl and Palmer, 1981, 1983; Smillie, 1992).

## Granodiorite Dikes, Plugs, and Stocks

The major plutons of foliated gray granitoids in the Dry Valleys area are intruded by swarms of granodioritic dikes and the odd plug or stock. These rocks were originally named Theseus Granodiorite by McKelvey and Webb (1962), who mapped their occurrence throughout much of the Asgard and Olympus Ranges adjacent to Wright Valley, and designated the north wall of Wright Valley between Mt. Theseus and Bull Pass as a type locality. The dikes were described as ramifying freely without consistent trend. Allen and Gibson (1962) noted similar dikes in the Victoria Valley area, but also found a coherent body of the granodiorite 650 m wide at Purgatory Peak north of Lower Victoria Glacier, and another stocklike body poorly exposed north of Lake Vashka. Haskell et al. (1965b) carried the Theseus Granodiorite southward into Taylor Valley. In addition to dikes, both ramifying through and concordant with foliation in the enclosing gneisses and metasediments, larger tabular bodies were recorded in the lower Taylor Valley at Mt. Coleman and Cresent Glacier. Allibone et al. (1991) mapped similar dikes, plugs, and stocks with particularly intense swarms adjacent to Taylor Glacier at Cavendish Rocks and between the Rhone and Stockling Glaciers, and in Wright Valley at the snout of Valhalla Glacier. Findlay (1985) mentions that numerous sills, dikes, and bosses of gray biotite granite in the Blue Glacier area may be equivalent to the Theseus Granodiorite of McKelvey and Webb (1962).

The mineralogy of the rocks just cited is quartz + plagioclase + orthoclase + biotite. They are nonfoliated and nonporphyritic, except for some gneissic texture at a few places mentioned by Allen and Gibson (1962). Ghent and Henderson (1968) described hornblende-bearing granodioritic gneisses that occur as pods and ramifying veins in Koettlitz Group in the Mt. Falconer area, eastern Asgard Range. They suggested that they are equivalent to the Theseus Granodiorite, but the presence of hornblende and the decidedly gneissic character suggest otherwise.

## Pink Granites

The work of Skinner and Ricker (1968), Skinner (1983b), and Findlay (1985) has shown that pink granites of southern Victoria Land cannot be conveniently lumped as the Irizar Granite of Gunn and Warren (1962) or the Vida Granite of McKelvey and Webb (1962). To be sure, there are a number of occurrences of discordant, nonfoliated to weakly foliated hornblende-bearing granites with pink to brick-red K-feldspar akin to the original descriptions, but these characteristics alone are not enough to designate a rock as being late in the intrusive sequence.

Skinner (1983b) notes that the occurrence of Vida Granite around Lake Vida contains biotite, but not hornblende, and also that the sheet of Vida Granite between the Kukri peneplain and basement sill as mapped in the upper Wright Valley by McKelvey and Webb (1962) is in fact Dais Granite and Olympus Granite Gneiss, the same as that found locally beneath the basement sill.

Haskell et al. (1965b) mapped Irizar Granite at several localities in the Taylor Valley area. In the western Kukri Hills they show Irizar Granite above, separated from Larsen Granodiorite below by the basement sill, in a fashion similar to the

Vida–Dais relationship mapped by McKelvey and Webb (1962) in upper Wright Valley. Allibone et al. (1991) map their Catspaw pluton in the western Kukri Hills occurring both above and below the sill. To the east the Catspaw pluton is intruded by their Hedley pluton, which continues southward across Ferrar Glacier cropping out in the western portion of Cathedral Rocks, where it was designated Irizar Granite (?) by Gunn and Warren (1962) and Vida Granite by Findlay (1985).

Rocks mapped as Irizar Granite by Gunn and Warren (1962) and Haskell et al. (1965b) around Taylor Glacier at Pearse Valley, Friis Hills, Solitary Rocks, Cavendish Rocks, and the western tip of the Kukri Hills comprise the Pearse pluton of Allibone et al. (1991). The alteration effects of the Ferrar dolerite on this rock at the western end of the Kukri Hills were studied by Craw and Findlay (1984).

On the north side of Taylor Valley between Canada and Commonwealth Glaciers another locality mapped as Irizar Granite by Haskell et al. (1965b) was the focus of a detailed study by Ghent and Henderson (1968; Ghent, 1970). The so-called Mt. Falconer pluton was mapped in detail, and a variety of chemical analyses were performed on both rocks and individual minerals. The succession of rocks in the area is Koettlitz Group intruded by granodiorite gneiss, intruded by microdiorite and granophyre dikes, intruded by the Mt. Falconer pluton, a quartz monzonite covering approximately 3 km$^2$ of outcrop area, intruded by camptonite dikes.

Smillie (1992) states that the Mt. Falconer pluton is identical in field relations, mineralogy, and geochemistry to his Rhone pluton around Rhone Glacier and to the Pearse pluton of Allibone et al. (1991). It should be noted that the Catspaw and Hedley plutons are part of Smillie's (1992) Dry Valley 1 Suite, and the Rhone pluton (and by implication the Mt. Falconer and Pearse plutons) are part of his Dry Valley 2 Suite, emphasizing the fact that the pink Irizar and Vida Granites of earlier workers should not be lumped together as one cogenetic group.

## Mafic and Felsic Dike Swarms

Intense swarms of dikes with both mafic and felsic compositions intrude considerable portions in Victoria, Wright, and Taylor Valleys (Allen and Gibson, 1962; Haskell et al., 1965b; Allibone et al., 1991) (Fig. 3.12). They continue southward into the Koettlitz–Blue Glaciers area, but there they are not so extensively developed (Blank et al., 1963). McKelvey and Webb (1962) named these rocks the Vanda Lamprophyre and Porphyry after the area in Wright Valley to the east of Lake Vanda where the dike rocks comprise up to 70% of the outcrop (Fig. 3.1). As much as 50% of the outcrop is dike rock in the central St. Johns Range north of Victoria Valley (Waters, 1991) (Fig. 3.13). Throughout the region the strike is consistently northeast. Wu and Berg (1992) report that most dikes are nearly vertical, 0.5–3 m long, and from a few to 10–20 km long. Keiller (1988) notes that east of Lake Vanda the dikes dip 30°–40°SE and in the vicinity of Mt. Loke vary from nearly vertical to gently southeast. Allen and Gibson (1962) report that in the Victoria Valley area at the Insel Range, Sponsers Peak, and west of Clark Glacier the rocks dip from vertical to 45°NW (Fig. 3.2).

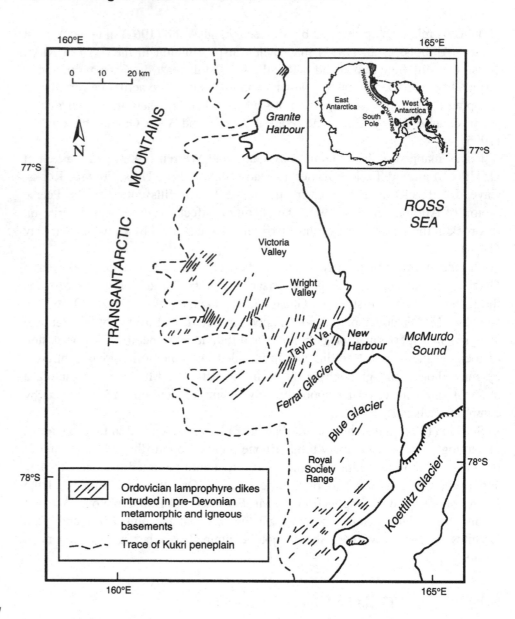

**Figure 3.12.** Schematic sketch map of distribution of lamprophyre dikes throughout southern Victoria Land. After Wu and Berg (1992). Used by permission.

In general the felsic dikes crosscut the mafic dikes, although the opposite situation also exists (Keiller, 1988; Allibone et al., 1991), indicating the close relationship of the two. This is reinforced by the observation of a composite dike with mafic borders and a felsic core in the vicinity of Mt. Loke (Keiller, 1988). The dikes typically have chilled margins that in some cases are more resistant to erosion than either the inner parts of the dikes or the surrounding country rock, producing miniature, paired ridges that outline the dikes as they cross the outcrop.

The mafic dikes have a variety of compositions and may be porphyritic or equigranular. The groundmass commonly contains plagioclase, hornblende, biotite, and interstitial alkali feldspar. Pyroxene is present in some cases. Alteration is common with brown hornblende rimmed by green hornblende, biotite partially gone to chlorite, and plagioclase sericitized and epidotized (Allibone et al., 1991). Plagioclase and brown hornblende phenocrysts are most common. Augite and biotite occur in some of the mafic porphyries, and quartz is a rare phenocryst.

**Figure 3.13.** Mafic dikes cutting through the southern portion of the St. Johns Range.

Wu and Berg (1992) undertook a major trace- and rare-earth-element geo-chemical study of the lamprophyres throughout the McMurdo Sound area. Most of the rocks are calc–alkaline, although some dikes in the Royal Society Range are alkaline or ultramafic in composition. Depletion of Nb/Ti relative to large-ion lithophile elements and light rare earth elements indicates the likelihood that these magmas were subduction-related. A paleomagnetic study on lamprophyre dikes from the Kukri Hills was conducted by Manzoni and Nanni (1977).

The felsic dikes apparently are universally porphyritic, with phenocrysts of quartz and feldspar, feldspar and hornblende, or feldspar alone set in a ground-mass of quartz, biotite, plagioclase, alkali feldspar, and amphibole (McKelvey and Webb, 1962; Allibone et al., 1991)

The Pearse pluton of Allibone et al. (1991) contains enclaves of felsic por-phyry similar to the felsic Vanda dikes, but it also is crosscut by similar dikes, suggesting a close relationship between this pluton and the Vanda dikes.

To date essentially nothing has been published on the kinematic or tectonic significance of these dike swarms. They received passing mention in a recent synthesis of information about the dike swarms of Antarctica (Sheraton et al., 1987). The bimodal suite implies extension. The northeasterly orientation implies a northwest–southeast orientation to extension, perhaps related to transtensional movement with a right-lateral component parallel to the orogen, or to southwest-erly oriented subduction, as suggested by Wu and Berg (1992).

## Suite Subdivision of Southern Victoria Land Granitoids

Recent efforts by New Zealand geologists to understand the plutonic evolution in southern Victoria Land (Allibone et al., 1991; Cox and Allibone, 1991; Smillie, 1992) have resulted in the abandonment of the descriptive, lithology-based classifications of previous workers. These workers' approach has been to map individual plutons, walking out the contacts where possible, with the intent of erecting a relative chronology of intrusion. Petrological descriptions are made for each pluton, which highlight not only the difference between plutons, but also internal variations. Finally, chemical analyses are performed on the various plutons and the phases within them, data are displayed on variation diagrams, and the rocks are grouped into petrogenetic suites in a manner following White and Chappell (1983) and Pitcher (1983, 1984).

The mapping by Allibone et al. (1991) at 1:50,000 covers an area from the western Kukri Hills across the central Asgard Range into a small portion of Wright Valley east of Lake Vanda. The sequence of plutonism that has been determined by these workers is shown in Figure 3.14. Cox and Allibone (1991) determined the geochemistry of the Calkin and Dun plutons and the orthogneisses. Smillie (1992) determined the geochemistry of the younger rocks. His work concentrated somewhat to the east of the area mapped by Allibone et al. (1991), and so bodies included in his sequence of plutonism vary from that of Allibone et al. (1991), mainly in the designation of dikes and small plutons in the upper part of the sequence (Fig. 3.14). A number of these plutons have been mentioned in the preceding sections.

The Calkin and Dun plutons as well as undifferentiated orthogneisses have been deformed along with Koettlitz Group during the $D_2$ episode of deformation, which produced upright, tight, northwest-trending folds. The Bonney pluton, the largest mapped plutonic body in the area, is compositionally and texturally heterogeneous. Compositions range between monzodiorite, quartz monzodiorite, granodiorite, and monzogranite. Mafic minerals include both biotite and hornblende. K-feldspar phenocrysts are common, with the largest up to 8 cm in length. Textures vary due to alignment and segregation of K-feldspar phenocrysts, xenoliths, and mafic enclaves. At the margins of the pluton, gneissic banding and alignment of xenoliths are prevalent, but drop off toward the interior of the pluton.

The Catspaw pluton is a small body of monzogranite composition. It differs from the Bonney pluton in that it is homogeneous and lacking in flow foliation. The Hedley pluton is a small body of granodiorite, lacking hornblende as a mineral phase, as well as mafic enclaves. The Valhalla pluton is similar in composition to the Hedley pluton. It should be noted that the Catspaw and Hedley plutons were previously mapped as Irizar Granite by Gunn and Warren (1962) and Haskell et al. (1965b), although they have now been shown to occur relatively early in the sequence of intrusions in the area (Allibone et al., 1991).

The Pearse pluton (Allibone et al., 1991), the Rhone pluton (Smillie, 1992), and the Mt. Falconer pluton (Ghent and Henderson, 1968) are all small, pink, biotite–hornblende-bearing bodies of variable composition that postdate most of

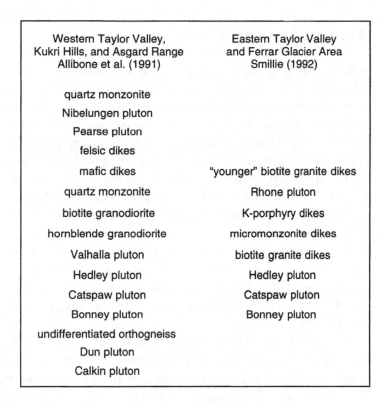

| Western Taylor Valley, Kukri Hills, and Asgard Range Allibone et al. (1991) | Eastern Taylor Valley and Ferrar Glacier Area Smillie (1992) |
|---|---|
| quartz monzonite | |
| Nibelungen pluton | |
| Pearse pluton | |
| felsic dikes | |
| mafic dikes | "younger" biotite granite dikes |
| quartz monzonite | Rhone pluton |
| biotite granodiorite | K-porphyry dikes |
| hornblende granodiorite | micromonzonite dikes |
| Valhalla pluton | biotite granite dikes |
| Hedley pluton | Hedley pluton |
| Catspaw pluton | Catspaw pluton |
| Bonney pluton | Bonney pluton |
| undifferentiated orthogneiss | |
| Dun pluton | |
| Calkin pluton | |

**Figure 3.14.** Sequence of plutonism in Dry Valleys determined by Allibone et al. (1991) and Smillie (1992).

the dike emplacement in the area, but are nonetheless crosscut by younger felsic dikes.

Geochemical data on the complete range of plutonism in Taylor Glacier area are presented by Cox and Allibone (1991) and Smillie (1992). Most of the analyses plot within the I-type fields of White and Chappell (1977, 1983). Typical linear trends are apparent on Harker diagrams as a function of varying silica composition. However, the diagrams show two distinct groupings or suites, in particular for variations of $K_2O$, rubidium, and vanadium, which divide the plutonism temporally. Smillie (1992) designates two suites, the older DV1, which includes the Bonney, Catspaw, and Hedley plutons and biotite granite dikes, and the younger DV2, which includes the Vanda dike swarms, Rhone pluton, and younger felsic dikes. Cox and Allibone (1991) analyzed the orthogneisses in the area and found that the Dun pluton was distinct from either the DV1 or DV2 Suite, but that the other rocks coincided with DV1 trends; they concluded that the DV1 Suite represents an episode of plutonism that spanned the time of deformation of Koettlitz Group.

Using various discrimination diagrams, Smillie (1992) concludes that the DV1 Suite is calc–alkaline and I-type typical of Cordilleran granitoids formed in continental magmatic arcs, whereas the DV2 Suite is more akin to Caledonian I-type granitoids as defined by Pitcher (1983), which are considered unrelated to subduction processes and formed in extensional settings. This is consistent with the interpretation of the Vanda dikes as having formed in an extensional mode, but at odds with the interpretation of Wu and Berg (1992) that they are subduction-related.

## Structural History

Deformation throughout southern Victoria Land occurred during multiple episodes, mainly at amphibolite grade of metamorphism, and involved early phases of the Granite Harbour Intrusives. Early workers recognized the general trend of bedding or compositional layering that strikes northwest in the area from Victoria Valley to Blue Glacier, and more variably between the Blue and Koettlitz, but they did not appreciate the aspect of multiple folding (Allen and Gibson, 1962; Gunn and Warren, 1962; McKelvey and Webb, 1962; Blank et al., 1963; Haskell et al., 1965b). Several groups undertaking detailed studies recognized the structural complexity in small areas, but did not develop a comprehensive regional picture (Smithson et al., 1970; Williams et al., 1971; Murphy, 1972). Findlay et al. (1984) structurally mapped the area from Koettlitz to Ferrar Glacier in fair detail and extended reconnaissance work northward into Taylor and Wright Valleys. Good sketch maps and line drawings of oblique structural views are presented by Findlay (1978), Mortimer (1981), Findlay et al. (1984), and Allibone (1987).

Throughout the Victoria, Wright, and Taylor Valleys the Koettlitz Group and associated orthogneiss generally dip steeply and strike northwest, parallel to the elongation of the metamorphic belts and the plutons that surround them. The most prominent deformation throughout this area produced tight to isoclinal folds with northwest-trending axes that are horizontal or gently plunging, with foliation or schistosity parallel to compositional layering on the limbs and axial-planar in the hinge regions. Working in a small area on the south side of Wright Valley, Murphy (1972) recognized an earlier set of isoclinal folds with northeast-trending, variably plunging axes, as well as a later set that were of a more open style and coaxial with the northwest-trending second generation.

In the middle Taylor Valley and Ferrar Glacier area, Allibone (1987) also recorded upright, tight to isoclinal, shallowly plunging folds with a northnorthwest trend. These followed an earlier folding episode that is seen in small, rootless isoclines paralleling the foliation, which involve the associated orthogneiss and are coaxial with the later folding. Allibone et al. (1991) mapped an antiform with steep eastern limb and shallow western limb on both sides of lower Taylor Glacier, and Haskell et al. (1965b) mapped three synforms in the lower part of Taylor Valley, at Nussbaum Riegel, and on either side of Crescent Glacier.

In a detailed study at Nussbaum Riegel, Williams et al. (1971) observed northwest-trending folds that contain small, rootless closures within the foliation. They also recorded extensive boudinage of quartzofeldspathic layers within the marble, concluding that in that area considerable shortening had occurred perpendicular to foliation and that extension had occurred in all directions parallel to it.

The upright northwest-trending structures north of Ferrar Glacier continue southward through the Con-Rod Hills, approximately to Salmon Valley (Findlay et al., 1984). Possible inversions of units occur due to the second generation of folding. A prominent mineral lineation and elongation of pebbles is parallel with fold axes in this area.

Between the Salmon and Miers Valleys, the Salmon Marble Formation has been mapped in a variety of complex outcrop patterns (Blank et al., 1963; Williams et al., 1971; Mortimer, 1981). Williams et al. (1971) recognized two generations of folding on the north side of Garwood Valley. The earlier is generally isoclinal and recumbent with gentle north plunge and the development of axial-plane cleavage. The second generation of folds is generally angular or boxlike with thick hinges and thin limbs, little or no axial-plane cleavage, and gentle plunge to the west. The complex patterns developed in Salmon Marble are interpreted as due to intersection of disparate (noncoaxial) fold trends (Williams et al., 1971; Mortimer, 1981). On the south wall of Miers Valley the Koettlitz Group strikes east–west and dips south, parallel to the axial plane of folds that plunge 20°–60°SE (Findlay et al., 1984). These rocks swing north at the head of the valley, parallel to the boundary with the granitic pluton that divides the northern belt of Koettlitz Group from the southern one in this area.

In the southern area around Walcott and Radian Glaciers, the structure is dominated by a third generation of folds, which are asymmetrical, even monoclinal, with steep axial planes (Findlay et al., 1984). Axes plunge shallowly and swing from a north–south trend on Radian Ridge through northeast to east–west on eastern Rucker Ridge and Dismal Ridge. Findlay (1978) interprets these folds as "drape" structures caused by local intrusion of granites. Two earlier generations are generally recognizable in Salmon Marble, but not in Hobbs Formation. The earliest generation produced attenuated isoclinal folds with angular hinges, compared with the second generation, in which the hinges are rounded with a tight to open chevron style. Axial directions of both of these generations are highly variable, and interference patterns are common in outcrop. Two shear zones with highly transposed schist and marble have been recognized, possibly due to recumbent folding and thrusting during the earlier deformations (Findlay et al. 1984).

Multiple deformation in Skelton Group rocks was recognized by Murphy et al. (1970) and Flory et al. (1971). The interpretation of deformation of the Skelton Group by Skinner (1982) and Rees et al. (1989a) varies considerably. The study by Skinner (1982) included outcrops on both sides of Skelton Glacier and at Teall Island (Fig. 3.8). Rees et al. (1989a) examined rocks in the area north of Cocks Glacier in detail (Fig. 3.9). Skinner (1982) discerned that Anthill Limestone (including rocks in the central part of Teall Island) had been deformed during three episodes, while Cocks Formation (including rocks at the northern and southern ends of Teall Island) had been deformed only twice. Rees et al. (1989a), however, found the structure north of Cocks Glacier only in part comparable to that described by Skinner (1982). Orientations of folds vary throughout the Skelton Glacier area, so that a coherent kinematic picture is not apparent. According to Rees et al. (1989a) the first two deformations preceded emplacement of small granite plutons north of Cocks Glacier, while the third deformation either followed or accompanied the granite intrusion.

## Isotopic Dating

In the decades since the IGY a number of isotopic studies have examined the timing of rock groups and events within southern Victoria Land. Some of the results can be readily interpreted, others cannot. The earliest ages were a shotgun blast of K/Ar determinations ranging from 533 to 433 Ma (Goldich et al., 1958; Angino et al., 1962; Pearn et al., 1963), with the pattern centering around 470–465 Ma in a study by McDougall and Ghent (1970) that focused on the Mt. Falconer pluton and adjacent rocks. Sampling for these studies was from a variety of rocks, both plutonic and metasedimentary, from Victoria to Taylor Valley. Both biotite and hornblende were analyzed by McDougall and Ghent (1970) with concordant results; the other studies analyzed only biotite. The first reported date from the Transantarctic Mountains was a single K/Ar age of 529 Ma on biotite from a gneiss sample from Gneiss Point (Goldich et al., 1958). This early date indicated that the orogenic belt had cooled below argon-retention temperatures of biotite within a Cambro-Ordovician time frame.

Dating of the late-stage plutonic rocks (Vida or Irizar Granite) has produced a consistent set of Rb/Sr ages of around $470 \pm 10$ Ma. Most recently Graham and Palmer (1987) presented a statistically tight whole-rock isochron for six samples from two plutons in Granite Harbour with an age of $473 \pm 8$ Ma (IR = 0.70883). Three mineral isochrons of $476 \pm 3$ $471 \pm 5$, and $469 \pm 12$ Ma reinforce the whole-rock data. Faure and Jones (1974) analyzed a suite of Victoria Intrusives (Vida Granite and Vanda Porphyry) from Wright Valley and found that they loosely fit a whole-rock isochron of $471 \pm 44$ Ma (IR = 0.7104). Working on core from the Dry Valley Drilling Project (DVDP) hole 6 in Victoria Valley, Stuckless and Ericksen (1975) produced a four-point whole-rock isochron of $475 \pm 14$ Ma (IR = 0.7098). These studies, and especially the one by Graham and Palmer (1987), confirm the assumption by Deutsch and Webb (1964) of an IR of 0.709 for the granite at Granite Harbour, on which they determined Rb/Sr model ages of $472 \pm 15$ Ma for biotite and $462 \pm 80$ Ma for feldspar.

In their Rb/Sr study Deutsch and Webb (1964) dated a range of samples, including Asgard Formation, Olympus Granite Gneiss, Dais Granite, Theseus Granodiorite, and Vanda Porphyry, all with an assumed IR of 0.709. Most of the dates were on biotite and ranged from 485 to 465 Ma. One dike of Vanda Porphyry produced an enigmatic set of ages: biotite + hornblende, $467 \pm 15$Ma; feldspar, $922 \pm 79$ Ma; and whole rock, $979 \pm 78$ Ma. As a check on this age Jones and Faure (1967) analyzed two grab samples of Vanda Porphyry from Wright Valley for whole-rock and feldspar Rb/Sr. The four analyses plotted on an isochron of $460 \pm 7$ Ma (IR = 0.7119), leading Faure and Jones (1974) to interpret the old apparent age of Deutsch and Webb (1964) as being due to contamination by [87]Sr from the country rock at the time of intrusion.

Faure and Jones (1974) also dated a suite of Wright Intrusives (Olympus Granite Gneiss and Dais Granite) from Wright Valley. The nine analyses give a loose whole-rock isochron of $488 \pm 43$ Ma (IR = 0.7109). Graham and Palmer (1987) hint that preliminary work on foliated granitoids in southern Victoria Land has produced ages of around 485 Ma, bearing out the data of Faure and Jones (1974), but this work has yet to be published. Jones and Faure (1969) had earlier combined Rb/Sr whole-rock data from Olympus Granite Gneiss, Dais Granite, and Vida Granite to produce an isochron of $480 \pm 14$ (IR = 0.7109).

Mapping by Cox and Allibone (1991) has shown that the Bonney pluton includes portions of Vida Granite, Dais Granite, and Olympus Granite Gneiss, from which five and possibly seven samples were collected for the study by

Faure and Jones (1974). By plotting these data separately, Cox and Allibone (1991) report whole-rock isochron ages of 512 Ma (IR = 0.7105) for five samples and 498 Ma (IR = 0.7101) for all seven samples, for the Bonney pluton, the main phase of their DV1 Suite.

U/Pb systematics have been applied only three times to the rocks of southern Victoria Land. In an early study, Deutsch and Grögler (1966) analyzed three size fractions of zircons from a sample of Olympus Granite Gneiss that spread along a chord intercepting concordia at the origin and at 588 Ma. The $^{207}Pb/^{206}Pb$ ages on the individual size fractions, from coarsest to finest, were $585 \pm 25$, $595 \pm 25$, and $576 \pm 25$ Ma. The authors argued that if inherited zircons were present the more heterogeneous fine fraction should have a higher $^{207}Pb/^{206}Pb$ age, or, stated oppositely, since the fine fraction is not older than the other fractions the total population of zircons either crystallized, was completely recrystallized, or suffered significant lead loss at around 588 Ma, in any case signaling a significant thermal event in the region.

The implication of these dates, which remained unacknowledged until the discussion of Skinner (1983b), is that if the Olympus Granite Gneiss is indeed Proterozoic, then the Koettlitz Group that it intrudes (or is derived from) is older, precluding correlation of Koettlitz Group with Cambrian sequences south of Byrd Glacier, as had been suggested by various authors (Gunn and Warren, 1964; Laird, 1981a, b; Skinner, 1982).

Vocke and Hanson (1981) analyzed four size fractions of zircons from Olympus Granite Gniess from the DVDP-6 hole for uranium and lead. The data were very discordant and plotted with a lower concordia intercept of $462 \pm 6$ Ma and a projected upper intercept of $2,555 \pm 330$ Ma. The $^{207}Pb/^{206}Pb$ age of the most discordant fraction (>80 mesh) was 947 Ma, suggesting that a component of the zircons is Proterozoic or older. Cathodoluminescence and microprobe analysis indicated that the zircons are zoned with low-uranium cores and high-uranium rims. Vocke and Hanson (1981) suggested that the 2555-Ma age may date the provenance age of detrital zircons or alternatively that the cores of the zircons formed in situ during an early metamorphic event. Either way, the authors state that due to the difference in uranium content, the difference in concordance between their samples and those of Deutsch and Grögler (1966) indicates that the zircons of the two studies experienced significantly different geological histories.

The conclusion that Olympus Granite Gneiss is Proterozoic in age was recently reinforced by Rowell et al. (1993b), who determined a U/Pb zircon date of $550.5 \pm 4$ Ma on a quartz syenite that crosscuts Skelton Group (both Cocks Formation and Anthill Limestone) north of Cocks Glacier. The granite itself is nonfoliated and is younger than at least two generations of folding in the Skelton Group, although its relationship to a third generation of folding recognized locally is uncertain.

Felder and Faure (1990) have dated the granitic rocks in the Brown Hills by both Rb/Sr and Ar/Ar techniques. A Rb/Sr whole-rock isochron on the Carlyon Granodiorite gave an age of $568 \pm 54$ Ma (IR = 0.7121), further demonstrating that magma generation had begun in the Neoproterozoic throughout a broad area of southern Victoria Land. Whole-rock data for the Mt. Rich Granite are more scattered, but suggest a similar age, though lower IR (= 0.7084). A whole-rock isochron age of $484 \pm 6$ Ma was obtained for dikes of leucocratic granite that crosscut the Carlyon Granodiorite and Mt. Rich Granite.

Hornblende and biotite were analyzed by incremental heating for $^{40}Ar/^{39}Ar$ (Felder and Faure, 1990). One hornblende gave an excellent plateau age of $534 \pm 5$ Ma. Several biotite plateau ages average $508 \pm 3$ Ma. An additional seven $^{40}Ar/^{39}Ar$ biotite ages for which total gas was released have a mean of $500 \pm 4$ Ma.

The chronology indicated for the granitic rocks in the Brown Hills is one of generation of magma around 568 Ma with cooling through approximately 550°C by about 534 Ma and through approximately 300°C by about 508 Ma, with subsequent intrusion of leucocratic dikes at about 484 Ma, which did not elevate the temperature of the surrounding rocks sufficiently to reset argon systems.

Although the timing of plutonism and cooling is beginning to emerge, the age of the metasedimentary rocks is poorly constrained. As already mentioned, an estimate of the $T_{DM}$ of the Skelton Group is 800–700 Ma, based on analyses of pillow lava found in the Cocks Formation (Rowell et al., 1993b).

Adams and Whitla (1991) undertook a Rb/Sr whole-rock study of the Asgard Formation collected from lower Wright Valley and upper Victoria Valley. They concentrated on "finer-grained protolithologies" with the assumption that those would be the best candidates for strontium homogenization during recrystallization. The data set from Wright Valley scatters within a long band whose bounding IRs are 0.707 and 0.713 and whose slope indicates an age of approximately 670 Ma. The authors offer the tentative conclusion that if the IRs are realistic, the 670-Ma age dates the time of metamorphism. If either interpretation of the origin of Olympus Granite Gneiss is correct – that is, that it was derived by metasomatism of Asgard Formation at the peak of regional metamorphism or that it intruded before or synchronously with the deformation and metamorphism of Asgard Formation, then the approximately 670-Ma date of Adams and Whitla (1991) is at odds with the U/Pb dating of Olympus Granite Gneiss (588 Ma) of Deutsch and Grögler (1966), but it nevertheless suggests thermal activity affecting Asgard Formation in the Neoproterozoic rather than the Cambro-Ordovician time frame of the traditional Ross orogeny.

Three of the samples analyzed by Adams and Whitla (1991) were marbles. These defined a whole-rock isochron of $840 \pm 30$ Ma (IR = 0.7081), close to the value of seawater during the Neoproterozoic. Arguing that because the abundance of strontium at the time of limestone deposition would far outweigh subsequent radiogenic strontium, the authors interpret the isochron age as indicating the time of sedimentation.

## Evolution of Southern Victoria Land

The earliest geological activity in southern Victoria Land is the deposition of shallow marine sediments of both arenaceous and calcareous affinity. If the correlation of Koettlitz and Skelton Groups is correct, this represents a single depositional cycle. Sedimentation had begun by 800–700 Ma, as indicated by pillow lavas in Skelton Group (Rowell et al., 1993), and may go back as far as about 840 Ma based on dating of marble in the Koettlitz Group (Adams and

Whitla, 1991). The pillow basalts most likely signal continental rifting at the time of their eruption.

Deformation, metamorphism, and plutonism began at a later time in the Neoproterozoic. The earliest intrusions, as exemplified by the Olympus Granite Gneiss, were intimately associated with the deformation, which dates perhaps around 588 Ma (Deutch and Grögler, 1966; Skinner, 1983b). Two episodes of deformation had affected Skelton Group by approximately 550 Ma, as dated by a crosscutting pluton (Rowell et al., 1993b). Deformation, however, continued after this time as well. The aforementioned pluton intruding Skelton Group accompanied or preceded a third episode of deformation locally, and many of the older gray granitoids in southern Victoria Land were also affected by deformation. Dating of these rocks falls in a range between approximately 510 and 490 Ma, and includes the DV1 Suite, which was likely to have formed due to subduction processes (Faure and Jones, 1974; Cox and Allibone, 1991; Smillie, 1992). By around 500 Ma portions of southern Victoria Land had cooled to temperatures sufficient to retain argon. Continuing to around 470 Ma were late-stage, posttectonic intrusions, including consistently oriented dike swarms and the DV2 Suite, likely to have formed in an extensional setting (Faure and Jones, 1974; Graham and Palmer, 1987; Smillie, 1992).

# 4 Central Transantarctic Mountains

## Geological Summary

The central Transantarctic Mountains are distinctive for the variety of geological units displayed across the region. These include a suite of upper-amphibolite-grade metamorphics, collectively known as the Nimrod Group, which crop out in the Miller and Geologists Ranges on the plateau side of the mountains and are deformed by a major, ductile shear zone. To the east of this and cropping out all the way to the ice shelf is a belt of lower-greenschist-grade graywackes and pelites called the Beardmore Group. Unconformably and tectonically overlying the Beardmore Group along its western portion is the fossil-dated Early Cambrian Shackleton Limestone, which crops out in a widening belt northward to Byrd Glacier. This is uncomformably overlain at places by clastic rocks of the Douglas Conglomerate and Starshot Formation, which together with the Shackleton Limestone comprise the Byrd Group. A variety of structural styles are represented in the various belts of the central Transantarctic Mountains, all of which are punctured by isolated plutons and small batholiths of the Granite Harbour Intrusives.

In the early interpretations of the tectonics of the region (Grindley and Laird, 1969; Grindley and McDougall, 1969) three orogenies were proposed: the Nimrod orogeny causing the ductile shearing of the Nimrod Group, before the Neoproterozoic Beardmore orogeny causing folding of the Beardmore Group, unconformably overlain by the Byrd Group, itself folded during the Cambro-Ordovician Ross orogeny. As will be discussed in a later section, the interpretations of orogenic events in the central Transantarctic Mountains have undergone considerable revision, in particular with regard to deformation of the Nimrod Group, for which there now is evidence that ductile shearing was active during the period 540–520 Ma (Goodge et al., 1993c).

## Chronology of Exploration

The central Transantarctic Mountains were first charted by the Polar Party of Scott's *Discovery* expedition. Their traverse took the party from Minna Bluff southward across the Ross Ice Shelf to 82°17′S. Towering in the distance was the twin-peaked Mt. Markham massif, named for Sir Clements Markham, president of the Royal Geographical Society and the acknowledged father of the expedition. An attempt to reach land at the group's end point was thwarted by a broad shear zone of crevasses, so no rock was examined.

**Figure 4.1.** View to northwest from Mt. Markham across Nimrod Glacier. Underlain by Nimrod Group, the Geologists Range marks the left distance. Underlain by Cobham and Goldie Formation, Gargoyle Ridge and the Cobham Range are the dark massif at the right rear. Kon Tiki Nunatak sits in Nimrod Glacier in front of Gargoyle Ridge. Hamilton Glacier drains the foreground, which is underlain by undifferentiated Goldie Formation and Shackleton Limestone and remains unvisited by geologists.

In 1908–9 Shackleton's Polar Party made its way to the polar plateau via the Beardmore Glacier. The group collected rocks at Buckley Island at the head of the glacier and from moraines at Granite Pillars and The Cloudmaker. David and Priestley (1914), Skeats (1916), and Mawson (1916) described the returned samples. Of particular note was the discovery of archeocyathid-bearing limestones in blocks from moraines and in outcrop at Buckley Island, which for 50 years remained the only in situ Cambrian fossils from the continent.

Retracing Shackleton's route, Scott's ill-fated Polar Party collected samples along Beardmore Glacier, which were recovered from the party's tent the following season. These samples were described by Debenham (1921) and Smith (1924). Among the collection were the first fossil plants from the continent, including *Glossopteris,* discovered in the Beacon sequence at Mt. Buckley and described by Seward (1914).

The next visit to the area took place during the IGY (1957–58) when J. H. Miller and G. Marsh, members of the New Zealand section of the Commonwealth Trans-Antarctic Expedition, sledged across from the tractor train bound for the Pole and touched ground at the southern tip of the Miller Range.

In 1959–60 members of the New Zealand Alpine Club Expedition climbed Mt. Kyffin near the eastern mouth of Beardmore Glacier. Geological descriptions were given by Oliver (1964, 1972). Between 1959 and 1966 a series of New Zealand geological and survey parties systematically mapped the areas between

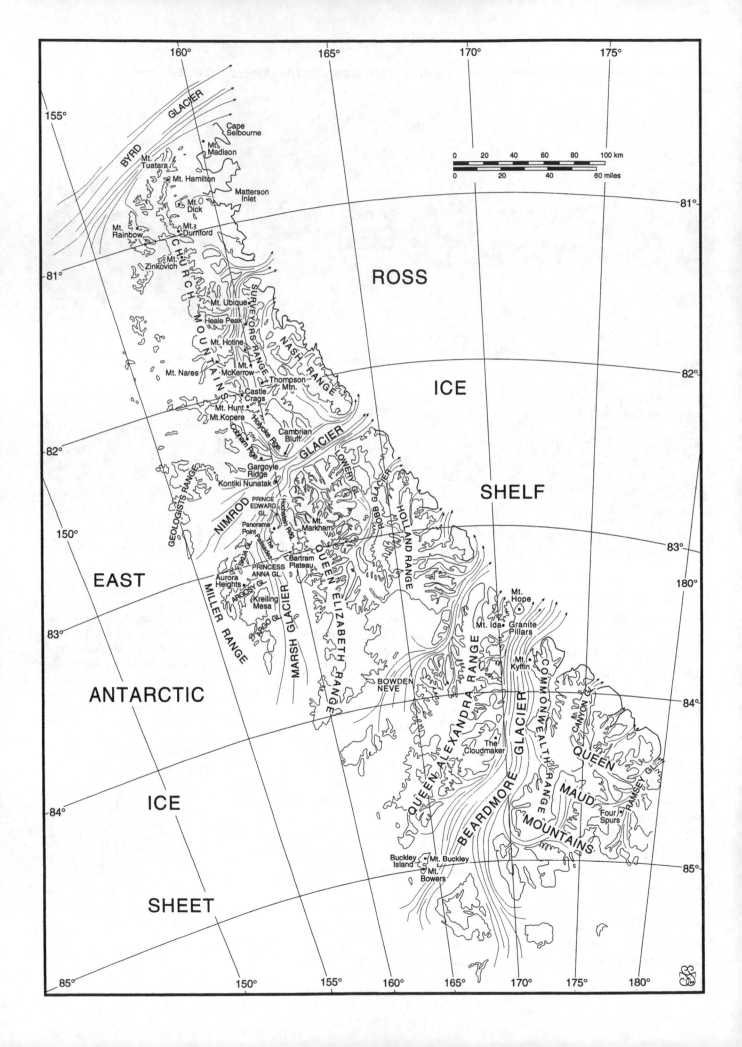

Byrd and Beardmore Glaciers (Gunn and Walcott, 1962; Grindley, 1963; Laird, 1963, 1964; Skinner, 1964, 1965; Grindley and Laird, 1969; Laird et al., 1971; Young and Ryburn, 1968).

A helicopter-supported party from Texas Tech University made observations in the Ramsey Glacier area in 1964–65 (Wade et al., 1965a, b; Wade and Cathey, 1986). Field parties from Ohio State University were active in the region between Nimrod and Ramsey Glaciers in 1967–68 (ground-based), and 1969–70 and 1970–71 (helicopter-supported) (Barrett et al., 1968, 1970; Elliot, 1970; Elliot and Coates, 1971; Barrett and Elliot, 1973; Lindsay et al., 1973; Elliot et al., 1974). New Zealand and U.S. parties operating from a helicopter-supported camp on Darwin Glacier visited outcrops south of Byrd Glacier in 1978–79 (Burgess and Lammerink, 1979; Stump et al., 1979; Stump, 1980). U.S. parties worked from a helicopter-supported camp on Bowden Névé in 1985–86 (Borg and DePaolo, 1986; Borg et al., 1986a; Rowell et al., 1986; Stump et al., 1986a). U.S. ground-based parties undertook detailed studies between Byrd and Nimrod Glaciers in 1984–85, 1986–87, and 1987–88 (Rees et al., 1985, 1987, 1988) and in the upper Nimrod Glacier area in 1989–90 and 1990–91 (Goodge et al., 1990, 1991b).

## Nimrod Group

Cropping out in the Miller and Geologists Ranges at the head of Nimrod Glacier are the metamorphic rocks of the Nimrod Group. In various accounts (e.g., Stump, 1973; Elliot, 1975; Grindley, 1981) these rocks have been considered the cratonic margin to the geoclinal deposits of the Ross orogen. From the field descriptions of Miller and Marsh from the southern tip of the Miller Range and their own petrographic observations, Gunn and Walcott (1962) named the Miller Formation, designated a type locality, and described gray and white granular marble and black amphibolite schist.

The Miller Range was subsequently mapped in reconnaissance fashion by members of the Northern Party of the 1961–62 New Zealand Geological and Survey Expedition (Grindley, 1963). The name "Nimrod Group" was proposed for all of the metamorphic rocks found therein and subdivided into four formations, in ascending order, the Worsley, Argosy, Aurora, and Miller (Grindley et al., 1964). Grindley and Laird (1969) later reported that the lower three formations are conformable, but that the Miller Formation structurally overlies the others across a major discontinuity that Grindley (1972) named the Endurance thrust. Grindley and Laird (1969) also added the name "Argo Gneiss" for occurrences of dioritic orthogneiss and eclogite that had been incorporated along the thrust. The Geologists Range was visited by a New Zealand party in 1964–65, but the findings were never published, except to be mapped as Nimrod Group by Grindley and Laird (1969).

In 1967–68 a U.S. party studied the Miller Range south of Argosy Glacier (Barrett et al., 1968). Gunner (1969) described in detail the petrography of the metamorphic rocks that he encountered, grouping them according to lithology rather than formation. The most detailed geological map to date of the Miller Range is that of Grindley (1972) (Fig. 4.4). Focused studies by U.S. parties in

**Figure 4.2** (*facing page*). Location map of the central Transantarctic Mountains.

the late 1980s and early 1990s have amplified, and to some degree modified, earlier conclusions (Borg et al., 1986a, 1987, 1990; Goodge et al., 1990, 1991a, b, 1993a, b). Goodge et al. (1991a) have demonstrated that shearing is distributed in a broad zone around the Endurance thrust throughout both the Miller and Geologists Ranges and have proposed the name "Endurance shear zone" for this feature. They also recognized that the Aurora Formation had a magmatic rather than a sedimentary protolith and accordingly used the name "Aurora or-thogneiss." Previously Borg et al. (1990) had called these rocks the "Camp Ridge Granodiorite." In their mapping of the Geologists Range, Goodge et al. (1990) found lithologies similar to those in the Miller Range, but mapped them by rock type rather than formation (Fig. 4.5). In fact, in their recent mapping of the Miller Range, Goodge et al. (1993c) have abandoned a formational subdivision, choosing instead to identify the units by lithology collectively termed Nimrod Group. Although in so doing Goodge et al. (1993b) are acknowledging their uncertainty of the stratigraphy and its coherency throughout the Miller and Geologists Ranges, the original subdivisions of Grindley et al. (1964) are useful in describing the major lithological associations, and so these are followed in the subsequent sections.

## Miller Formation

The Miller Formation is a layered sequence of polydeformed gneisses and schists, with lesser micaceous quartzite, marble, calc–silicate, and amphibolite. Cropping out in the small ridges at the head of Argosy Glacier and throughout much of the exposure south of Argo Glacier, it is structurally the highest unit in the Miller Range, in contact with the Aurora Orthogneiss across the Endurance shear zone. The Miller Formation is thought to be the oldest unit of the Nimrod Group, based on its thrust relation with the other formations, its higher metamor-phic grade, and the possibility that it underwent an episode of deformation before the others (Goodge et al., 1991a).

The metamorphic grade of the Miller Formation is characterized by upper amphibolite to transitional granulite-facies assemblages (Goodge et al., 1991a). Gneisses contain hornblende + plagioclase + biotite ± garnet ± epidote ± clinopyroxene. Pelitic schists include kyanite ± sillimanite. Calc–silicate rocks contain calcite + quartz + scapolite + clinopyroxene ± epidote ± tremolite. Geothermometry and geobarometry indicate peak metamorphic temperatures of 680°–750°C and pressures of 8–14 kb (Goodge et al., 1991a, 1993a).

## Worsley Formation

The marbles and calc schists cropping out at the eastern end of Aurora Heights and northward toward the head of Skua Glacier were designated the Worsley Formation by Grindley et al. (1964). These rocks lie stratigraphically beneath the Argosy Formation. Lacking good stratigraphic control, Borg et al. (1990) chose to lump the Worsley as part of a composite Argosy Formation; however, Goodge et al. (1991a) retained Grindley et al.'s (1964) original nomenclature.

The Worsley Formation contains tremolite-bearing marble beds 10–80 m thick and garnet–biotite schist units up to 150 m thick, with a total thickness of the

**Figure 4.3** (*facing page*). Location map for figures in Chapter 4.

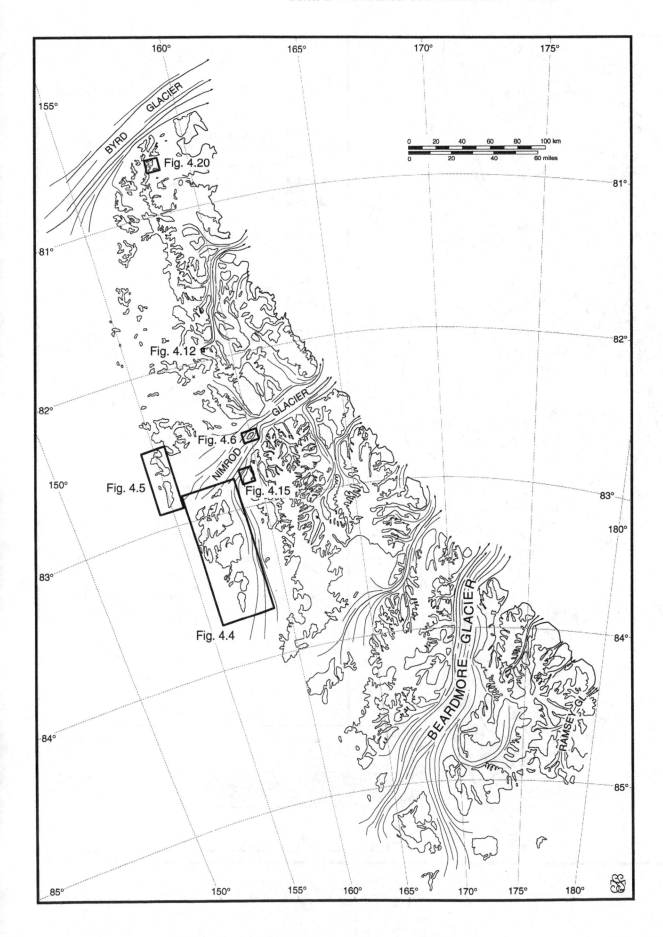

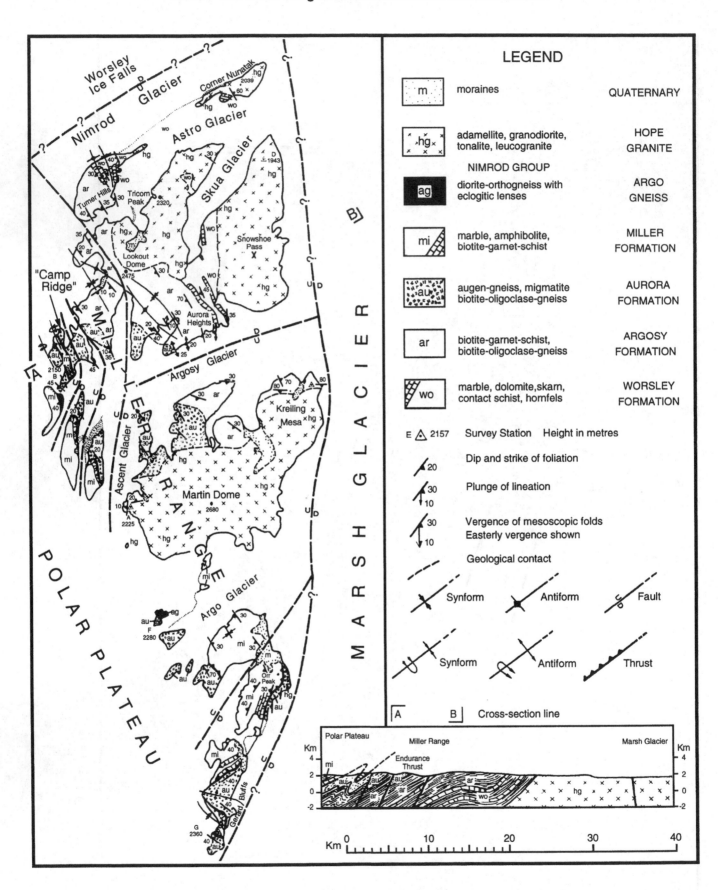

LEGEND

| | | |
|---|---|---|
| m | moraines | QUATERNARY |
| hg | adamellite, granodiorite, tonalite, leucogranite | HOPE GRANITE |

NIMROD GROUP

| | | |
|---|---|---|
| ag | diorite-orthogneiss with eclogitic lenses | ARGO GNEISS |
| mi | marble, amphibolite, biotite-garnet-schist | MILLER FORMATION |
| au | augen-gneiss, migmatite biotite-oligoclase-gneiss | AURORA FORMATION |
| ar | biotite-garnet-schist, biotite-oligoclase-gneiss | ARGOSY FORMATION |
| wo | marble, dolomite, skarn, contact schist, hornfels | WORSLEY FORMATION |

E △ 2157    Survey Station    Height in metres

⟋ 20    Dip and strike of foliation

↑ 30    Plunge of lineation
↓ 10

↑ 30    Vergence of mesoscopic folds
↓ 10    Easterly vergence shown

– –    Geological contact

⟋ Synform    ⟋ Antiform    Fault

⟋ Synform    ⟋ Antiform    Thrust

A ⌐    B ⌐    Cross-section line

formation indicated at about 700 m (Grindley et al., 1964). The formation is intruded to the east and the north by a pluton of the Granite Harbour Intrusives, in *lit-par-lit* fashion at places, which has resulted in contact metamorphism producing diopside and cummingtonite in the marbles and andalulite–sillimanite in the schists.

## Argosy Formation

The Argosy Formation crops out on the south side of Argosy Glacier at Greene Ridge, and on the north side from Aurora Heights to Turner Hill. The formation is predominantly a garnet–biotite schist sequence with intermittent beds of psammitic gneiss. Minor lithologies include quartzite, calc–silicate schist, marble, and amphibolite. Relict cross-bedding, laminations, and possible relict graded bedding are pictured by Gunner (1969, pp. 20, 21) from occurrences on Kreiling Mesa. Grindley et al. (1964) state that the thickness of the Argosy Formation is approximately 1,000 m.

Although previously the whole of the Nimrod Group was thought to have been metamorphosed to amphibolite grade (e.g., Grindley, 1972; Gunner, 1976), Goodge et al. (1991a) have recognized greenschist-facies assemblages in at least part of the Argosy and Worsley Formations. South of Argosy Glacier, pelitic schists and phyllites contain muscovite + biotite ± chlorite ± garnet, and farther to the west typical schists contain muscovite + biotite + garnet ± epidote ± chloritoid.

## Aurora Orthogneiss

The Aurora Formation, first described as feldspathized, lenticular-banded, biotite gneiss, was considered by Grindley et al. (1964) to be of metasedimentary origin and to have been deposited stratigraphically above the Argosy Formation, although they did recognize "diorite-*ortho*-gneiss, in places verging towards granodiorite" (p. 209) in part of the unit. Later treatments of the Nimrod Group, however, stressed the sedimentary protolith of the Aurora (e.g., Grindley and Laird, 1969; Grindley, 1981; Adams et al., 1982). Borg et al. (1986a) recognized the Aurora to be a mylonitized gneiss derived from granodiorite that was intruded into adjacent schists. In order to emphasize the reinterpreted protolith, Borg et al. (1990) renamed this unit the Camp Ridge Granodiorite, while Goodge et al. (1991a) chose to modify the original name to Aurora Orthogneiss, the nomenclature followed herein. However, as discussed later there now appears to be a distinction between the orthogneiss found at "Camp Ridge" and those found elsewhere in the Miller Range.

## Mafic and Ultramafic Tectonic Blocks

Enclosed within sheared portions of the Nimrod Group are mafic and ultramafic tectonic blocks, the cores of which contain eclogitic mineral assemblages (Goodge et al., 1993a). These rocks were named the Argo Gneiss by Grindley and Laird (1969). Mineralogical data from the mafic blocks by Goodge et al. (1993a) indicate that the eclogitic metamorphism took place at temperatures of

**Figure 4.4** (*facing page*). Geological map and cross section of the Miller Range. Detail at the head of Argosy Glacier shows folding of Endurance shear zone. Compare with Figure 4.15. After Grindley (1972). Used by permission.

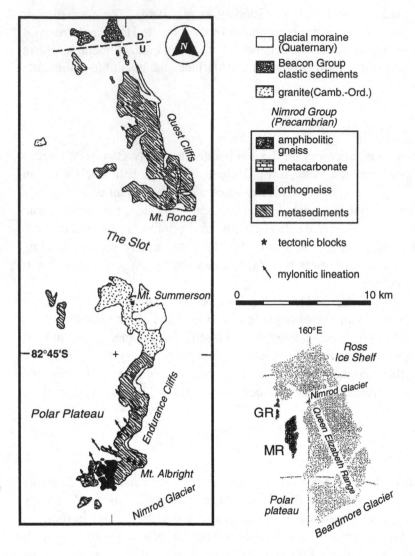

**Figure 4.5.** Geological sketch map of the Geologists Range, with subdivision of Nimrod Group by lithology. Mylonitic tectonites with gently northwest-plunging lineations are widespread. After Goodge et al. (1990). Used by permission.

approximately 600°C and pressures of 12–25 kb. The eclogitic blocks are surrounded by a coarser-grained rim with a (garnet–)amphibolite assemblage characterized by hornblende + plagioclase ± garnet ± biotite, which was probably formed during the upper amphibolite to granulite-facies metamorphism (somewhat higher temperature and lower pressure) of the enclosing gneisses and schists (Goodge et al., 1993a). The blocks are subspherical to ellipsoidal, with sizes from 0.5 to 50 m. Foliation of the surrounding sheared rocks wraps around the blocks, indicating their relative rigidity at the time of ductile shearing.

## Isotopic Determinations

Gunner and Faure (1972) undertook a Rb/Sr study of the Nimrod Group. The results showed that these rocks were of greater antiquity than any others known in the Transantarctic Mountains. Their best-guess estimate of 1,980 Ma for partial homogenization of the samples (either diagenesis or metamorphism) was based on whole-rock data that scattered so widely on an isochron diagram, and depending on the grouping of samples gave such varying initial strontium ratios, that the significance of the postulated age is questionable.

Sm/Nd isotopic determinations on samples from the Miller and Argosy Formations, and a sample of orthogneiss from "Camp Ridge" that is closely associated with Miller Formation, all have $T_{DM}$ ages of 2.7–2.9 Ga (Borg et al., 1990). Discordant U/Pb data from zircons were interpreted by Gunner (1982; Gunner and Mattinson, 1975) as indicating source material for the Nimrod Group possibly as old as 2.8 Ga.

A widely quoted personal communication from V. C. Bennett (Borg et al., 1990; Goodge et al., 1991a) suggests that the orthogneiss at "Camp Ridge" has single-crystal Pb/Pb zircon ages of ~1.7 Ga, dating the time of magmatic crystallization. However, samples of several other orthogneisses and plutonics with various degrees of shearing, some of which were mapped as Aurora Orthogneiss, have U/Pb zircon ages of 541–521 Ma (Goodge et al., 1993c). If both sets of dates are correct, then two portions of what has been called Aurora Formation and Aurora Orthogneiss are distinct in terms of age and tectonic history. Acknowledging this distinction, Goodge et al. (1993c) speak of the "Camp Ridge orthogneiss" in reference to the locality dated as 1.7 Ga. Also, due to uncertainties in discerning a meaningful stratigraphy, they map all rocks in the Miller Range by their lithologies under the single name "Nimrod Group."

In a footnote to an earlier work, Goodge et al. (1991a) discuss the uncertainty of whether the orthogneiss on "Camp Ridge" intrudes Miller Formation as discussed by Borg et al. (1990) or Argosy Formation by their own reckoning, and also whether what they call Argosy Formation on "Camp Ridge" is the same as the Argosy Formation as mapped by Grindley (1972) throughout the rest of the Miller Range.

Regardless of the confusion over the integrity of the various units, the important conclusions of Goodge et al. (1993c) are that movement of the Endurance shear zone was occurring during emplacement of plutonic rocks in the interval 541–521 Ma and that it had begun at some time before this due to the lesser development of shear fabric in the plutonics than the surrounding metamorphic rocks. Metamorphic monazite from pelitic schist of the Nimrod Group, interpreted to have grown during the shearing, was U/Pb-dated at 525–522 Ma. In addition, U/Pb dating of zircon from an undeformed pegmatite indicated that the shearing had ceased by about 515 Ma.

Early K/Ar studies of Nimrod Group gave dates ranging from 1,000 to 450 Ma, with the oldest dates postulated to record the Nimrod orogeny (McDougall and Grindley, 1965; Grindley and McDougall, 1969). Further K/Ar investigation revealed that the broad spread before approximately 570 Ma was due to excess argon contamination (Adams et al., 1982a); however, these authors postulate, essentially from no data, that the Beardmore orogeny was occurring in the interval 650–580 Ma.

An Ar/Ar study by Goodge and Dallmeyer (1992) affirms that the older K/Ar dates on Nimrod Group are due to excess argon. Release spectra indicate hornblende cooling ages of 525–487 Ma and muscovite cooling ages of 508–486 Ma. This range of ages overlaps the K/Ar dates on plutonic rocks from the Miller Range and elsewhere in the central Transantarctic Mountains.

## Beardmore Group

The Beardmore Group was named by Gunn and Walcott (1962) for metagraywackes and argillites found between Beardmore and Nimrod Glaciers (Goldie Formation), and gneisses and marbles at the southern end of the Miller Range (Miller Formation). Subsequently, the Miller Formation was assigned to the Nimrod Group (Grindley, 1963), and the Cobham Formation, a sequence of schist, marble, and quartzite in the Cobham Range, was included in the Beardmore Group (Laird et al., 1971). Several other formations, cropping out in the Queen Maud Mountains, have been assigned to the Beardmore Group (Stump, 1982) (see Chapter 5), and so too the Patuxent Formation of the Pensacola Mountains by some authors' correlations.

## Cobham Formation

Cobham Formation occurs at a single locality along the western margin of the Cobham Range between Mt. Kopere and Gargoyle Ridge, and perhaps at the western end of Kon Tiki Nunatak, where it has been observed only from the air (Goodge et al., 1991b) (Fig. 4.6). Laird et al. (1971) designated a type locality as the western end of Gargoyle Ridge and measured a 510-m section. The top of the formation, marked by the highest marble bed in the sequence, is conformably overlain by Goldie Formation (Fig. 4.7). The base of the Cobham Formation is not exposed. A high-angle reverse fault at the southwestern end of the Cobham Range duplicates a small portion of the type section.

Lithologically, the Cobham Formation is a sequence of schists, calc schists, marbles, and quartzites (Fig. 4.8). Alternating pelitic schist and quartzite occur throughout the sequence. Quartzite is the most common rock type in the lower

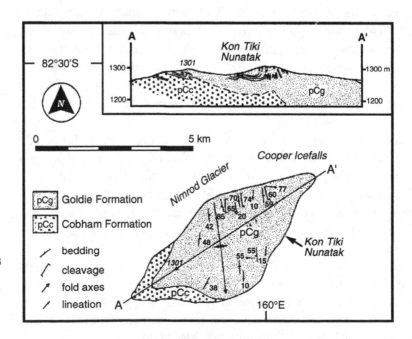

**Figure 4.6.** Geological map and cross section of Kon Tiki Nunatak. Cobham Formation viewed from the air. After Goodge et al. (1991b). Used by permission.

**Figure 4.7.** Southwest corner of the Cobham Range at its intersection with Gargoyle Ridge. Cobham Formation extends to the summit. Contact with Goldie Formation is set at the top of the last marble in the Cobham Formation, indicated on photo. A high-angle reverse fault cuts across the face as indicated. The location of the recumbent fold pictured by Laird et al. (1971, Figure 19, p. 464) is indicated by an arrow.

300 m of the section, whereas marble is more prevalent in the upper portion. Bedding and fine lamination are the only sedimentary structures observed. The bedding ranges from several centimeters to 15 m. Units toward the top of the section are up to 45 m thick, but bedding within them may have been obscured by recrystallization.

Regional metamorphism accompanying deformation has produced biotite–almandine in the pelites and tremolite in the calc schists of the Cobham Formation. Later contact metamorphism, presumably associated with emplacement of the Granite Harbour Intrusives, created a granoblastic texture throughout much of the formation. The environment of deposition of the Cobham Formation was probably shallow-water marine at the inner margin of a continental shelf (Laird et al., 1971).

**Figure 4.8.** Calc schists in Cobham Formation at southern end of the Cobham Range.

## Goldie Formation

Cropping out for more than 400 km between Starshot and Ramsey Glaciers, the Goldie Formation is one of the most areally extensive units in the basement of the Transantarctic Mountains. The first specimens, fragments of gray, green, and purple slate and phyllite, collected from the western moraine of Beardmore Glacier by Shackleton's Polar Party, were later described by Mawson (1916).

Throughout its exposure, the Goldie Formation exhibits great uniformity of lithology, with alternating beds of graywacke and pelite, in general metamorphosed to the extent that fine-grained portions may be described as phyllite, argillite, or slate. The graywacke units vary in thickness from several tens of centimeters to about 2 m, while the pelites are generally less than a meter thick. The rocks are typically medium to dark gray, in many places weathering to brown; however, purple and green slates have also been reported (Mawson, 1916; Grindley, 1963).

The graywackes typically are poorly sorted and matrix-supported. In some cases the matrix is a minor enough component for the rocks to be arenites or quartzites. Clasts in the graywackes are subangular to subrounded. Quartz predominates, with plagioclase, microcline, and muscovite as minor constituents. Trace detrital minerals include apatite, sphene, zircon, epidote, tourmaline, and hornblende (Gunn and Walcott, 1962; Edgerton, 1987). Rock fragments are conspicuously absent, except for Gunner's (1976) report of minor fragments of mica schist, granite, and chert at several localities. Nowhere have volcanic or pelitic rock fragments been seen. Source rocks for the Goldie Formation appear to have been crystalline, distant from any area of tectonic activity.

A common feature of the Goldie Formation from throughout its area of exposure is the presence of calcite within the matrix of the graywackes. In some cases it is abundant enough to warrant calling the rock a calcarenite. In addition, rare limestone beds have been reported by several authors (Gunn and Walcott, 1962; Laird et al., 1971; Oliver, 1972; Gunner, 1976). Fine conglomerates with rounded clasts of quartz or calcite or both have also been reported from a few localities (Laird et al., 1971; Oliver, 1972; Gunner, 1976).

The most common sedimentary structure in the Goldie graywackes is graded bedding, observed by various authors (Laird et al., 1971; Oliver, 1972; Gunner, 1976; Stump et al., 1991). Although the graywackes superficially appear to be massive, subtle grading is quite common. Also, the tops of graywacke beds commonly grade up into the overlying pelite, whose top is in sharp, or even scoured, contact with the overlying graywacke (Stump et al., 1991). Laminations typically occur in the pelites. Other sedimentary structures of limited occurrence include ripple marks, ripple drift lamination, and cross-bedding. It is generally agreed that the Goldie Formation was deposited through the action of turbidity currents (Laird et al., 1971; Stump, 1976a).

Diamictite occurs at four horizons within the Goldie Formation at Panorama Point on the northern end of Cotton Plateau (Stump et al., 1988). The thickest of these (about 6 m) lies directly below a thick pile of pillow basalt. The other three units are thinner (70 cm to 1 m) and are several tens of meters downsection.

The thickest unit is mainly structureless, but does have some parallel-laminated portions. Sparse pebbles and cobbles are found throughout the unit, and pebbly pods occur locally, bound in a pelitic matrix. Clasts range in size from fine sand to boulders up to 40 cm in diameter. Clast lithologies include quartzite, calcareous quartzite, dolomite, and quartz–muscovite schist. The three thinner units are structureless and contain smaller clasts and sandier matrix than the thick unit. Definitive criteria such as striated clasts and dropstone sags are lacking, possibly due to the high degree of shearing suffered by the uppermost unit, but the overall lithological characteristics suggest possible glaciogenic affinity (Stump et al., 1988).

Igneous rocks associated with the Goldie Formation have been reported from three localities. At Four Spurs in the Ramsey Glacier area the sediments are intruded by three sills of basalt, described as varying somewhat in composition, but composed of plagioclase (andesine–labradorite), augite, chloritized hypersthene, and opaques (Wade and Cathey, 1986). Whether the basalts are similar in age to the Goldie Formation or were emplaced during later periods of igneous activity (Paleozoic or Mesozoic) was not suggested by the authors. Oliver (1972) reports "a small amount of felsite or acid lava" on the eastern side of Beardmore Glacier, but no further description is given.

Both metabasalt and rhyolite crop out in limited occurrences at Cotton Plateau interbedded with Goldie sedimentary rocks (Stump et al. 1991). Yellow-weathering metarhyolite occurs about 50 m downsection from the metabasalt at Pan-

**Figure 4.9.** Pillow basalt in Goldie Formation at Panorama Point.

orama Point. It is approximately 5 m thick and directly underlies the lowermost of the thin diamictite units. In thin section the rock is an intergrowth of finely recrystallized quartz, plagioclase, muscovite, and garnet. Sprinkled sparsely throughout are crystal fragments of quartz (up to 1 mm), suggesting an explosive origin for this rock. High sodium and low potassium contents of the rhyolite indicate that it was probably spilitized.

The metabasalt occurs at two locations: (1) along the crest and upper slopes of the Palisades around Panoroma Point, where it occupies the core of a syncline and is the highest unit stratigraphically in the Goldie, and (2) on the western and headward walls of Prince Edward Glacier. There a contact with bedded Goldie Formation at glacier level was observed through binoculars, giving the appearance of a fault.

As discussed later, at The Palisades some portions of the metabasalt are highly sheared, some portions are not. An area on the upper slopes just south of Panorama Point preserves excellent pillow structure (Fig. 4.9). Pillow breccia was also observed. Toward the southern end of the outcrop area within the zone of shearing, coarse-grained gabbros are associated with the basalts. They appear to intrude the basalts in an irregular fashion, but the poorly preserved condition of the rock precluded a sure interpretation.

The metabasalts are composed of a metamorphic assemblage of calcium–amphibole–clinozoisite–biotite–plagioclase. Some portions contain altered euhedral plagioclase phenocrysts with relict albite twinning; other portions are non-porphyritic. Calcite occurs in the interstices between some grains. Amygdules are of limited occurrence.

The gabbro contains plagioclase and clinopyroxene with a hypidiomorphic granular texture and considerable variability of grain size (Borg et al., 1990). The plagioclase is somewhat altered to epidote, and an alteration rim of amphibole surrounds the pyroxene. Some calcite is also present in the interstices and along fractures.

Sm/Nd analysis of whole-rock basalt and gabbro, plus plagioclase and pyroxene, separates from the gabbro plot to give an isochron age of $746 \pm 82$ Ma (Borg et al., 1990). The whole-rock gabbro value falls somewhat off the isochron. Borg et al. (1990) argue that it may be rejected owing to the presence of calcite and other alteration material in the whole rock. When this is done, the three-point isochron gives a value of $762 \pm 24$ Ma, which the authors interpret as the best date for the magmatism. This is an important number, for it is the only direct date on the age of deposition of the Goldie Formation. It compares with the date of (700–800 Ma) on the pillow basalt in the Cocks Formation (Skelton Group) (Rowell et al., 1993b). $\varepsilon_{Nd}$ values on the basalt and gabbro at The Palisades are around $+6.8$, indicating an oceanic affinity for these rocks (Borg et al., 1990). $T_{DM}$ ages dating the provenance of Goldie graywacke from near the mouth of Nimrod Glacier average 1.65 Ga (Borg et al., 1990).

Metamorphism of Goldie Formation is most conspicuously of a thermal type, produced posttectonically throughout the area during intrusion of Granite Harbour Intrusives. The development of slatey cleavage during deformation of Goldie Formation was accompanied in some places by the growth of sericite in a preferred orientation. No further reports of regional metamorphism have been made, however, except in the area of Cotton Plateau. There muscovite parallels cleavage, and two generations of biotite are present, the older with a spongy texture and preferred orientation, presumably dynamically produced, and the younger, randomly oriented and euhedral due to thermal effects (Edgerton, 1987).

Thermal metamorphism of Goldie Formation is widespread (Gunn and Walcott, 1962; Laird, 1963, 1964). Spotted slates are reported from many localities. Assemblages in the graywackes and the pelites include biotite $\pm$ muscovite $\pm$ chlorite. In some cases chlorite is seen to replace biotite. Quartz–feldspar–calcite spots are also reported (Gunn and Walcott, 1962). These lower-grade thermal effects are recorded as far as 30 km from outcrops of igneous rocks, leading to the conclusion that much of the area must be underlain by plutonic rocks at reasonably shallow depths. Close to granitic plutons higher-grade hornfels assemblages are found. Hornblende is common in the metagraywackes, andalusite $\pm$ cordierite in the pelites, and diopside in some places in calcareous rocks. K/Ar dating of two whole-rock samples of Goldie slate gave ages of 477 and 465 Ma, indicating the time of cooling following thermal metamorphism (Adams et al., 1982a).

On the basis of the similarity of calcareous units in their lower portions passing upward into predominantly pelitic sequences, Stump et al. (1991) have suggested that the Cobham and lower Goldie Formations are correlative with the Worsley and Argosy Formations of the Miller Range. Laird et al. (1971) also suggested possible correlation between the Cobham Formation and rocks in the Geologists Range. However, the $T_{DM}$ ages of Argosy Formation (2.72 Ga) and Goldie Formation (1.65 Ga) preclude the two having a common source (Borg et al., 1990). The Pb/Pb age of 1.7 Ga on the "Camp Ridge orthogneiss" indicates that it is distinctly older than the Goldie Formation dated at approximately 760 Ma by the gabbro at Cotton Plateau. If the "Camp Ridge orthogneiss" is intruding the Argosy Formation, then the correlation is precluded; however, if it intrudes Miller Formation, a correlation is allowable. Although there is uncertainty as to what unit the "Camp Ridge orthogneiss" intrudes (see footnote in Goodge et al., 1991a, p. 61), the difference in $T_{DM}$ ages alone is enough for the author to back away from his suggested correlation and consider the Nimrod and Beardmore Groups as having been deposited during different episodes.

## Flattened Conglomerate

A poorly sorted conglomerate with clasts ranging in size up to 1 m in diameter crops out at the southern end of Cotton Plateau and on the northern end of Bartram Plateau (Laird et al., 1971). The clasts are highly flattened (aspect ratios as high as 20:1) in the more easterly exposures (Fig. 4.10) and become less so to the west, where they can be seen to be rounded to subrounded and clast-supported (E. Stump, D. Edgerton, and R. Korsch, unpub. data). The lithology of the clasts includes quartzite, calcarenite, and mafic volcanics. The matrix is composed of quartz, albite, sericite, biotite, muscovite, chlorite, actinolite, and unstrained calcite.

The stratigraphic affinity of this conglomerate is uncertain. Laird et al. (1971) suggested that it is a member of either the Goldie or Cobham Formation. This is a reasonable assumption given the metamorphic grade and the fact that all of the clasts could be derived from recognized lithologies in the Cobham and Goldie Formations. The lithology is different from the Douglas Conglomerate insofar as clasts of silicic plutonic rocks and abundant limestone are not represented. The deformation of this rock will be discussed in the following section.

## Byrd Group

The name "Byrd Group" was proposed by Laird (1963) for the strongly folded limestones and lesser conglomerates, sandstones, and shales cropping out from Byrd Glacier to south of Nimrod Glacier and at an isolated exposure at the head of Beardmore Glacier. The most extensive formation is the Shackleton Limestone, named by Grindley (1963) in honor of Ernest Shackleton, whose Polar Party discovered the archaeocyathid-bearing limestone in outcrop south of Buckley Island during their bid for the Pole. The rock was described by Skeats (1916). For more than 50 years this stood as the only in situ fossil-dated rock of Cambrian age from the continent. Conglomerates containing archaeocyathid-bearing clasts were also collected by Shackleton's party from moraines at Granite Pillars and The Cloudmaker, and later described by Taylor (in David and Priestley, 1914).

**Figure 4.10.** Flattened conglomerate at southern end of Cotton Plateau.

These conglomerates have been recognized as analogous to the Douglas Conglomerate (Rowell and Rees, 1989), a formation named by Skinner (1964) for occurrences in the northern Churchill Mountains. Skinner (1964) also named the Dick Formation for a sequence of argillites and sandstones with grit lenses in the same vicinity. The remaining member of the Byrd Group is the Starshot Formation, named by Laird (1963) for a sequence of calcareous sandstone, shale, grit, and conglomerate exposed on the eastern side of Starshot Glacier.

## Shackleton Limestone

Although Grindley (1963) named the Shackleton Limestone, Laird (1963) mapped a much more extensive area of exposure and designated the type locality as a site on the east side of the central Holyoke Range. Laird et al. (1971) reported thicknesses in the Holyoke Range of 8,300 m west of Mt. Hunt and 4,400 m at Cambrian Bluff. Skinner (1964) reported a thickness in excess of 3,660 m south of Mt. Hamilton in the northern Churchill Mountains. Due to structural complications, both folding and faulting, as well as discontinuous exposure due to snow cover, Rees et al. (1985; Rowell et al., 1988b; Rowell and Rees, 1991) have questioned the validity of these extreme thicknesses, suggesting instead that the thickness of the Shackleton Limestone in the Holyoke Range is probably between 1,000 and 2,000 m. Likewise, Burgess and Lammerink (1979) suggest a thickness of 2,000 m south of Mt. Hamilton. Stump et al. (1979) measured a 1,940-m section of Shackleton Limestone in continuous outcrop at Mt. Tuatara.

Shackleton Limestone contacts Goldie Formation at five known localities: in the Holyoke Range on the east side of Cambrian Bluff and west of Mt. Hunt, in the vicinity of Cotton Plateau at the Palisades, at Mt. Lowe, and north of the mouth of Princess Anne Glacier. Laird et al. (1971) consider all of them to be unconformities. Rowell et al. (1986), however, interpret the contacts to be faults where they have examined them, in the Holyoke Range and around Mt. Lowe. In addition, Laird et al. (1971) register uncertainty about the contact north of Princess Anne Glacier.

At The Palisades a syncline of Shackleton Limestone truncates folded Goldie Formation (Laird et al., 1971, Fig. 20; Stump et al., 1991) (Fig. 4.11). Beneath the horizontal basal beds of Shackleton at the trough of the syncline, the Goldie Formation is folded into tight, recumbent folds. The exposed section of Shackleton Limestone gives the impression of being the basal portion of the formation, for the lowermost beds are micaceous sandstones, and sandstones are interbedded throughout the lower 160 m of the formation. Above this the Shackleton is pure limestone. The contact is marked by a yellow clay-rich zone about 1 m thick that could be fault pug. However, immediately adjacent to this, the basal beds of Shackleton show no evidence of shearing. On the western limb of the Shackleton syncline the contact between Shackleton Limestone and Goldie Formation was highly sheared, probably during formation of the syncline. Nevertheless, the relations at the trough of the syncline point to this contact as an angular unconformity, with the folding of Goldie Formation followed by erosion and deposition of the Shackleton Limestone. The interbedded quartzites at the base of the section at The Palisades are missing in the limestones at Mt. Lowe, which adds support to Rowell et al.'s (1986) interpretation that that locality is a fault.

The Shackleton Limestone is typically gray, although pink, cream, white, and black shades occur at places. The formation is well bedded, exhibits a great variety of textures, and, except for the interbedded quartzite layers in the basal section at The Palisades, its lithology is almost entirely calcareous or dolomitic. A number of conglomeratic or brecciated horizons originally included in the Shackleton Limestone (Laird, 1963; Laird et al., 1971) have been reinterpreted as infaulted packets of Douglas Conglomerate (Rees et al., 1985; Rees and

**Figure 4.11.** Unconformity between Shackleton Limestone and Goldie Formation. Shackleton Limestone is horizontally bedded at the trough of a syncline. The Goldie Formation beneath it is recumbently folded, as seen in Figure 4.16. Located midway down The Palisades below Panorama Point.

Rowell, 1990). Likewise, a 200-m-thick unit of megabreccia west of Mt. Hamilton, interpreted as a slideblock within Shackleton Limestone (Burgess and Lammerink, 1979), has been assigned to Douglas Conglomerate (Rees et al., 1988).

The Shackleton Limestone is strongly folded throughout its extent but is essentially unmetamorphosed, except in the vicinity of Cotton Plateau and in the vicinity of Byrd Glacier where portions of the limestone are recrystallized to white marble (Laird et al., 1971). At Cotton Plateau anhedral biotite and zoisite are found in the basal quartzites at The Palisades (Edgerton, 1987), and wollastonite rosettes occur in marble adjacent to a dolerite sill at Mt. Lowe (Laird et al., 1971). Adjacent to Byrd Glacier the north face of Mt. Tuatara is recrystallized to a coarse-grained marble. In the northern Churchill Mountains Skinner (1965) reported very minor chlorite within the limestones. At Mt. Madison amphibolite-grade metamorphism was reached in the marbles and schists of the Selbourne Marble (Skinner, 1964), but whether this formation is part of the Byrd Group is equivocal, as discussed later.

The sedimentary environment of the Shackleton Limestone has been interpreted as a shallow-water carbonate shelf, emergent in part and populated with archeocyathid bioherms (Laird et al., 1971; Burgess and Lammerink, 1979; Rees et al., 1989b). Laird et al. (1971) and Burgess and Lammerink (1979) called for tectonic instability and penecontemporaneous erosion and slumping to account for conglomeratic and brecciated units within the formation. As mentioned earlier, these coarse-grained units were in all cases interpreted by Rees et al. (1985, 1988) as belonging to the Douglas Conglomerate, leaving only lithologies indicative of a tectonically quiescent basin of low relief comprising the Shackleton Limestone.

The most detailed sedimentological study of the Shackleton Limestone was undertaken by Rees et al. (1989b), concentrating primarily on sequences in the Holyoke Range. These authors recognize four principal environment associations: (1) intertidal, (2) carbonate sand shoal, (3) low-energy shallow subtidal shelf, and (4) archaeocyathan–microbial reef. Both low- and high-energy components are recognized.

The intertidal association is typically dolomitic, with fenestrally laminated lime mudstones and parallel-laminated and ripple cross-laminated grainstones. It is relatively rare in the Holyoke Range, but appears to be better developed in the northern exposures of the formation, where evaporite collapse breccias and red carbonate mudstones are reported (Burgess and Lammerink, 1979). The sedimentary structures are indicative of a low-energy tidal flat environment. Flat-pebble conglomerates represent rip-up deposits from episodic storms.

The carbonate sand shoal association is represented by 10- to 50-m-thick carbonate sands composed primarily of ooids with lesser coated grains, oncoids, and bioclasts. Some beds are parallel-stratified; others are planar-cross-stratified. Thin rudstone with ooid clasts and bioclastic and peloidal grainstones are interlayered with the ooid sands.

The most widespread association is the shallow subtidal shelf, of which burrow-mottled limestone is the most common lithology, indicating quiet shelf conditions below wave base. Interbedded with the mottled limestones are grainstones with bioclasts, peloids, and/or intraclasts indicative of current action on the shelf, perhaps due to storm waves. Rare oncoidal packstone beds also exist.

Situated within the subtidal shelf deposits are numerous archaeocyathan–microbial reefs, some up to 50 m thick in the northern Holyoke Range. They are composed of the microbial microfossils *Epiphyton, Girvanella,* and *Renalcis.* Archaeocyathans make up as much as 30% of some reefs, but are usually less than 10% and may be absent altogether. The reefs have a simple mound shape. They are typically 0.5–3.0 m in diameter and 0.3–2.0 m in thickness, though complexes up to 50 m thick occur in the northern Holyoke Range. Judging from the geometry of enclosing limestone beds, the reefs were not more than a few tens of centimeters in relief.

The occurrence of Shackleton Limestone at the head of Beardmore Glacier is of special note both from a historical standpoint and because it extends the basin of deposition 200 km beyond the main outcroppings to the northwest. The first specimens of the limestone were collected by Frank Wild of Shackleton's Polar Party in 1908–9. The locality was described as Buckley Island by David and Priestley (1914, p. 241). Grindley (1963) located the locality at several tiny nunataks south of Mt. Bowers, adjacent to the crevasse-free corridor used by both Shackleton's and Scott's parties on their way to and from the Pole. Young and Ryburn (1968) returned to the locality and mapped Buckley Island, Mt. Bowers, and Mt. Darwin, finding a few more archaeocyathids. Most recently, Rowell and Rees (1989) examined the locality, recovering in situ trilobites and mollusks but no archaeocyathids. Shackleton Limestone at the nunataks south of Mt. Bowers includes massive and mottled limestone, some peloidal packstones, and some fossiliferous grainstones.

As already mentioned, the first fossils found in the Shackleton Limestone were poorly preserved archaeocyathids brought back by Shackleton's party, which determined a Cambrian age for the formation. The first well-preserved archaeo-cyathids were collected in the Nimrod Glacier area (Laird and Waterhouse, 1962). Systematic descriptions are contained in Hill (1964a, b, 1965) and Debrenne and Kruse (1986, 1989).

Trilobites were first noted by Burgess and Lammerink (1979). Well-preserved trilobites, as well as inarticulate brachiopods, hyoliths, mollusks and kennardiids, were described by Rowell et al. (1988a, b), Rowell and Rees (1989), and Evans and Rowell (1990). All of the fossils within the Shackleton Limestone have been referred to the Lower Cambrian. The bulk of the archaeocyathid material is correlated with the Botomian Stage of Siberia, with a horizon from Mt. Egerton possibly as young as Toyonian (Debrenne and Kruse, 1986). The trilobites appear to correlate with the Atdabanian and Botomian (Rowell et al., 1988b).

## Post-Shackleton Clastic Sequences

Clastic rocks, in large part conglomeratic, occur in numerous exposures throughout the outcrop area of the Shackleton Limestone. Originally thought to represent deposition synchronous with or following on the Shackleton, conglomerates are now known to rest with angular unconformity on a karsted surface of Shackleton Limestone in the headreaches of Starshot Glacier northwest of Castle Crags (Rees et al., 1985; Rowell et al., 1988b) (Fig. 4.12). In addition, clasts of folded Shackleton Limestone are included in the conglomerates, themselves deformed, possibly as many as three times (Rees et al., 1987). Lacking body fossils and nowhere intruded by Granite Harbour Intrusives, the rocks are constrained only to be post–late Early Cambrian Shackleton Limestone and pre-Devonian Beacon Supergroup. Rowell et al. (1988b) suggested a probable Ordovician age, but more recently have favored "Middle or possibly early Late Cambrian" (Rees and Rowell, 1991).

Three formations were named by the early workers: Douglas Conglomerate and Dick Formation for sequences in the northern Churchill Mountains (Skinner, 1964) and Starshot Formation for sequences best exposed in the Surveyors Range on the east side of Starshot Glacier (Laird, 1963). Rowell et al. (1986) extended the occurrence of Douglas Conglomerate to outcrops in the Holyoke Range and also to numerous erratics found in moraines along Beardmore Glacier. It was

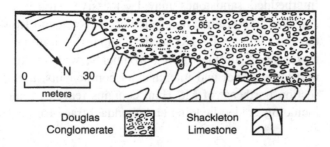

**Figure 4.12.** Detailed geological sketch map (not cross section) showing Douglas Conglomerate un-conformably overlying Shackleton Limestone at northern tip of Holyoke Range. Folds are diagrammatic. After Rowell et al. (1988b). Used by permission.

from limestone clasts in the conglomerates from the Beardmore that the best archaeochathine material was recovered by Shackleton's party (Taylor, in David and Priestley, 1914).

More recently, Rees and Rowell (1991) have reinterpreted all of the conglomeratic or brecciated portions of the Shackleton Limestone, including a 200-m-thick horizon at the type locality, as being Douglas Conglomerate, either in faulted or in unconformable contact.

The Dick Formation crops out at Mt. Dick and the peak to its north (spot height $1,710 \pm$ m), as well as on a pair of nunataks on the south side of Entrikin Glacier, which are designated the type locality (spot height 841 m and nunatak 5 km to its southwest) (Skinner, 1964). The formation consists of slatey argillite, indurated siltstone and sandstone with grit lenses, and occasional muddy limestone (Skinner, 1965). At its type locality the sandstones are rippled and cross-bedded, and the argillites have rare mudcracks (Rees et al., 1988). Dick Formation is overlain conformably by Douglas Conglomerate at spot height 841 m, which is also designated the type locality of the latter (Skinner, 1964).

A 9-m-thick spilite flow, composed of chlorite and sericitized plagioclase phenocrysts, was reported by Skinner (1964, 1965) as interbedded with Dick Formation northeast of Mt. Dick. Rees et al. (1988) state that these volcanics are pillowed basalts that at two localities are snow-bound. They present a K/Ar date of $586 \pm 20$ Ma for the basalts and assert that the volcanics must be considered unrelated to the Byrd Group. In light of the facts that no analytical data are presented with the date, that the basalts are highly altered, and that no other outcrops of pre–Byrd Group rocks are exposed anywhere in the vicinity, this interpretation must be viewed with caution. Since the post–Shackleton Limestone clastics were deposited in a tectonically active setting, accompanying volcanism is not unreasonable.

The Douglas Conglomerate consists of a highly varied assemblage of polymict conglomerate and breccia, with lesser sandstone and shale. Whereas the Douglas Conglomerate overlies finer-grained clastics of the Dick Formation at its type locality, northwest of Castle Crags more than 500 m of coarse conglomerates directly overlie karsted Shackleton Limestone with approximately 30 m of relief (Rees et al., 1985; Rowell et al., 1988b). Pebbles and boulders are predominantly Shackleton Limestone and quartz. Some of the limestone clasts were folded before deposition; some contain archaeocyathids (Rees and Rowell, 1991). Additional clast types include argillite, siltstone, quartzite, arkosic sandstone, diorite, two-mica granite, gneissic granite, hematitic gneiss, and amygdaloidal volcanics (Skinner, 1965).

Many of the conglomerates are massive or crudely stratified clast-supported units. Others show cross-stratification. In some cases the clasts are imbricated. Interbedded sandstones may be massive, stratified, or cross-stratified. Thinly stratified fine-grained rocks also occur. Rees and Rowell (1991) have interpreted the association of rocks in the Douglas Conglomerate and Dick Formation as representing alluvial fan environments, with proximal, mass-flow components, mid-fan channel and interchannel components, and distal fan components. One section northeast of Mt. Hunt contains features indicative of marine deposition, including graded bedding and abundant trace fossils (Rees and Rowell, 1991).

Starshot Formation is composed of calcareous sandstones and shales, with subsidiary portions of conglomerate (Laird, 1963; Laird et al., 1971). Its type locality was designated as Mt. Ubique. The conglomerates, generally not coarser than pebble size, are predominantly limestone and quartz, with lesser quantities of argillite and graywacke and sparse intermediate volcanic rock (Laird, 1964). Sedimentary structures include ripple marks, cross-bedding, load casts, and graded bedding. Several bands of intensely altered silicic volcanic rock crop out on Mt. Ubique, and two bands of altered mafic igneous rock (chlorite and albitized plagioclase) occur on Mt. Heale. Owing to lithological similarities, Laird et al. (1971) suggest that the Starshot Formation is correlative with the conglomeratic rocks as the type locality of the Shackleton Limestone (i.e., Douglas Conglomerate of Rees and Rowell, 1991).

## Granite Harbour Intrusives

A suite of discrete plutons crops out in the area from Starshot to Ramsey Glacier, where they intrude Goldie Formation and the metamorphic rocks of the Miller Range. The contacts with Goldie Formation are sharp and steep, and have hornblende–hornfels aureoles out several tens of meters, whereas in the Miller Range the contacts show more assimilation and *lit-par-lit* effects (Grindley, 1963). The plutons intruding Goldie Formation are posttectonic, whereas those in the Nimrod Group are both synchronous with and postdate the Endurance shear zone. A small pluton intrudes Shackleton Limestone at the northern end of Bartram Plateau (Laird et al., 1974), and several dikes of quartz porphyry intrude Dick Formation and Douglas Conglomerate between Starshot and Byrd Glaciers (Skinner, 1965).

The main outcroppings were compiled by Grindley and Laird (1969) from mapping by Laird (1963) in the Nash Range, by Gunn and Walcott (1962) between Nimrod Glacier and the Holland Range, by Grindley et al. (1964) in the Miller Range, by Laird et al. (1971) in the Cotton Plateau area, by Grindley (1963) in the Queen Alexandra Range, and by Oliver (1972) southeast of the mouth of Beardmore Glacier. Ground mapping and helicopter reconnaissance by Gunner (1971, 1976; Barrett et al., 1970; Lindsay et al., 1973) turned up several additional plutons between the Miller Range and Canyon Glacier.

The first description of granites from the area was by Mawson (1916), who reported on samples collected by Shackleton's Polar Party near Mt. Hope at the mouth of Beardmore Glacier. These were fine-grained and porphyritic biotite granites. The fine-grained sample contained minor muscovite and the porphyry trace amounts of sphene. Gunn and Warren (1962) introduced the name "Hope Granite" for the suite of plutons between Nimrod Glacier and the Beardmore. This nomenclature was followed by other workers for the next decade (Grindley, 1963; Grindley et al., 1964; Laird, 1963; Laird et al., 1971; Gunner, 1976), although Gunner (1971) introduced the name "Ida Granite" for a single pluton around Granite Pillars that intrudes the Hope Granite body at Mt. Hope and the surrounding area. However, Borg et al. (1986a) recommended abandoning the term "Hope Granite," recognizing that the rocks to which this term had been applied range widely in composition and are not related to a single parent magma. Borg et al. (1991) went further to suggest that even "Granite Harbour Intrusive Complex" may be inappropriate since the wide span of Rb/Sr and K/Ar dates (570–460 Ma) found throughout the Transantarctic Mountains may encompass more than a single tectonic event, and the rocks around Granite Harbour may have had a different tectonic setting and petrogenesis than those south of Byrd Glacier.

Petrography of the plutonic rocks is well described in the earlier studies in the area, wherein a limited variety of rocks were identified. The most recent study by Borg et al. (1991), emphasizing isotopic compositions, recognized a plutonic suite ranging from gabbro to granite, but did not describe their petrology systematically. What was called Hope Granite by the early workers is a medium- to coarse-grained biotite granite, typically with phenocrysts of microcline, but also with some equigranular phases, typically gray but with a pink blush in some cases. Gunner (1976) recognized that the most common rock type of the Hope Granite is quartz monzonite, with lesser amounts of granite, granodiorite, and quartz diorite. Earlier descriptions of plutonics other than granite sensu stricto are granodiorites and quartz monzonite at Cape Laird, Deverall Island, and the southwestern Nash Range (Laird, 1964), granodiorite adjacent to Robb Glacier (Gunn and Walcott, 1962), tonalite at Bartram Plateau (Laird et al., 1971), and granodiorite and quartz diorite east of the mouth of Beardmore Glacier (Oliver, 1972).

Muscovite was recognized in specimens of Hope Granite from throughout the area (Gunn and Walcott, 1962; Grindley, 1963; Grindley et al., 1964; Laird, 1965; Laird et al., 1971). Gunn and Warren (1962) noted that in specimens from the coastal plutons that they collected, muscovite is almost entirely within the boundaries of microcline, but they nevertheless felt that it was probably primary. Hornblende was also found as an accessory in some of these plutonics (Laird, 1964; Gunner, 1976). Since primary muscovite and hornblende are mineralogical earmarks of S- and I-type granites, respectively, the lumping of all the plutonics into a single suite by the early workers was obviously an oversimplification. The plutonic rocks in the Miller Range, in particular, are tourmaline-bearing, two-mica granites (Borg et al., 1986a), but elsewhere if both muscovite and hornblende were reported from the same rocks, the muscovite is more likely of secondary origin.

The Ida Granite of Gunner (1971) contrasts with "typical" Hope Granite in that it is finer-grained, nonporphyritic, and with less calcium and magnesium and more silicon. At different places its contact with the Mt. Hope pluton is both sharp and diffuse, and it has been interpreted as a differentiate probably intruded before total solidification of the parent body.

Aplitic and pegmatitic dikes are found in association with plutons and nearby country rock throughout the region (Gunn and Walcott, 1962; Grindley et al., 1964; Laird, 1964; Gunner, 1976), but they are not abundant, particularly in contrast to the dike swarms of southern Victoria Land. Tourmaline, garnet, and/or muscovite are typically constituents of these dikes. Also, a swarm of lamprophyre dikes of this vintage has been noted in the Miller Range (Grindley et al., 1964).

## Geochronology

The earliest dating of the plutonic rocks in the central Transantarctic Mountains, K/Ar determinations on biotite and muscovite by McDougall and Grindley (1965), gave cooling ages between 480 and 465 Ma, though no errors were fixed to these numbers. The plutons sampled were from the Miller Range and the Mt. Hope area. The dating added to the emerging picture of plutonism throughout the Transantarctic Mountains occurring in a Cambro-Ordovician time frame.

Further K/Ar determinations by Adams et al. (1982a) affirmed this cooling interval. A set of nine samples from the granites and associated xenoliths in the Miller Range gave ages of 495–474 Ma. Two of these were nearly concordant biotite and hornblende pairs, indicating that cooling of the plutons was fairly rapid. Samples of hornblende-bearing quartz monzonite and tourmaline–muscovite pegmatite intruding Goldie Formation in the Nash Range had, respectively, biotite and muscovite ages of $447 \pm 3$ and $449 \pm 3$ Ma, indicating a later cooling in the coastal area and possibly a younger age for the intrusions there.

Additional cooling data are two Rb/Sr whole-rock–plagioclase dates obtained by Borg et al. (1990): 470 Ma for a diorite from the Campbell Hills near the mouth of Nimrod Glacier and 460 Ma for a diorite on the north side of Ramsey Glacier.

Gunner and Mattinson (1975) reported a Rb/Sr whole-rock study of plutonic rocks collected in the Miller Range and in the coastal ranges between Nimrod and Beardmore Glaciers. Data from the two areas plotted on two distinct isochrons. The one from the Miller Range gave an age of $488 \pm 34$ Ma (IR = 0.734) and the one from the coastal ranges an age of $463 \pm 12$ Ma (IR = 0.710). These results accord with the K/Ar determinations, indicating that magma emplacement and subsequent cooling occurred earlier in the Miller Range than in the region to the east.

In fact, recent U/Pb dating of zircons by Goodge et al. (1993c) pushes the onset of plutonism in the Miller Range back to approximately 541 Ma. Plutonics generated in the time span 541–521 Ma were syntectonic, overlapping the later stages of development of the Endurance shear zone, which imparted them with tectonite fabrics. An undeformed pegmatite in the vicinity was U/Pb zircon dated at around 515 Ma, giving a bound on the timing of cessation of shearing.

Isotopic studies on Sm/Nd, Rb/Sr, and oxygen systems in the plutonic rocks of the area by Borg et al. (1990) show systematic changes across the central Transantarctic Mountains. Because the Transantarctic Mountain front cuts obliquely across the structural trends in the sedimentary and metamorphic rocks, it is useful to view the samples collected in the coastal exposures from Nimrod to Ramsey Glacier as projected onto a continuation of a transect down Nimrod Glacier from the Miller Range and out onto the Ross Ice Shelf. $\varepsilon_{Nd}$ values vary from $-1.5$ to $-12$ in a westerly direction, whereas $\varepsilon_{Sr}$ increases from $+40$ to $+540$ (Fig. 4.13). $\delta^{18}O_{WR}$ varies from $+8.9$ to $+12.7\%$, generally increasing to the west but with highest values around the mouth of Nimrod Glacier. Borg et al. (1990) note that a marked shift in $\varepsilon_{Nd}$ and $\varepsilon_{Sr}$ occurs across Marsh Glacier separating the Miller Range from the rest of the central Transantarctic Mountains and, on this basis, suggest separate crustal blocks for the plutonic source areas:

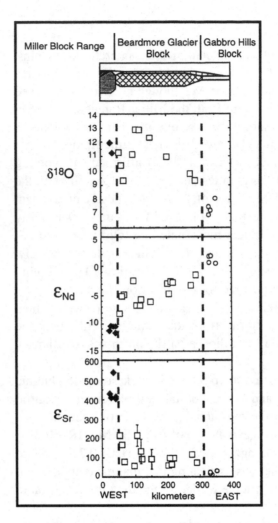

**Figure 4.13.** Borg and DePaolo's (1991) interpretation of lower crustal provinces in the central Transantarctic Mountains defined by neodymium, strontium, and oxygen isotopic compositions of ~500-Ma granites. Diamonds: peraluminous granodiorite to granite; squares: metaluminous to slightly peraluminous granodiorite to granite, occasionally diorite; circles: gabbro, diorite, tonalite, and metaluminous granodiorite. Used by permission.

the Miller block to the west and the Beardmore block to the east. $T_{DM}$ ages indicate that the source of the granites in the Miller block was 2.0-Ga crust. $T_{DM}$ ages for intrusive rocks of the Beardmore block range from 1.8 to 1.3 Ga, which they generalize to 1.7 Ga coincident with the time of generation of the Aurora Orthogneiss.

## Deformation

### Deformation of Nimrod Group

Pervasive shearing and multiple phases of folding at amphibolite conditions of metamorphism characterize the Nimrod Group. The rocks have a prominent compositional layering or foliation, which for the most part probably mimics bedding. Except around the margins of the plutons that intrude Nimrod Group, where considerable variation occurs, the layering strikes fairly consistently to the northwest and dips at moderate angles to the southwest. An elongate mineral lineation trends northwest–southeast within the layering. Mesofolding is fairly pervasive throughout the Miller Range; however, the predominant structure is

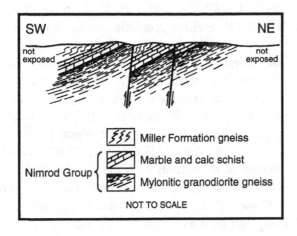

**Figure 4.14.** Interpretive sketch of Endurance shear zone offset by reverse faults. From Borg et al. (1986a). Used by permission.

the Endurance thrust or shear zone (Grindley, 1972) that crops out in the ridges at the head of Argosy Glacier and in the southern end of the range south of Argo Glacier. Intensity of deformation is greatest adjacent to the Endurance shear zone and drops off somewhat in exposures to the east, for example, along Aurora Heights.

Overall, shearing has effected a structural thickness of 12–15 km in both the Miller and Geologists Ranges (Goodge et al. 1991a, 1993a). The shear zone is characterized by porphyroclastic mylonitic gneiss, with a southeast-trending, elongate-mineral lineation that focuses on a zone of ultramylonite approximately 5 m wide. The mylonitic gneisses are intimately interleaved with calc–silicate tectonites. At places eclogitic blocks (Argo Gneiss) are incorporated along the shear zone. Within the main part of the shear zone the mylonitic layering is planar, but both above and below, the layering becomes irregularly folded. At the head of Argosy Glacier the shear focuses on a single zone; however, at least three zones of high strain occur south of Argo Glacier. Grindley (1972) interpreted the situation south of Argo Glacier as being due to repetition of the Endurance thrust by asymmetric folding (Fig. 4.4). Borg et al. (1986a), however, interpreted the repetition as being due to high-angle reverse faulting (Fig. 4.14), and Goodge et al. (1991a) suggested the additional possibility that the three zones were splays of the single zone at the head of Argosy Glacier. A consistent sense of shear with tops to the southeast, parallel to the mineral elongation, is indicated by a number of structures including S–C fabrics, rotated porphyroclasts and boudins, and asymmetrical folds (Goodge et al., 1991a).

Mesoscopic folding is widely developed throughout the Nimrod Group, with the predominant axial trend northwest–southeast parallel to the mineral elongation direction. Grindley (1972) designated five phases of folding ($F_1$–$F_5$), the first three of which he associated with the contractions that produced the Endurance shear zone. Goodge and Dallmeyer (1992) suggested three tectonothermal events ($D_1$–$D_2$ and $M_1$–$M_3$), with the third a thermal recrystallization due to intrusion of local plutons of posttectonic Hope Granite. The $D_2$ of Goodge and Dallmeyer (1992) is associated with the main movements of the Endurance shear zone. Refolded mesofolds and rootless isoclines within schists and gneisses may represent an earlier, separate episode of deformation ($D_1$) overprinted by $D_2$ or simply relics of the early stages of shear deformation ($D_2$) (Goodge and Dallmeyer,

1992). Grindley's (1972) $F_1$–$F_3$ correspond to $D_2$ of Goodge and Dallmeyer (1992). His earliest generation of folds ($F_1$) are intrafolial and isoclinal, are of limited occurrence, and are best developed in lower portions of the Aurora Orthogneiss. They have a northwest–southeast axial trend. The second generation of folds ($F_2$) are the most widespread of the Nimrod Group. They are coaxial with $F_1$, with axes plunging less than 10° from horizontal, and have axial planes dipping 45°–60°SW. Grindley (1972) interpreted $F_2$ as due to a northeasterly vergence along the Endurance shear zone, whereas Goodge and Dallmeyer (1992), with the additional weight of a variety of sense-of-shear indicators, interpreted these folds as being due to ductile deformation with a southeasterly sense of movement. Gunner (1969, Fig. 9) pictures an $F_1$ fold refolded by $F_2$.

Grindley's (1972) third generation of mesoscopic folds ($F_3$) are similar in style to $F_2$, but their axes have a southwest trend and plunge directly downdip on the $F_1$ foliation. Vergence on these folds is toward the southeast. They are best developed at the head of Argosy Glacier.

Grindley (1972) also recognized several postmetamorphic open flexures ($F_4$) in the area between Nimrod and central Argosy Glaciers. Small, irregular folds ($F_5$) and associated schistosity, as well as local doming of foliation, were produced during intrusion of the posttectonic granite plutons in the Miller Range, particularly in the vicinity of Skua Glacier (Grindley, 1972).

## Deformation of Beardmore and Byrd Groups

A paradoxical situation exists in the central Transantarctic Mountains, exclusive of the Miller and Geologists Ranges, in that deformation apparently increases in intensity and complexity in successively younger formations. Goldie Formation appears to have the simplest structure, which throughout most of its outcrop area is represented by a single generation of upright chevron-style folds. The structure of the Shackleton Limestone is described by most authors as "complex," with two generations of folding recognized at places (Burgess and Lammerink, 1979; Rowell et al., 1986). In Douglas Conglomerate three episodes of deformation have been described (Rees et al., 1987). Bear in mind that each of these formations is thought to be unconformable on the former.

Except within the westernmost exposures of Beardmore Group, where two episodes of folding are documented in the Cobham and Goldie Formations, deformation is simple and extremely uniform. Throughout most of its outcrop area, Goldie Formation appears to have been folded once with a tight chevron style, approaching isoclinal according to some authors (Gunn and Walcott, 1962; Grindley, 1963; Laird et al., 1971). Bedding strikes consistently in a north–south direction except, notably, at the northernmost exposures around Dickey Glacier, where it swings to the northeast. Dips of bedding are usually greater than 45°, and often greater than 60°.

In general, the folds in Goldie Formation are of large amplitude without the accompanying involvement of mesofolds. Broad distances of uniformly dipping bedding, separated by sharp fold closures, appear to be the norm. Examples are pictured by Gunn and Walcott (1962, p. 412) and Gunner (1976, pp. 14, 15). Fold axes within Goldie Formation are oriented within 10° of north–south and are generally subhorizontal. However, Linder et al. (1965) report a synform south of the mouth of Nimrod Glaicer with an axial plane oriented 40°NE with steep SE dip and fold axis plunging 60°NE. Oliver (1972) reports mesoscopic folds in the Commonwealth Range whose axes plunge at various angles between horizontal and vertical. Axial-plane cleavage, generally dipping subvertically, is developed at many places in the pelites of Goldie Formation, though by no means everywhere. Where bedding is steep enough the cleavage is oriented subparallel to it.

Beardmore Group has suffered two episodes of folding along the western margin of its exposure at Cotton Plateau: the Cobham Range and possibly Kon Tiki Nunatak. Isoclinal recumbent folds overprinted by a second generation of mesofolds were reported by Laird et al. (1971) from the Cobham Formation in the Cobham Range. These authors also mapped a folded unconformity at Cotton Plateau where folded Goldie Formation is unconformably overlain by Shackleton Limestone, itself folded into an open syncline whose trough and eastern limb are exposed on the Palisades. This locality demonstrates that Beardmore Group was deformed pre–Early Cambrian (Botomian) and was the *a priori* evidence used to designate the Beardmore Orogeny as an event distinct from the Cambro-Ordovician Ross orogeny (Grindley and McDougall, 1969).

Stump et al. (1986a, 1991; Edgerton, 1987) conducted a structural study to characterize the two deformations of western exposures of Beardmore Group. Their observations are summarized in Figure 4.15 and the following account.

Goldie Formation crops out continuously from the steep, west-facing slopes of The Palisades across the northern end of Cotton Plateau. East of Panorama Point from the northwestern corner of Cotton Plateau, Goldie Formation is basically a homoclinal sequence, dipping steeply (60°–80°) and younging to the northwest, although at places attitudes scatter widely due to mesoscopic folding. At Panorama Point the formation is folded tightly around a north-trending syncline cored by the metabasalt unit already described. The main bed of diamictite crops out on both sides adjacent to the metabasalt.

A vertically dipping shear zone cuts through portions of the metabasalt, as well as the eastern limb of the Shackleton syncline. The syncline in Goldie Formation was formed before the syncline in Shackleton Limestone and is not coaxial with it. The locus of the shear zone is the hinge surface of the Goldie syncline, but it appears to have formed during folding of the Shackleton syncline, causing attenuation and overturning of the latter's eastern limb. A small pluton of Granite Harbour Intrusives intrudes Goldie Formation north of Panorama Point where it crosscuts the shear zone.

The fold in Shackleton Limestone has an open cylindrical geometry with a subhorizontal beta axis trending approximately 40°NW. It appears to be primarily a flexural structure lacking in cleavage except in the sheared eastern limb. One can follow a systematic change in bedding orientation in Goldie Formation where it crops out along the unconformity adjacent to the Shackleton Limestone from west of the shear zone to the trough of the Shackleton syncline. When the Shackleton syncline is "unfolded" around its axis, orientations in the Goldie Formation trace an east-vergent anticline.

Two generations of cleavage and of mesofolds are developed in Goldie Formation at the northern end of Cotton Plateau. East of Panorama Point the mesofolds are all upright structures. The older set is east-vergent, with subverti-

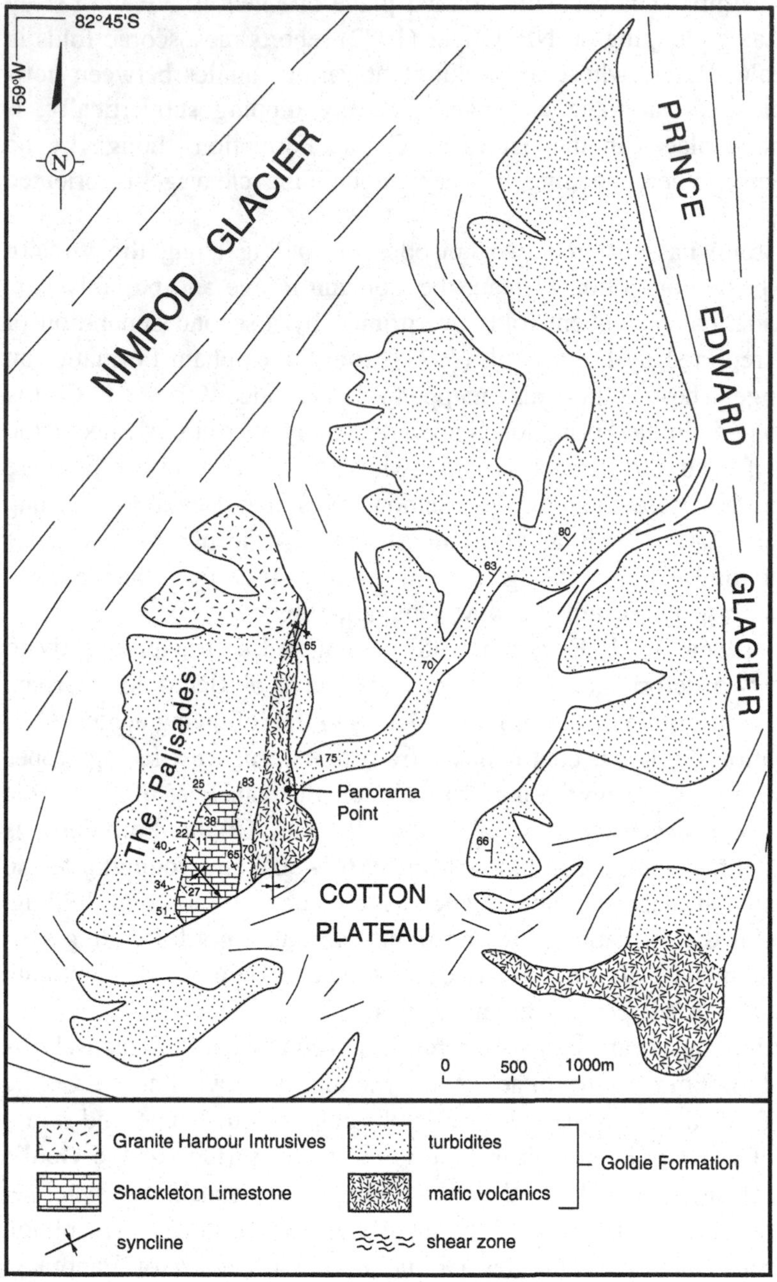

**Figure 4.15.** Geological map of northern Cotton Plateau. Modified from Edgerton (1987).

cal cleavage striking north–south, consistent with vergence on the limb of the Goldie syncline. The younger mesofolds are west-vergent, with cleavage oriented northwest–southeast, parallel to the axial surface of the Shackleton syncline, and crosscutting the earlier cleavage. On the lower slopes of The Palisades beneath subhorizontal Shackleton Limestone, Goldie Formation crops out in a series of recumbent or reclined folds, which at places are overprinted by more upright mesofolds and steep cleavage (Fig. 4.16). Orientations of the two sets of folds and cleavage are complex and cannot be easily generalized.

The northern end of Cotton Plateau appears to straddle a distinct structural boundary in Goldie Formation, marked by the transition from recumbent to

**Figure 4.16.** Recumbently folded Goldie Formation beneath Shackleton Limestone at The Palisades.

upright folds across the Palisades and the shear zone that cuts through their upper slopes. An interpretation of events consistent with the available field data is that Goldie Formation was folded around a north–south axis, possibly with an east-vergent movement direction. It was eroded and overlain by Shackleton Limestone. Renewed compression produced northwest–southeast mesostructures in Goldie Formation and a fold of the same orientation in Shackleton Limestone, focusing along a shear zone parallel to the hinge surface of a syncline in Goldie Formation. Movement along this shear zone was vertical to high-angle reverse with east side up, causing upturning and attenuation of Shackleton Limestone.

**Figure 4.17.** Recumbent and upright folding in Goldie Formation exposed on the northern flank of Kon Tiki Nunatak.

In the flattened conglomerate at the southern end of Cotton Plateau, the plane of flattening and a cleavage in the matrix have a northnorthwest strike and steep westerly dips. Tension gashes within some of the clasts indicate movement both with west-up and east-up senses of motion. Because of the considerable strain these rocks have suffered and the approximate structural alignment with The Palisades, it is tempting to equate the movement producing the flattened conglomerate with the shear zone to the north.

The structural pattern in Beardmore Group as seen at the northern end of Cotton Plateau, of recumbent structures in westerly exposures becoming upright to the east, is repeated at both Kon Tiki Nunatak and the Cobham Range.

The relationship is exposed on an uninterrupted cliff face on the north side of Kon Tiki Nunatak (Fig. 4.17). Stump et al. (1991) interpreted the relationship to be the result of two episodes of deformation with upright folding overprinting recumbent folding. Goodge et al. (1991b), however, explain the geometry as being due to a single west-vergent tectonic event. Both groups would agree that the west-vergent deformation was related to the Ross orogeny (i.e., post–Shackleton Limestone in age).

At the southwestern corner of the Cobham Range recumbent folding occurs in the lower portion of the Cobham Formation (Laird et al., 1971, Fig. 19). Eastward along Gargoyle Ridge the Cobham and overlying Goldie Formation form a homoclinal sequence, dipping 50°–70°E, which is folded at a synclinal hinge on the eastern side of Tarakanov Ridge, adjacent to Prince Philip Glacier (Laird et al., 1971).

Two episodes of mesofolding and cleavage can be discerned in the Cobham Range. The earlier cleavage, characterized by aligned growth of biotite, is fairly common but not pervasive. The later cleavage is developed to a limited extent as fine crenulations of the earlier cleavage.

The earlier set of mesofolds plunge broadly toward the southeast. Poles to axial surfaces spread through a great circle, suggesting refolding. The vergence of these folds, if apparent, is toward the east or northeast. The younger set of mesofolds is more widespread. Axial surfaces are fairly consistent with north–south strikes and steep to vertical dips. Fold axes generally have a gentle to moderate plunge to the northnortheast. Some, however, are distributed throughout a great circle approximately parallel to the dominant orientation of axial surfaces, suggesting that bedding was not planar at the time of the second folding due to the effects of the earlier episode. Where vergence is indicated, it is consistently toward the west. This is consistent with the direction of movement on the high-angle reverse fault that offsets Cobham Formation at the southern end of the Cobham Range (Fig. 4.7).

The Shackleton Limestone is folded throughout its exposure. In general the structural trend is in a north–south direction, but at a number of localities bedding and fold axes are oriented east–west. South of Nimrod Glacier the Shackleton Limestone is preserved in three elongate strips: (1) along The Palisades, (2) around Mt. Lowe and Hochstein Ridge, and (3) underlying the Markham Plateau and extending north from it. At The Palisades, Shackleton Limestone is folded into a syncline with an overturned and sheared eastern limb, as described earlier. At Mt. Lowe and Hochstein Ridge, Grindley and Laird (1969) have mapped an anticline–syncline pair. Rowell et al. (1986, p. 49) picture steeply dipping, irregularly-folded Shackleton Limestone north of Mt. Lowe, where they interpret the contact with Goldie Formation as a fault.

From aerial observation and photo interpretation, Grindley and Laird (1969) map a syncline of Shackleton Limestone straddling Markham Plateau and extending northward from it, and pinching out south of Nimrod Glacier. Stump et al. (1986a) examined steeply dipping Shackleton Limestone north of Mt. Markham (Fig. 4.18). From that vantage the limestone appeared to continue all the way to Nimrod Glacier. The contact with Goldie Formation in this area has not been visited.

Owing to opposing dips and tops indicators in outcrops on the eastern and western margins of the Holyoke Range, Laird et al. (1971) mapped a major north-trending syncline throughout the range. At Cambrian Bluff at the southern tip of the Holyoke Range, the Shackleton Limestone is complexly deformed with both recumbent and upright folds. Various views of these structures are pictured by Laird (1963, pp. 476, 477) and Rowell et al. (1986). The relationship of this complexity to the larger syncline is unclear.

Laird (1963) mapped several small folds with an eastnortheast trend at Mt. Hotine and Mt. McKerrow on the east side of Starshot Glacier, while 3 km south at Thompson Mountain an anticline trends southeast. Along the steep spurs of the central Churchill Mountains between Mt. Zinkovich and Mt. Nares, tightly

**Figure 4.18.** Vertically dipping Shackleton Limestone overlain by Beacon Supergroup at the northern edge of the Mt. Markham massif below Mt. Korsch.

folded Shackleton Limestone with nearly vertical bedding strikes north–south (Fig. 4.19).

In the northern Churchill Mountains, the structural pattern becomes more complicated. Macrofolds occur in a variety of orientations with plunges up to 50°, and mesofolds are more prevalent than farther to the south. At places Douglas Conglomerate and Dick Formation are interspersed with Shackleton Limestone. A number of faults have been mapped or postulated.

In the outcrops along Byrd Glacier, marking the northernmost exposures of the Byrd Group, folds can be traced with a fairly consistent northeast–southwest trend parallel to the glacier. South and east of this, however, folds are oriented in all quadrants. At Mt. Durnfom the north–south trend of the central Churchill Mountains is joined by an east–west set of folds (Burgess and Lammerink, 1979). The east–west trend predominates northward as far as Mt. Egerton, but from there northward to beyond Mt. Hamilton northeast–southwest and north-west–southeast folds occur (Skinner, 1964). Burgess and Lammerink (1979) reported that the east–west-trending set was overprinted by the northeast–southwest-trending set, and subsequently Rees et al. (1987) observed that around Mt. Hamilton the northeast–southwest-trending folds are older and overprinted by the northwest–southeast set, which in turn are overprinted by northeast–southwest kink folds (Fig. 4.20). The present author has observed cross-folding to the southwest of Mt. Tuatara, as shown in Figure 4.21.

Skinner (1964) mapped several major faults in this northern area. Outcrops are on widely spaced ridgelines and nunataks, which did not deter him from connecting structures across broad areas of snow cover. Two north–south-trending structures, which he named the Albert Markham and Rainbow faults, may well be the result of uplift of the present-day Transantarctic Mountains. However, a northeast–southwest-trending fault at Mt. Rainbow (Kiwi fault of Skinner, 1964) was observed by Burgess and Lammerink (1979) not to offset the Beacon Supergroup at that locality.

Reverse faults with northeast–southwest trends were observed by Skinner (1964) in nunataks at the head of Matterson Inlet and at Constellation Dome, and south of Mt. Hamilton. In a detailed mapping of the area around Mt. Hamilton, Rees et al. (1987) found Shackleton Limestone bounded by reverse faults and Douglas Conglomerate. They draw this pair of faults arcing from an east–west trend south of Mt. Hamilton to a northeast trend east of it and relate the faulting to the northwest–southeast compression that also produced the older northeast–southwest-trending folds mentioned earlier (Fig. 4.22).

Throughout most of its outcrop the Byrd Group shows essentially no metamorphic effects, although calcite veining is common in many places where deformation has been strong. However, on the north side of Mt. Tuatara the author has observed that Shackleton Limestone changes to a coarse-grained white marble. Bedding, which dips and youngs steeply to the southsoutheast toward the core of a syncline mapped by Skinner (1964), is crosscut by a crude, steeply dipping foliation marked by darker variations in the marble (Fig. 4.20).

At Mt. Madison the rocks are marbles, schists, and calc schists, named the Selbourne Marble by Skinner (1964). They have been metamorphosed to the amphibolite facies and contain assemblages with such minerals as muscovite, biotite, phlogopite, scapolite, hornblende, anthophyllite, and garnet. Skinner (1964) mapped macrofolds with both northeast–southwest trends and northwest–southeast trends.

Layering within the schists and calc schists is irregular and uneven, possibly representing transposed bedding. Where observed by Stump (1980), this layering contains small fold closures and is overprinted by a crenulation cleavage and accompanying folds with an east–west trend.

On the basis of metamorphic grade, Skinner (1964) suggested that the Selborne Marble is Precambrian in age and correlative with the Nimrod Group. In this case the rocks at Mt. Madison could be exposed basement to the Byrd and Beardmore (?) Groups. Since the lithologies of the Selborne Marble (schist, calc schist, marble) could have been derived from the Dick Formation pelites and Shackleton Limestone, Stump (1980) suggested that it was a high-grade correlative of the Byrd Group. It now seems to this author that either suggestion is possible.

The structural attributes of the Byrd Group are in many ways characteristic of a foreland fold and thrust belt (Stump, 1992b). These include multiple episodes of nonmetamorphic deformation. The Douglas Conglomerate may be interpreted as a clastic wedge generated by erosion from an early stage of deformation, then caught up in subsequent movements (Panttaja and Rees, 1991). The tectonic nature of most, if not all, of the observed contacts between Byrd Group and Goldie Formation, and the apparently simple style of folding of the Goldie, suggest that the unconformity may have served as a *décollement* during some of

**Figure 4.19.** Folded and faulted Shackleton Limestone unconformably overlain by Beacon Supergroup on spurs of Mt. Coley.

the movement. However, the pattern in the southern part of the area of in-faulted or in-folded strips of Shackleton Limestone suggests that Goldie Formation was to some extent involved in the movement of the Shackleton.

The pattern of north–south-trending folds in Byrd Group throughout its exposure from the central Churchill Mountains southward, giving way northward to variably oriented folds and finally to northeast–southwest folds parallel to Byrd Glacier, may be interpreted as resulting from oblique movement whose principal component was perpendicular to the main structures, with a lesser component parallel to the orogen resulting in the oblique trends at the northern end of the area. The reverse faulting around Mt. Hamilton is consistent with such movement.

The marked contrast in rock types on opposite sides of Byrd Glacier has been noted by various authors. Right-lateral strike–slip movement was postulated by Grindley (1981), while Davey (1981) suggested left-lateral. Stump (1976a) postulated that the crystalline rocks north of Byrd Glacier were not displaced by transverse faulting, but rather were an abutment to movement of the Byrd Group from the south. In addition, Borg and DePaolo (in press) report that $T_{DM}$ ages in plutonic rocks north of Byrd Glacier are the same as those to the south, implying that there has been no large strike–slip movement along the boundary. The discontinuity across Byrd Glacier may have occurred during the postulated sepa-

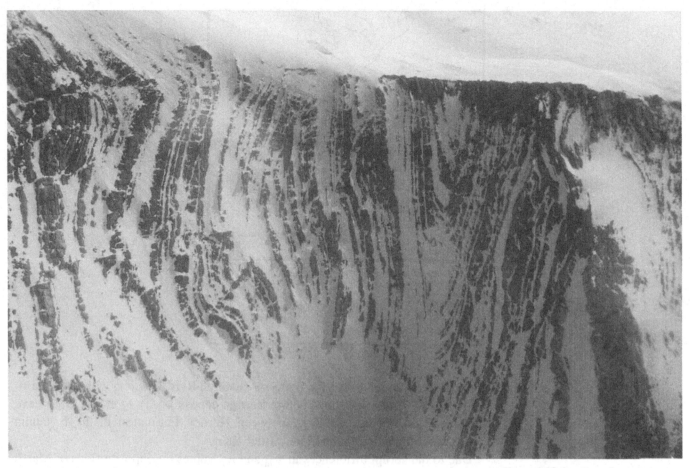

**Figure 4.19** *(continued).*

ration of Laurentia from Antarctica, from movement along a transfer fault in which continental crust to the south was more attenuated, and later defined the northern boundary for deposition of the Byrd Group.

## Evolution of the Central Transantarctic Mountains

Parts of the central Transantarctic Mountains lend themselves to straightforward geological interpretation, but the relationships of the parts and the evolution of the whole are not so readily apparent. One of the important issues is the relationship of the Miller and Geologists Ranges to the rest of the central Transantarctic Mountains, that is, the relationship of the Nimrod Group to the Beardmore and Byrd Groups. Common to both of these areas are posttectonic Granite Harbour Intrusives, ranging from 450 to 495 Ma, with those in the Miller Range apparently cooling during the earlier part of this interval. The different crustal levels exposed in the two areas indicate considerably greater exhumation of the Miller and Geologists Ranges than areas to the east before deposition of the Beacon Supergroup. This implies that by the end of the Ross orogeny a greater thickness of continental crust existed west of Marsh Glacier than to the east. The same

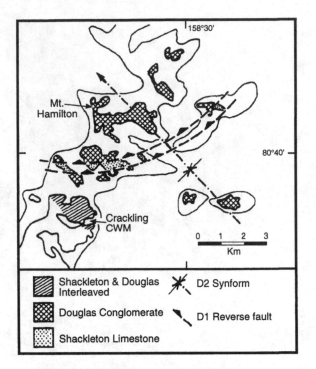

**Figure 4.20.** Generalized geological map around Mt. Hamilton, northern Churchill Mountains, showing thrust faulting of Shackleton Limestone both above and below Douglas Conglomerate. After Rees et al. (1987). Used by permission.

argument can be made for the opposite sides of Byrd Glacier, with the area to the north being more exhumed before Beacon deposition. Borg et al. (1989) have called attention to the similarity between Horney Formation in the Britannia Range and Miller Formation in the Miller Range.

Due to the abrupt differences in $\varepsilon_{Nd}$ and $\varepsilon_{Sr}$ values on opposite sides of Marsh Glacier, Borg et al. (1990) postulated that it marks a major lower crustal boundary, possibly a suture, designating the Miller Range block to the west and the Beardmore Glacier block to the east.

On the basis of the similarity of calcareous units in their lower portions passing upward into predominantly pelitic sequences, Stump et al. (1991) suggested that the Cobham and lower Goldie Formations are correlative with the Worsley and Argosy Formations of the Miller Range. To the contrary, considering the different neodymium model ages of the Argosy and Goldie Formations, and the different crystallization ages of the "Camp Ridge orthogneiss" and the gabbro in Goldie Formation at Cotton Plateau, this author reconsidered the suggestion of a correlation between Nimrod Group and lower Beardmore Group. However, the deformational and metamorphic history of the Nimrod Group remains problematic, since it has now been shown that the Endurance shear zone, which developed under amphibolite-grade metamorphic conditions, was active at least in its later stages during the interval 541–521 Ma, as dated by syntectonic plutons (Goodge et al., 1993c).

The 1.7-Ga Pb/Pb date on the "Camp Ridge orthogneiss" indicates emplacement during the Paleoproterozoic, though not the metamorphic and structural conditions of the host rock at the time. Goodge and Dallmeyer (1992, p. 95) recognized a thermotectonic episode before the Endurance shearing "based largely on cryptic petrologic and structural evidence," which "may have involved regional high-grade metamorphism and concomitant deformation."

Considering all factors, it seems most likely that deposition and subsequent deformation and metamorphism of the Nimrod Group occurred in the Paleoproterozoic (with the Miller Formation having distinctly older source material than the Argosy and Worsley Formations), before deposition of the Beardmore Group. The establishment of the Beardmore basin may have been associated with rifting, as indicated by the approximately 760-Ma basalt and gabbro in Goldie Formation. This rifting is postulated to have occurred due to the separation of Laurentia from Gondwanaland (Moores, 1991; Stump, 1992b). Initial foundering resulted in the shallow marine deposits of the Cobham Formation, followed by the deeper-water turbidites of the Goldie Formation.

Constraints on the timing of deformation of the Endurance shear zone are that it was younger than 1.7-Ga orthogneiss, had begun before 541-Ma orthogneiss emplacement, and had ended by approximately 515-Ma pegmatite emplacement (Goodge et al., 1993c). Constraints on the initial deformation of the Beardmore Group are that it was younger than 760-Ma gabbro and older than Atdabanian (middle Lower Cambrian) Shackleton Limestone. Constraints on the main deformation of the Shackleton Limestone are that it was post-Toyonian (upper Lower Cambrian) and before posttectonic Granite Harbour Intrusives (~500 Ma).

Because the pre–Shackleton Limestone folding of Goldie Formation at Cotton Plateau was east-vergent and the sense of shear on the Endurance shear zone was top to the southeast, Stump et al. (1991) suggested that both structures resulted

**Figure 4.22.** White coarsely crystalline Shackleton Limestone on the north face of Mt. Tuatara, above Byrd Glacier. The dark layer exhibits a coarse foliation axial-planar to eastnortheast-trending folds in the area.

from the same compressional event (the Beardmore orogeny of Grindley and McDougall, 1969). Borg et al. (1990) presented a model in which this resulted from the suturing of a terrane or microcontinent (Beardmore Glacier block) to the craton (Miller Range block) in the vicinity of Marsh Glacier. Because of the oceanic character of the basalt at Cotton Plateau, they suggested this terrane was allochthonous, with an ocean of considerable breadth between. To this author a better scenario is a basin formed by rifting, but not significant separation, of continental crust and the production of minimal oceanic material, followed by collapse and suturing of the basin along its western margin. The basalts and gabbro at Cotton Plateau are not the foundation of the Goldie Formation, but rather occupy the highest stratigraphic position in the exposures at Cotton Plateau. The fact that recumbent folding is found only in the westernmost exposures of the Beardmore Group lends support to the idea that a collision or suturing occurred along the western boundary of the Beardmore basin. Since it is uncertain whether the contacts between Goldie Formation and Shackleton Limestone outboard of Cotton Plateau are unconformities, and because only one episode of deformation has been recognized in Goldie Formation in these areas, it is possible that the initial folding of Beardmore Group occurred only along its western margin adjacent to the zone of suturing.

In this case Shackleton Limestone could have been deposited unconformably on Goldie Formation in the west, but conformably farther outboard. The interval of deposition of Shackleton Limestone was fairly brief, with most of the several thousand meters of deposition occurring in the Botomian. Whether or not folding of Goldie Formation occurred in its outboard areas before deposition of Shackleton Limestone, it must have been involved in the Shackleton deformation because of the distribution of in-faulted or in-folded strips of Shackleton Limestone south of Nimrod Glacier. Not to have been folded before Shackleton deposition would explain its apparently simple structure. To have been folded before Shackleton deposition suggests that deformation of Shackleton Limestone was coaxial with the earlier Goldie folding, or alternatively that the deformation of Goldie Formation associated with Shackleton Limestone was focused into relatively narrow zones that are yet to be recognized, except in the shear zone at Cotton Plateau.

Distinguishing the deformation episodes that affected Shackleton Limestone before and after deposition of Douglas Conglomerate is equally problematic. The evidence that Shackleton was deformed before deposition of Douglas Conglomerate is their unconformable relationship and the presence of folded clasts of Shackleton in Douglas.

The picture that emerges from the foregoing discussion is that compressional deformation was an ongoing feature east of Marsh Glacier probably from the uppermost Proterozoic until after deposition of the Douglas Conglomerate, with brief intervals during which deposition prevailed. West of Marsh Glacier movement on the Endurance shear zone apparently began before 541 Ma and waned through the interval 541–521 Ma, ceasing by around 515 Ma. Following the Bowring et al. (1993) timescale for the Cambrian, the emplacement of magma in the Miller Range and movement on the Endurance shear zone directly coincided with the deposition of Shackleton Limestone.

Goodge et al. (1991a, 1993a, b, c) have developed a model of left-lateral oblique convergence along the continental margin in which shallow-level crustal components were deformed primarily by shortening perpendicular to the orogen, as can be seen in folds throughout much of the Transantarctic Mountains, while deeper crustal levels were deformed by ductile shearing in a top-to-the-southeast sense, as typified by the Endurance shear zone in the Miller Range.

This model of different crustal levels behaving in kinematically different ways may be used as a rationale for the apparently continuous nature of deformation in the Miller and Geologists Ranges and the episodic nature of deposition and deformation to the east. Furthermore, it appears that deformation of the Byrd Group continued as a foreland fold and thrust belt in upper crustal levels while deformation in deeper crustal levels to the west had waned. The change in orientation of structures in Byrd Group as it approaches Byrd Glacier is consistent with kinematics in which there is a left-lateral component to the convergence.

Posttectonic plutonism is widespread both in the Miller and Geologists Ranges and in the Goldie Formation. A pluton intruding Shackleton Limestone at Bartram Plateau indicates that deformation of that formation had also been completed at least in the southern area before the final intrusion. This plutonism marks the end of tectonic activity of the Ross orogen, which was being rapidly exhumed by about 480–475 Ma in the Ordovician.

# 5 Queen Maud and Horlick Mountains

## Geological Summary

The Queen Maud and Horlick Mountains are dominated by a composite batholith of proportions rivaling any in the world, with plutonic rocks continuous from east of Shackleton Glacier to the Ohio Range, a distance of 550 km. Because field parties were working at opposite ends of this batholith in the early days of mapping, two different names came to be applied: to the west the "Queen Maud batholith" (McGregor, 1965) and to the east the "Wisconsin Range batholith" (Faure et al., 1968). The area where the nomenclature impinges is Scott Glacier, the widest exposure of the batholith throughout its length. In preference to a new term, the name "Queen Maud–Wisconsin Range batholith," albeit cumbersome, is probably the best designation for this plutonic body.

Although there are some exceptions, in general the batholith separates distinct suites of sedimentary and volcanic rocks on the outboard Ross Ice Shelf side and inboard polar plateau side of the Transantarctic Mountains. The oldest unit of the inboard suite is the La Gorce Formation, a folded sequence of graywacke and pelite, probably equivalent to the Goldie Formation (Beardmore Group) of the central Transantarctic Mountains. This is followed by the Wyatt Formation, a massive, silicic porphyry with both extrusive and hypabyssal phases that is known to have intruded La Gorce Formation after its folding. At one locality in the La Gorce Mountains the Wyatt Formation is conformably overlain by a clastic sequence known as the Ackerman Formation. Dating of the Wyatt and Ackerman Formations has been problematic, but it is likely that they are correlative, respectively, with the Thiel Mountains porphyry and Mt. Walcott Formation of the Thiel Mountains, which have been dated by the Rb/Sr whole-rock method at about 500 Ma (Pankhurst et al., 1988).

The outboard suite includes two distinct sequences of metamorphic rocks that occur in scattered outcrops along the Ross Ice Shelf. One of these is the Liv Group, characterized by bimodal volcanic rocks, quartzites, and marbles. The marbles have been fossil-dated as both Early and Middle Cambrian. The other sequence contains pelitic and calc–silicate schists of the Duncan Formation (Duncan Mountains) and the Party Formation (Leverett Glacier area). These units are thought to be older than the Liv Group and have previously been correlated with the Beardmore Group, but they lack fossils, and the only known contacts with Liv Group are faults. So whether they are Neoproterozoic, like the Beardmore Group, has not been proved.

Deformation of the outboard suite produced considerably more mobility than the inboard suite. Intrusion of the main phase of the Queen Maud–Wisconsin

**Figure 5.1.** View from Mt. Sletton to the southeast across Scott Glacier to Mt. Zanuck and Grizzly Peak. Area underlain entirely by Granite Harbour Intrusives.

Range batholith was mainly posttectonic, with cooling ages similar to those of the Granite Harbour Intrusives elsewhere in the Transantarctic Mountains.

## Chronology of Exploration

The Queen Maud Mountains were encountered by Roald Amundsen's party en route to the South Pole in 1911, extending the known length of the Transantarctic

**Figure 5.2.** Queen Maud Mountains viewed to the northwest from Mt. Griffith. High point on the horizon is Mt. Fridtjof-Nansen. At that point the Transantarctic Mountains narrow to about 50 km. The polar plateau marks the left skyline and the Ross Ice Shelf the right. All outcrops in foreground are Granite Harbour Intrusives.

**Figure 5.3** (*facing page*). Location map from Queen Maud to Horlick Mountains.

Mountains an additional 300 km beyond Shackleton's route up Beardmore Glacier. The only landfall of the Norwegians' entire venture was at Mt. Betty, a tiny nunatak near the mouth of Strom Glacier, close to the cache of supplies left before the ascent of Axel Heiberg Glacier and the crossing of the polar plateau. Amundsen (1912, p. 39) recounts that he and Bjaaland skied over to the outcrop, where they "photographed each other in 'picturesque attitudes' [and] took stones for those who had not yet set foot on bare earth." However, the collection failed to impress the others in camp. Amundsen writes, "I could hear such words as 'Norway – stones – heaps of them,' and I was able to piece together and understand what was meant."

The rocks were returned to Norway and later described by Schetelig (1915). The specimens numbered "about twenty" and consisted of medium- to fine-grained light-gray granites, aplites, fine-grained orthogneisses, and one sample of mica schist.

While approaching the mountain front on the journey south, at about 84°S, Amundsen noted the appearance of land to the east, which seemed to veer to the southwest and connect in a great semicircle to the Queen Maud Mountains. The intersection of the two land masses was mapped as the southern apex of the Ross Ice Shelf (correctly), and the eastern highlands were named Carmen Land.

The next visit to the area was made two decades later by a ground party of the First Byrd Antarctic Expedition, led by geologist Laurence M. Gould. The party

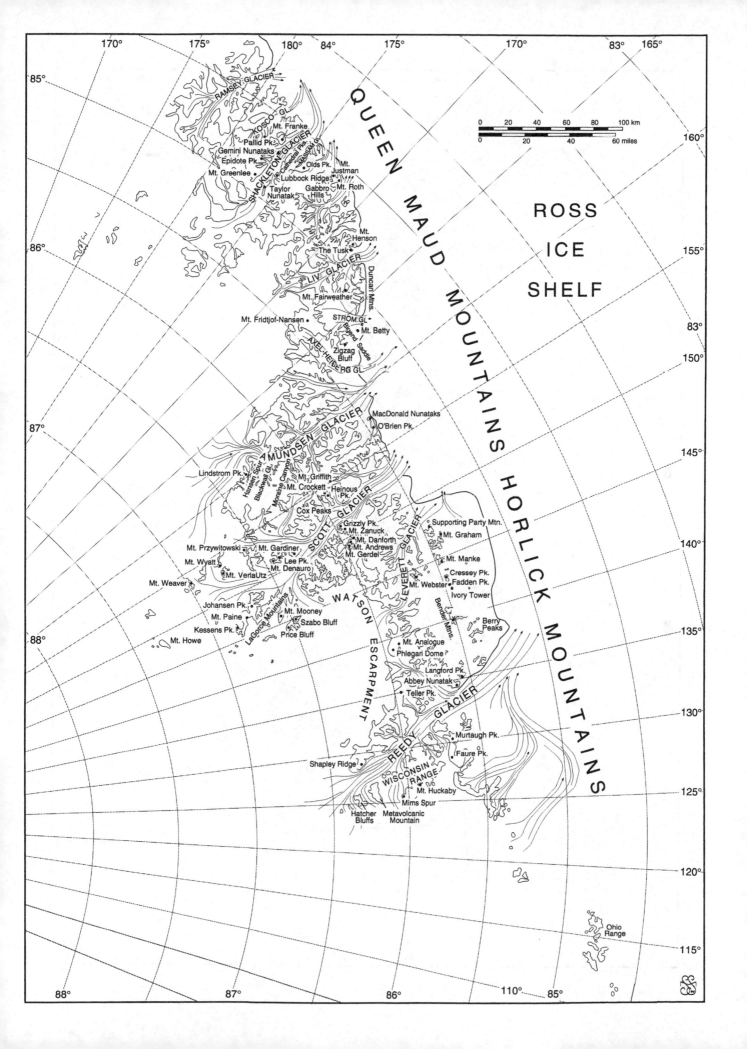

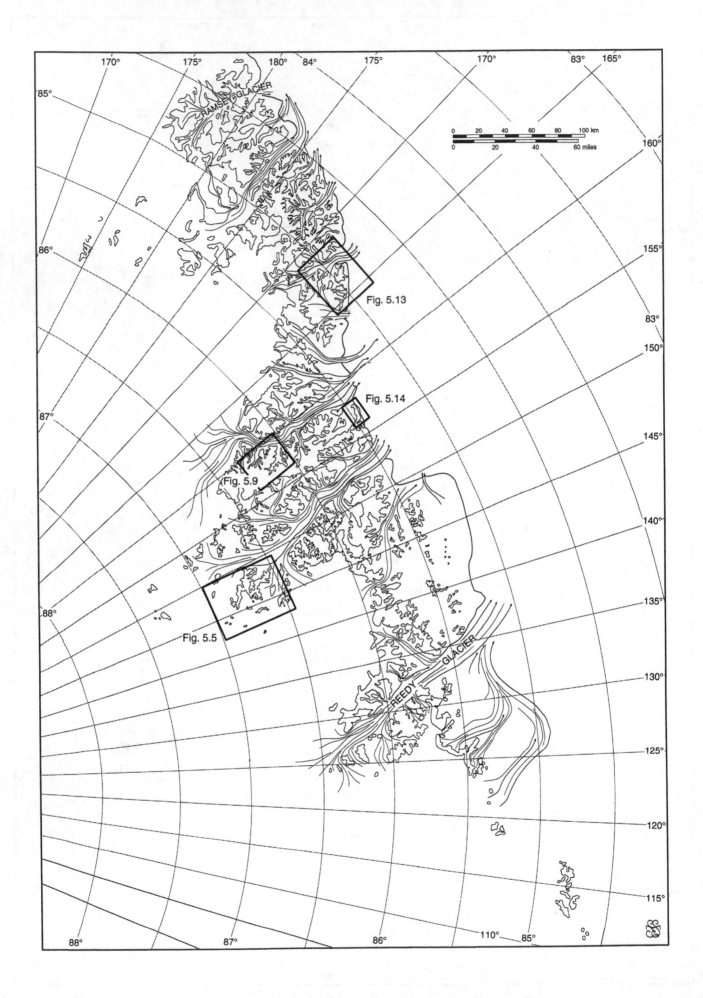

retraced Amundsen's route to the mountains, where they gave weather reports for Byrd's polar flight up Liv Glacier. With the flight successfully completed, Gould climbed up into Beacon rocks at the foot of Mt. Fridtjof-Nansen; then the party headed eastward along the mountains in search of Carmen Land, making landfalls at several localities and recording granitic and metamorphic rocks (Gould, 1931). The easternmost station was at Supporting Party Mountain, from which they could see conclusively that Carmen Land did not exist. In addition to the initial descriptions of the geology of the Queen Maud Mountains, Gould (1935) extended David and Priestley's (1914) idea of the "Great Antarctic Horst" to this part of the Transantarctic Mountains. Returned samples from this expedition were described by Stewert (1934a, b, c).

In 1934–35 a ground party of the Second Byrd Antarctic Expedition, led by geologist Quin A. Blackburn, returned to Supporting Party Mountain and from there forged a route to the polar plateau up Scott (then Thorne) Glacier. Blackburn (1937) recorded gneisses and pegmatitic veins in the nunataks on the eastern side of the mouth of the glacier, observed that for most of its length the glacier was bounded by mountains of granitic composition, and noted that the line of undulating hills on Ackerman Ridge was composed of "black, porphyritic material." Like Gould, however, Blackburn was most interested in the Beacon stratigraphy, and his principal geological effort was directed toward collecting and measuring a section at Mt. Weaver.

The area between Beardmore and Liv Glaciers remained unsighted until 1940, when a reconnaissance flight of the U.S. Service Expedition, based at Little America, closed the gap. Onboard the plane was the scientific leader of the expedition, geologist F. Alton Wade.

In the decade following the IGY, a number of field parties visited the Queen Maud and Horlick Mountains. Nevertheless, by 1969 with the production of the American Geographical Society's Antarctic Map Folio Series, the one significant area remaining undifferentiated at the 1:1,000,000 scale throughout the Transantarctic Mountains was in the coastal region between Reedy and Axel Heiberg Glaciers.

In November 1958 an overland geophysical traverse passed within 45 km of the eastern end of the Wisconsin Range. William Long and Fred Darling hiked over to the range and collected granites in situ and Beacon rocks from a moraine. The following month the traverse party camped adjacent to the central Ohio Range, where a number of rocks and fossil specimens were collected (Long, 1959), justifying further geological studies. During the 1960–61 and 1961–62 seasons, the United States fielded parties to the Ohio Range. Their primary focus was the stratigraphy and paleontology of the Beacon sequence; however, two reports covered the igneous and metamorphic rocks of the basement (Treves, 1965; Long, 1965a).

Geological observations were made at several coastal localities along the Queen Maud Mountains in 1961–62 during an airborne geophysical study of glaciers tributary to the Ross Ice Shelf (Linder et al., 1965; Craddock et al., 1964). During the same season a New Zealand party, working primarily along the polar plateau in the headreaches of Beardmore Glacier, traversed southward and retraced Amundsen's route down Axel Heiberg Glacier (Herbert, 1962; Grindley et al., 1964).

**Figure 5.4** (*facing page*). Location map for figures in Chapter 5.

Two U.S. field parties worked in the Queen Maud Mountains in 1962–63. One, under the leadership of Al Wade, went to a part of the area he had sighted in 1940, undertaking a ground traverse along the northern side of Shackleton Glacier. The other was based at the foot of Mt. Weaver, where Blackburn's (1937) observations were confirmed (Doumani and Minshew, 1965). The use of helicopters during part of this season permitted reconnaissance investigations throughout the upper Scott Glacier area (Minshew, 1965, 1966, 1967). Also during this season, the first helicopter flight to the South Pole was accomplished, with its departure from Mt. Howe, the earth's southernmost outcropping of rock (Anon., 1963).

In 1963–64 a New Zealand party traversed between Axel Heiberg and Shackleton Glaciers, examining rocks throughout the coastal areas (McGregor, 1965). The same season a U.S. party worked from a base camp on the western side of Nilsen Plateau (Long, 1965b; Mirsky, 1969).

In 1964–65 studies were continued in the Shackleton Glacier area using helicopters from a camp on McGregor Glacier (Wade et al., 1965a, b; Wade, 1974; Wade and Cathey, 1986). Helicopters were also used from a base camp in the Wisconsin Range, from which investigations were conducted throughout the Reedy Glacier area, with a single landing made at Mt. Webster in the Leverett Glacier area (Minshew, 1967; Murtaugh, 1969; Faure et al., 1968, 1979; Eastin, 1970).

Since 1969 several other parties have undertaken studies in the basement of the Queen Maud Mountains. In 1969–70 a New Zealand party traversed Scott Glacier from its headreaches to its mouth (Katz and Waterhouse, 1970a, b). In 1970–71 a U.S. party occupied two helicopter-supported base camps, one on McGregor Glacier and the other on the west side of Nilsen Plateau, from which they covered the entirety of the Queen Maud Mountains (Elliot and Coats, 1971; Stump, 1974, 1976a, b, 1985; Burgener, 1975).

Subsequently, all work on basement rocks in the region has been done by U.S. ground parties. In 1974–75 a party examined the Duncan Mountains (Stump, 1981a). This study was followed by others in the Leverett Glacier area in 1977–78 (Stump et al., 1978; Heintz, 1980; Lowry, 1980), on the western side of upper Scott Glacier in 1978–79 (Stump et al., 1979; Borg, 1980), and on the eastern side of upper Scott Glacier in 1980–81 (Smit, 1981; Stump, 1983; Stump et al., 1985, 1986a, b; Smit and Stump, 1986). In 1986–87 a party worked in the area of the Gabbro Hills (Borg et al., 1987a, 1990), and in 1987–88 a party examined rocks in the central Scott Glacier area in connection with a fission-track dating study (Stump and Fitzgerald, 1988). In 1990–91 a party worked in the Horlick Mountains (Borg et al., 1991; Borg and DePaolo, 1994) and in 1992–93 another studied rocks north of Leverett Glacier (Rowell et al., 1993).

## Inboard Formations

### La Gorce Formation

**Distribution.** The La Gorce Formation is a folded sequence of alternating metagraywacke and metapelite that crops out in scattered localities from the western side of Nilsen Plateau to the eastern side of Reedy Glacier. It is medium

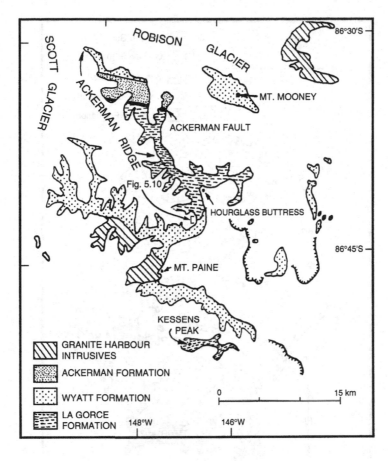

**Figure 5.5.** Geological map of La Gorce Mountains. After Stump et al. (1986a). Used by permission.

to dark gray and black, and typically weathers brown. Minshew (1967) informally designated the La Gorce Mountains as the type locality (Fig. 5.5). A considerable portion of Ackerman Ridge in the La Gorce Mountains, originally mapped from helicopter as La Gorce Formation (Minshew, 1967; Mirsky, 1969), was later found to be composed of Wyatt Formation (Katz and Waterhouse, 1970a).

Known localities of La Gorce Formation occur on the northwestern side of Nilsen Plateau at Hansen Spur and the head of Blackwall Glacier (Stump, 1985) (Fig. 5.6), on the western side of Scott Glacier at Lee Peak and Mt. Denaro (Stump et al., 1979) and at Cox Peaks (Katz and Waterhouse, 1970a), in the La Gorce Mountains at the southern end of Ackerman Ridge and at Kessens Peak (Stump et al., 1985), and in the Reedy Glacier area on the spur northeast of Teller Peak, on spurs north of Faure Peak and Murtaugh Peak, and around Mt. Huckaby (Mirsky, 1969; Murtaugh, 1969).

**Characteristics.** Neither the top nor the bottom of the La Gorce Formation is known. Due to the repetition of bedding by folding, the thickness is uncertain. Minshew (1967) estimated in excess of several thousand meters in the La Gorce Mountains, and Stump (1985) estimated in excess of 2,000 m at Hansen Spur. La Gorce Formation is older than Wyatt Formation, as demonstrated by intrusive contacts of hypabyssal phases of the latter at the head of Blackwall Glacier (Stump, 1985) and in the La Gorce Mountains (Stump et al., 1986b).

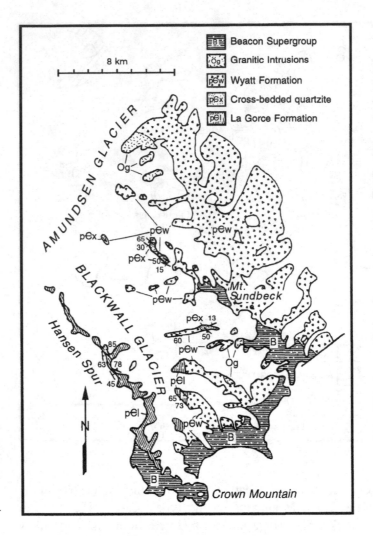

**Figure 5.6.** Geological map of Blackwall Glacier area, west of Nilsen Plateau. From Stump (1985). Used by permission.

At most localities the formation has been folded into nearly vertical attitudes (Fig. 5.7). Axial-plane cleavage is developed at places. The rocks have been metamorphosed to greenschist facies with quartz + feldspar + biotite ± muscovite ± chlorite assemblages typical in the pelitic fractions. Locally in the La Gorce Mountains cordierite and andalusite are present (Smit and Stump, 1986).

The most detailed sedimentological study of the La Gorce Formation is that of Smit (1981; Smit and Stump, 1986), who identified three types of sequences in the La Gorce Mountains based on bedding thicknesses and internal sedimentary structures of the graywackes. These include (1) graywacke and mudstone sequences, (2) Bouma sequences, and (3) laminated sandy siltstones.

In the graywacke and mudstone sequences, graywacke beds are typically 1–1.5 m thick, but range from 20 cm to 3 mrs. Mudstones are 5–20 cm thick. In the majority of cases the graywacke beds are massive; however, an appreciable number of beds are internally laminated, in some cases showing channels and cross-bedding, and a small percentage of the graywacke beds are internally graded. Contacts between graywackes and mudstones are often sharp, but many graywacke beds show a delayed graded transition into overlying mudstones. In rare occurrences, tops of mudstones contain scour marks.

**Figure 5.7.** Steeply dipping La Gorce Formation at Mt. Denaro.

Of secondary abundance are relatively thinner sequences (15–30 cm) that correspond to the A, C, and E divisions of Bouma's (1962) turbidite model. Sandy A divisions comprising 60% or more of most sequences grade sharply upward into silty cross-bedded C divisions generally 2–5 cm thick. These grade upward into pelitic E divisions at the top, which vary in thickness and are often eroded by subsequent turbidity currents.

Of minor importance are laminated sandy siltstones, 0.5–1 m in thickness, which vary from the other sequences by virtue of the irregularity of lamina thickness, absence of mudstone tops, and lack of grading.

Smit and Stump (1986) have interpreted the graywacke and mudstone sequences to be the result of high-density gravity flows and the Bouma sequences to be the result of turbidity currents. The laminated silty sandstones are perhaps the result of low-density bottom currents operating between gravity-flow depositional events. The environment of deposition was interpreted to have been middle-fan, depositional lobe, in a particularly broad shelf-rise setting.

A unique lithology of La Gorce Formation crops out at Cox Peaks, where massive conglomeratic units occur within beds of metagraywacke and argillite (Stump and Fitzgerald, 1988). Bedding within the conglomerates was not observed, but one unit is at least several hundred meters in thickness. Clasts in the conglomerate are derived primarily from the La Gorce Formation itself. They are flattened and elongate parallel to bedding in surrounding metagraywacke, but appear to have been angular to subrounded. In addition, there are rare, rounded clasts of vein quartz. The conglomerate is clast-supported with very little interstitial matrix.

**Petrography.** Petrographically, metagraywackes of the La Gorce Formation contain clasts of quartz and plagioclase set in abundant matrix. By the classification of Pettijohn et al. (1972) the rocks are all feldspathic graywackes. The quartz grains are subrounded to angular, with maximum grain sizes generally around 0.35 mm and the greatest observed size being 0.9 mm. Quartz extinction is slightly to highly undulose, and in some of the more metamorphosed samples the quartz is polygonized. Plagioclase grains are smaller than the coarsest quartz in a given sample. They vary in degree of alteration from practically clear to highly altered. The majority of the plagioclase grains exhibit albite twinning.

Except for rare composite grains of quartz and plagioclase, no other rock fragments have been observed. Persistent trace minerals include garnet, muscovite, epidote, clinopyroxene, sphene, and zircon. Apatite, rutile, corundum, kyanite, and tourmaline have also been observed (Smit and Stump, 1986). The detrital suite indicates that the La Gorce Formation was derived at least in part from a crystalline complex of relatively high-grade metamorphic and felsic plutonic rocks.

Minshew (1967) notes rare limestones and common calcite veins and vugs in the La Gorce Formation, although he is not specific as to their distribution. He also mentions a "calcareous–argillaceous facies" of the La Gorce at "Chlorite Hills" (presumably Langford Peak and Abbey Nunatak west of the mouth of Reedy Glacier). The present author has observed amphibolite and two marble beds at Langford Peak, which appear more closely related to Leverett Formation as seen in the Bender Mountains than to La Gorce Formation in the upper Scott Glacier area. Except for some calcite veining close to faults in the La Gorce Mountains, Stump (1985) and Smit and Stump (1986) have observed no calcite in the La Gorce Formation. Only on this point does the La Gorce Formation appear to differ from the bulk of the Goldie Formation in the Nimrod to Ramsey Glaciers area (see Chapter 4).

**Deformation.** La Gorce Formation is steeply dipping throughout its outcrop area. Nearly vertical axial-plane cleavage is developed to various degrees as well, more so in the pelitic beds and in regions of fold closures, while in metagraywacke beds in some areas it does not occur at all. Murtaugh (1969) reported one fold closure in the Reedy Glacier area, but other than that folds have not been observed outside of the La Gorce Mountains. This is probably more a matter of exposure, however, than of greater development in the latter area.

Stump et al. (1986b) summarized folding in the La Gorce Mountains where zones of closely spaced fold closures are separated by broad zones of steep, uniformly dipping beds. Folds are generally tight to isoclinal, with a chevron style, narrow hinge zones, and planar limbs (Figs. 5.8 and 5.9). Folds with concentric geometry also exist, although they are not as numerous. Asymmetrical folds are more common, but transitions to symmetrical forms exist. Vergence is

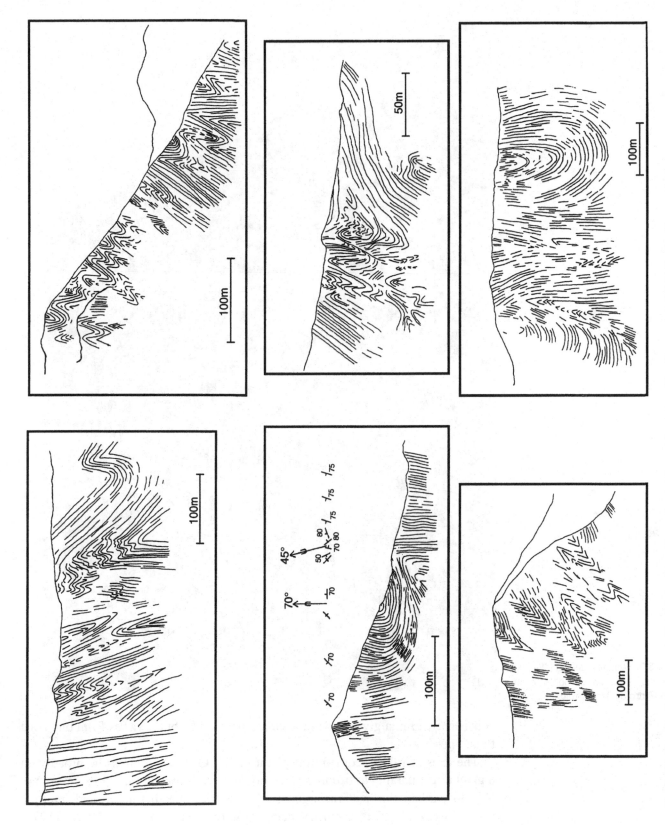

**Figure 5.8.** Sketches of exposures of folds in La Gorce Formation, La Gorce Mountains. After Stump et al. (1986b). Used by permission.

**Figure 5.9.** Steeply plunging folds in La Gorce Formation, La Gorce Mountains. Photo by Jerry Smit.

consistent within a given zone of asymmetrical folds, but it varies from one zone to another.

The axes of folds recorded throughout the La Gorce Mountains lie broadly on a great circle parallel to bedding-plane orientation. Stump et al. (1986b) were not able to explain this pattern, arguing against both sheath folding and multiple folding episodes. This geometry may be of regional extent, for Oliver (1972) notes similar co-planar fold axes in Goldie Formation near the mouth of Beardmore Glacier.

Other than local kinking associated with the Ackerman fault (see later), only one generation of folds has been observed in La Gorce Formation. No intersecting cleavages or oppositely vergent mesofolds, as in Goldie Formation at Cotton Plateau, are known to occur. In this, the folding of La Gorce Formation is similar to that of Goldie Formation as observed throughout the remainder of its outcrop.

The time of folding of the La Gorce Formation is not well constrained. Wyatt Formation, probably Late Cambrian in age (see later), intrudes across the folding, and the Wyatt itself was sheared to various degrees and faulted against La Gorce Formation presumably during the Ross orogeny. It has generally been assumed that folding of La Gorce Formation occurred at the same time as the Goldie Formation in the central Transantarctic Mountains and the Patuxent Formation in the Pensacola Mountains, that is, before their being unconformably overlain by Cambrian limestones.

## Wyatt Formation

**Distribution.** The Wyatt Formation is a silicic porphyry that crops out intermittently throughout approximately the same area as the La Gorce Formation. The informal type of locality was designated by Minshew (1967) as Mt. Wyatt on the western side of upper Scott Glacier. Wyatt Formation is known from the western side of Nilsen Plateau at Moraine Canyon, Blackwall Glacier, Lindstrøm Peak, and Crack Bluff (Stump, 1985) (Fig. 5.6), on the west side of Scott Glacier from Mt. Gardiner to Mt. Wyatt and Mt. Verlautz (Stump et al., 1979; Borg, 1980), in the La Gorce Mountains at the northern end of Ackerman Ridge and in the vicinity of Johansen Peak, as well as at Mt. Mooney (Stump et al., 1985) (Fig. 5.5), in roof pendants in the central Scott Glacier area at Mt. Danforth, Cox Peaks, Heinous Peak, and Mt. Griffith (Katz and Waterhouse, 1970a; Stump and Fitzgerald, 1988). Wyatt Formation also crops out in the Reedy Glacier area at Metavolcanic Mountain, Shapley Ridge, and east of Mt. Huckaby (Murtaugh, 1969).

Both volcanic and hypabyssal intrusive phases of the Wyatt Formation are known. Intrusive contacts with La Gorce Formation occur at Blackwall Glacier (Stump, 1985) and in the La Gorce Mountains (Stump et al., 1986b) (Fig. 5.10). Although no base of the eruptive phase has been seen, the porphyry is overlain conformably by sediments of the Ackerman Formation in the La Gorce Mountains (Stump, 1983). The Wyatt porphyry has been intruded by Granite Harbour Intrusives at a number of localities. In most instances the contacts are sharp; however, at Mt. Paine there is considerable intermixing of granite and porphyry. Also, on the northern side of Moraine Canyon, at Mt. Griffith, and at Shapley Ridge there are broad transition zones in which clearly identifiable Wyatt Formation gradually changes to coarse-grained granitic rock (Murtaugh, 1969; Stump, 1985; Stump and Fitzgerald, 1992).

**Field Appearance.** In outcrop the Wyatt Formation is a medium gray to black fine-grained porphyry with conspicuous phenocrysts of quartz and feldspar (Fig. 5.11). Throughout its exposure it varies considerably in degree of alteration and development of metamorphic foliation. At most localities the porphyry is

**Figure 5.10.** Wyatt Formation (upper right) intruding La Gorce Formation south of Hourglass Buttress in the La Gorce Mountains.

massive, showing no internal layering or other variation. One exception to this is at Lindstrøm Peak, where layering is apparent from greater and lesser concentrations of phenocrysts and lenses of chertlike material lacking phenocrysts altogether.

Also, in the area between Mt. Gardiner and Mt. Przytowski, most of the outcrops of Wyatt Formation are interspersed with light-colored felsic lenses, which define a foliation in the rock (Borg, 1980) (Fig. 5.12). The lenses are generally 1–2 cm thick and several centimeters long, with a maximum size of 5 cm by several tens of .entimeters. In the plane of foliation the lenses are roughly circular. The phenocryst content of the felsic lenses is identical to that of the surrounding groundmass, although the size of the crystals in the lenses appears to be slightly larger. In thin section it is apparent that K-feldspar is more abundant in the groundmass of the lenses than in the surrounding groundmass of the rock.

Throughout its exposure the foliation defined by the felsic lenses is remarkably constant, with strike varying from 125° to 150° and dip from 25°S to 50°S. Due to their regular shape and consistent orientation, Borg (1980) has interpreted the lenses as a relic of an original bedding feature, probably flattened pumice fragments. Similar features are pictured from the Thiel Mountains Porphyry by Ford (1964, p. 437).

**Figure 5.11.** Breccia in Wyatt Formation west of Mt. Farley. Scale: Phil Colbert, field assistant extraordinaire.

**Petrography and Geochemistry.** Stump et al. (1986b) published five major-element analyses of Wyatt Formation from the La Gorce Mountains and five analyses of volcanic porphyry from the Ackerman Formation. All are peraluminous dacites with $SiO_2$ contents between 65.5 and 69.8%, except for one sample of Ackerman Formation with $SiO_2$ of 72.9%. Three analyses on Wyatt Formation from the west side of Nilsen Plateau show similar results (Stump, 1976a).

Petrographically, the Wyatt Formation is a crystal-rich porphyry with a fine-grained, recrystallized groundmass. Phenocrysts vary in size from a few millimeters to 1 cm, and in abundance from 20 to 70% although 35–50% is typical. Quartz and plagioclase are plentiful in all samples. Biotite is appreciable in most. Minor amounts of K-feldspar occur in many of the samples, and hypersthene has been found in the vicinity of Mt. Wyatt (Borg, 1980).

Quartz phenocrysts vary from euhedral (hexagonal) to rounded, with most grains being embayed. Plagioclase crystals, in many cases broken, usually have oscillatory zoning and well-developed albite twinning. Depending on the degree of alteration of the rock, the plagioclase varies from nearly clear to completely sericitized or saussuritized. Biotite crystals commonly have opaque inclusions and sometimes are completely replaced by them. In some cases the biotite books appear to have been bent around other crystals. K-feldspar typically occurs as angular crystal fragments that are unzoned and untwinned.

**Figure 5.12.** Wyatt Formation porphyry interspersed with felsite lenses, interpreted to be flattened pumice fragments. Locality north of McNally Peak.

Hypersthene crystals are rounded and in a few cases embayed. They show various degrees of reaction with the groundmass, from none at all to complete replacement by brownish green fibrous amphibole. Opaque inclusions are common (Borg, 1980).

The groundmass of the Wyatt Formation is a granoblastic intergrowth of quartz, plagioclase, and usually K-feldspar. Biotite, sericite, chlorite, and/or epidote also occur, indicating that these rocks were metamorphosed under lower-greenschist conditions. The fine-grained micas may be dispersed throughout the groundmass, as well as the plagioclase, or be segregated into mottled or foliated patterns. At Mt. Mooney and the northern tip of Ackerman Ridge, the groundmass has a granophyric texture (Stump et al., 1986b).

**Nature of Emplacement.** Recrystallization of the groundmass is such that no original textures indicative of volcanic processes are preserved. The extrusive nature of portions of the Wyatt Formation is suggested by several features. Fragmented phenocrysts indicate explosive eruption. Bent biotite suggests movement during compaction. Felsic lenses have been interpreted as flattened pumice fragments. And rare layering is indicative of extrusive origin. Most geologists who have worked on the Wyatt Formation have interpreted it as originating, at least in part, by explosive volcanic processes (Minshew, 1967; Murtaugh, 1969; Borg, 1980; Stump et al., 1986b).

However, the intrusive nature of portions of the Wyatt Formation is affirmed by the contacts in the La Gorce Mountains and at Blackwall Glacier, and suggested by the granophyric textures at places. Overall, the Wyatt Formation appears to represent the remains of a silicic volcanic complex in which the bottom of the volcanic pile, as well as portions of the surrounding country rock (La Gorce Formation), were intruded by hypabyssal phases of the magma. Owing to the very considerable extent of the porphyries, it has been suggested that they originated in a caldera complex (Stump et al., 1986b).

## Ackerman Formation

A unique succession of sedimentary and volcanic rocks, named the Ackerman Formation, conformably overlies Wyatt Formation on Ackerman Ridge in the La Gorce Mountains (Stump, 1983). The base of the formation is marked by the appearance of a green shale unit. Approximately 2,000 m of section are exposed, with the top of the formation truncated by a fault with La Gorce Formation.

The volcanic rocks are similar to the Wyatt Formation in color, phenocryst content, and chemistry. One unit contains rip-up clasts in its base. Faint layering due to variations in phenocryst concentration also occurs in a few units. The extrusive nature of these porphyries is indisputable owing to their being interlayered with the sedimentary rocks.

The sedimentary units are mainly gray to green interbedded sandstones and shales or phyllites. Bedding varies from a few centimeters to 1 m. Although sedimentary structures are not plentiful, fine, planar lamination, ripple-drift lamination, rip-up clasts, and faint tabular cross-bedding have been recorded. Several sandstone units contain sparse, rounded cobbles.

The sandstones are composed primarily of quartz, some of which is embayed, but plagioclase and volcanic rock fragments are also appreciable. The source for the sandstones was, at least in part, local volcanic rocks. The shales and the matrix of the sandstones have been extensively sericitized, and chlorite is developed in some samples. Calcite is also present in minor amounts in a few units.

The Ackerman Formation represents shallow marine deposition of sediments around active volcanic islands or at a volcanically active basin margin (Stump, 1983). It appears similar to the Mt. Walcott Formation of the Thiel Mountains (Storey and Macdonald, 1987).

## Age Relationships

The evolution of isotopic dating of the Wyatt Formation and other basement units in the region is a classic example of the extent to which ideas about the tectonics of an area can become entrenched. The first Rb/Sr whole-rock study in the Transantarctic Mountains was conducted by Faure et al. (1968) on basement rocks of the Horlick Mountains. Two similar dates were published, $620 \pm 13$ Ma for the Wyatt Formation and $613 \pm 22$ Ma for the rapakivi phase of the Wisconsin Range batholith. A date of $450 \pm 16$ Ma was published for the La Gorce Formation, but because it was suspected from field relations to be older than Wyatt Formation, the date was interpreted to represent a time of metamorphic homogenization during the Ross orogeny.

Grindley and McDougall (1969) seized on $620$–$613 \pm$ Ma as well as similar ages from the Thiel Mountains Porphyry generated by the lead-$\alpha$ method (Ford et al., 1963) to date their newly proposed Beardmore orogeny, known to be pre–Early Cambrian because of field relations in the Nimrod Glacier area. The timing of the Beardmore orogeny was widely accepted by subsequent authors summarizing regional geology in the Transantarctic Mountains (e.g., Stump, 1973, 1976a, 1981b; Elliot, 1975; Gunner 1976, 1982).

In 1979 Faure et al. published a new study of Rb/Sr whole-rock geochronology from basement rocks between Nilsen Plateau and the Pensacola Mountains. New dates on the Wyatt Formation from Nilsen Plateau and the Horlick Mountains were $466 \pm 9$ Ma (IR = .7157) and $543 \pm 47$ Ma (IR = 0.7098), respectively. A new date for the rapakivi granites in the Horlick Mountains was $496 \pm 23$ Ma (IR = 0.7157). All of these dates were interpreted as representing resetting during the Ross orogeny, and the assumption continued that they were associated with the Beardmore Orogeny, "best estimated" by a new date of $646 \pm 77$ Ma (IR = 0.7069) from the Thiel Mountains Porphyry (Faure et al., 1979).

In their study of the La Gorce Mountains, Stump et al. (1986b) recognized that the folding of the La Gorce Formation was preintrusion of the Wyatt Formation. They also observed that the pluton of granite at Mt. Paine appeared to have intruded Wyatt Formation before its complete solidification. On this point, Stump et al.'s (1986b) interpretation was that the pluton was a Precambrian phase of the Granite Harbour Intrusives widely developed during the Cambro-Ordovician. A new Rb/Sr whole-rock isochron of $545 \pm 59$ Ma (IR = 0.7114, MSWD = 23.5) on Wyatt Formation from Mt. Wyatt (dating by R. J. Pankhurst) was interpreted by Stump et al. (1986b) to be a reset date following magmatism in the Neoproterozoic.

In 1988 Pankhurst et al. published four Rb/Sr whole-rock isochrons from different magmatic suites in the Thiel Mountains with dates of $495 \pm 6$, $493 \pm 24$, $491 \pm 12$, and $480 \pm 18$ Ma. These data supersede previous dates from the Thiel Mountains and require reinterpretation of the dates from the Wyatt Formation, given the likely correlation of Wyatt Formation and Thiel Mountains Porphyry (see Chapter 6). The dates on Wyatt Formation from Faure et al. (1979) and Stump et al. (1986b) fall barely within error of the dates from Pankhurst et al. (1988).

For the Wyatt Formation to be Cambrian or earliest Ordovician in age is reasonable in light of the other silicic volcanism of Cambrian age in the Queen

Maud–Horlick Mountains (Liv Group). It is also reasonable that Wyatt volcanism should closely precede the extensive plutonic magmatism of the Granite Harbour Intrusives that engulfs it. This reassignment of the age of the Wyatt Formation in no way negates the existence of the Beardmore orogeny, which is demonstrated by the angular unconformity of Shackleton Limestone on Goldie Formation (Chapter 4), but it removes the original basis upon which the orogeny was absolutely dated.

## Deformation

The most conspicuous deformation of Wyatt Formation is cleavage, developed in some of the unit throughout its regional extent. All degrees may be found, from faint alignment of micas in the groundmass to extensive cataclasis of phenocrysts and anastamosing growth of micas. The cleavage strikes fairly consistently in a northwest southeast direction, parallel to the regional structural grain of the mountains. The cleavage can be shown to be pre–Granite Harbour Intrusives in the Mt. Wyatt area where foliated xenoliths of Wyatt Formation in granite are randomly oriented adjacent to the foliated wall rock from which they have been derived (Borg, 1980).

Without planar markers, it cannot be said whether the Wyatt Formation has been folded. At three localities, however, the Wyatt or the Ackerman Formation is in fault contact with La Gorce Formation, on Ackerman Ridge, at Lee Peak, and in the Cox Peaks. All three faults strike in a similar eastsoutheast direction, but each has distinctive characteristics.

In the La Gorce Mountains the Ackerman fault is surrounded by a broad zone of mylonitic foliation in both the Ackerman and La Gorce Formations (Stump et al., 1986b). Cleavage overprints bedding for more than 500 m into the La Gorce Formation and completely overprints it for about 50 m. Cleavage is developed in Ackerman Formation throughout the 600-m-thick volcanic unit adjacent to the fault and occurs in the phyllitic units throughout the formation. A mineral lineation occurs downdip in the foliation (dipping 70°NE), suggesting a high-angle reverse sense of movement on this fault. Late-stage kink folds are developed in the cleavage of both the Ackerman and La Gorce Formations.

At Lee Peak the fault is marked by a 1-m-wide zone of yellow gouge. No cleavage is developed in the adjacent rocks, and no sense of movement was apparent.

At Cox Peaks cleavage is developed adjacent to the fault in the Wyatt Formation for 5–10 cm, and in the La Gorce Formation for 1–3 m. Quartz veining in steeply inclined tension gashes indicates a right-lateral sense of movement for this fault. Since granitic rocks engulf the Cox Peaks roof pendant, which shows no sign of offset along its boundary, the faulting must be preintrusion.

## Isolated Sedimentary Successions

Sedimentary sequences with uncertain affinities have been recorded at two places in the region, one at Blackwall Glacier (Stump, 1985) and the other at the head of Reedy Glacier (Murtaugh, 1969).

At Black Rock Glacier a nearly horizontal 150-m section of sandstone and shale, with a 2-m basal bed of sheared white marble, overlies massive Wyatt Formation. Whether the contact is sedimentary, intrusive, or a fault could not be determined. The sandstones are tan to gray, with some faint cross-bedding. The shales are black and in part laminated. No volcanic rocks occur in the sequence.

These rocks are distinctly different from the La Gorce Formation, which is exposed nearby. But neither do they bear much similarity to the Ackerman Formation in color or in the absence of volcanic units.

At the head of Reedy Glacier a section of clastic rocks is exposed in an isolated nunatak. The section dips about 50°NW, with cleavage developed irregularly in it. The sequence is formed primarily of argillaceous quartzite and siltstone, which contain some ripple marks. Again no volcanic units are present. Murtaugh (1969) states that the rocks are distinct from the La Gorce Formation and that their stratigraphic age is problematic. Bedding varies from a few centimeters to 1 m.

## Outboard Formations

Scattered in discontinuous outcrops along the Ross Ice Shelf from Kosco Glacier to Reedy Glacier are two distinct sequences of metamorphic rocks. Pelitic and calc–silicate schists, represented by the Duncan and Party Formations, are thought to be the older and have previously been correlated with rocks of the Beardmore Group. The other sequence consists primarily of bimodal volcanics, quartzites, and marbles that contain rare Cambrian fossils. These rocks belong to the Liv Group, including the Greenlee, Taylor, Fairweather, and Leverett Formations.

Because the two sequences have had common metamorphic and deformational histories during the Ross orogeny, with different manifestations at different places, they will be discussed according to the areas in which they are found, rather than chronologically.

### Shackleton Glacier Area

Along the northern half of Shackleton Glacier, rocks of the Taylor and Greenlee Formations straddle the western end of the Queen Maud–Wisconsin Range batholith. Wade et al. (1965a) named the Taylor Formation and informally designated Taylor Nunatak as the type locality. Wade (1974) later subdivided the Greenlee Formation from the Taylor Formation, for the fine-grained clastic rocks cropping out on the lower portions of Mt. Greenlee and Epidote Peak. Stump (1985) published a measured section of Taylor Formation from the northern half of Taylor Nunatak.

The strike of the Taylor and Greenlee Formations is roughly north–south parallel to Shackleton Glacier, with steep dips away from the glacier on both sides (Stump, 1985). The probable structure for the area is that of a large anticline breached along its crest by Shackleton Glacier, although two fault blocks dipping in opposite directions are also possible. The anticline may be cored by a gneiss dome and granitic pluton. A highly mobilized zone of migmatized Greenlee Formation crops out at the northern tip of Mt. Greenlee and at the bottom of the spurs east of Epidote Peak. North of that migmatites are exposed at Gemeni Nunataks and in the vicinity of Mt. Franke. On the east side of Shackleton Glacier a granitic pluton sharply intrudes Taylor Formation on the lower spurs of Lubbock Ridge and Cathedral Peaks. Along the western side of Massam Glacier from Nilsen Peak to Mt. Orndorff the silicic porphyries of the Taylor Formation are cut by a subvertical zone of myolinites (McGregor, 1965). Mineral lineation, particularly in the quartz phenocrysts, plunges directly down-dip, indicating nearly vertical movement, but no sense-of-shear indicators have been observed (Stump, 1976a).

Coherent stratigraphies can be demonstrated in three areas of Shackleton Glacier, but their relationship to one another, although suggestive, is not certain. The presumed lowest part of the section is on the west side of Shackleton Glacier, where the nonvolcanic Greenlee Formation is overlain by volcanic and clastic units of Taylor Formation. Upsection, high on Epidote Peak, marble makes its first appearance.

On the east side of Shackleton Glacier the rocks at Taylor Nunatak, with their large proportion of calcareous units, are distinct from the sequence exposed to the north from Lubbock Ridge through Waldron Spurs, which is predominantly quartzite interbedded with felsic volcanics. If no faulting exists between Lubbock Ridge and Taylor Nunatak, then by virtue of its topographic position and the orientation of bedding, the Taylor Nunatak section is stratigraphically beneath the section to the north.

## Greenlee Formation

The Greenlee Formation, exposed at the northern end of Mt. Greenlee and on the eastern slopes of Epidote Peak, consists of phyllites and fine-grained feldspathic quartzites. Some portions are slightly calcareous. Bedding is even, ranging in thickness from 5 cm to 1 m. Fine laminations are the only observed sedimentary structure. The formation appears to become more coarse-grained upsection, with a preponderance of phyllites at the bottom and quartzites at the top (Wade and Cathey, 1986).

Wade and Cathey (1986) were uncertain about the placement of the contact between Greenlee and Taylor Formations, due to a broad zone of brecciation and hydrothermal alteration. They site it about 6 km south of the northern tip of Mt. Greenlee. They also note the occurrence of ashfall tuffs and rare basalts near the top of the formation. In order to have a clear division, Stump (1985) defined the base of the Taylor Formation at the first appearance of volcanic rocks above the wholly sedimentary Greenlee Formation, with the contact occurring approximately 3 km south of the northern tip of Mt. Greenlee.

On the basis of the interpretation that the Greenlee Formation makes up the southern limb of an east–west-trending anticline, Wade and Cathey (1986) estimated the thickness of the Greenlee Formation to be at least 13,000 m. Mapping by Stump (1985) showed that the northwest–southeast trend of Greenlee Formation at glacier level on Mt. Greenlee swings to north–south on the higher slopes of Epidote Peak, making the exposed thickness of Greenlee Formation approximately 1,000 m.

The lack of significant graywacke, the lack of alternating fine- and coarse-grained units, and the lack of sedimentary structures other than laminations distinguish the Greenlee Formation from the turbidite sequences of the Beardmore Group. The fine-grained nature of the formation indicates very low-energy conditions of deposition. Wade and Cathey (1986) have suggested a shallow-water, near-shore environment. Both they and Stump (1982) have precluded correlation with the Goldie Formation.

## Taylor Formation

The Taylor Formation is characterized by a highly varied suite of volcanic, clastic, and calcareous rocks. The volcanics are predominantly dark-colored rhyolite porphyries, but basalts also occur low in the sequence on the west side of Shackleton Glacier. Many of the rocks on the west side of the glacier are highly sheared, so much of their original character is lost. On the other hand, the sequences on the east side of Shackleton Glacier are largely intact, and in some places preserve very delicate textures that belie the modes of eruption or deposition of the rocks. A zone of considerable shearing does, however, exist from Nilsen Peak southward along the west side of Massam Glacier.

A Lower Cambrian age of the Taylor Formation has been suggested on the basis of the enigmatic fossil *Cloudina*? found in a limestone breccia on Taylor Nunatak, along with trilobite fragments (Yochelson and Stump, 1977). Rowell and Rees (1989) have questioned the certainty of a Lower Cambrian age, posing the question of whether the fossil may have ranged into the Middle or Upper Cambrian. Grant (1990) was unable to be more positive than Yochelson and Stump (1977) in assigning the tube fossil to *Cloudina,* and he further questioned the identification of trilobite fragments.

In the lowermost portion of the Taylor Formation on Mt. Greenlee and Epidote Peak, altered basaltic lavas, some containing amygdules, are interbedded with quartzite and phyllite akin to the sediments of the Greenlee Formation beneath. Typical mineral assemblages of these metamorphic rocks are plagioclase–actinolite–sphene–chlorite–quartz and plagioclase–actinolite–epidote–chlorite. The plagioclase appears to be primary, occurring both as fine, aligned microlites and as phenocrysts up to 3 mm in length.

The basalts give way higher in the section to dark-colored silicic volcanics, both porphyritic and nonporphyritic, that are highly cataclastized. Several units of white, coarsely crystalline marble occur high on the ridge to the north of Epidote Peak. This first appearance of calcareous rocks in the section may tie across to the less deformed and metamorphosed limestone units on Taylor Nunatak.

Taylor Nunatak has two elongate parts, separated by ice cover. The southern half is composed entirely of volcanic and volcaniclastic rocks with highly variable attitudes, while the northern part contains similar rocks on its lower slopes that are overlain on the upper slopes by a 200-m sequence of limestones, volcanics, and volcaniclastics, striking north–south and dipping 70°E. This is the only locality within the outcrop of Taylor Formation in which such a thick limestone section is preserved. Northward from Lubbock Ridge to Nilsen Peak the Taylor Formation is characterized by rhyolitic porphyries interbedded with sedimentary units.

The most widespread volcanic rocks of the Taylor Formation are massive porphyritic rhyolites that are typically dark gray to black, though light gray and tan varieties do exist. The rhyolites contain phenocrysts of quartz, plagioclase, and in some cases orthoclase. The quartz ranges from euhedral to rounded and is often embayed. The plagioclase is typically euhedral, contains albite or Carlsbad twinning, and in some cases has been partially or wholly altered to sericite or epidote. In many specimens a portion of the phenocrysts have been fragmented. Volcanic rock fragments, both silicic and mafic, are present in small amounts in most samples. It is notable that no phenocrysts of ferromagnesian minerals occur.

The groundmass of the porphyries is usually a featureless, microcrystalline mesostasis of quartz, plagioclase, K-feldspar, sericite, and fine opaque minerals. Only in the area bounding the west side of Massam Glacier do the crystals show a preferred orientation.

In a few specimens of the porphyry, however, original groundmass textures have been preserved. These include both welded and unwelded shard structure, axiolitic structure, flow banding, spherulites, and perlitic cracks. These features demonstrate that the porphyries were erupted both as ash flows and as lavas. Ash-fall tuffs are also indicated by very fine-grained, finely laminated clastic rocks in the section at Taylor Nunatak. Accretionary lapilli has been found at a few localities (Stump, 1976b).

Extremely fine-grained silicic rocks with the appearance of chert occur at places in the Taylor Formation, in particular on Taylor Nunatak and around Mt. Orndorff. The rocks are shades of dark green or gray. They are generally bedded at 3- to 50-cm intervals and contain in some cases cross-beds, scours, and laminations. Rare conglomeratic horizons with rounded or subangular clasts of the same composition as the matrix have been recorded. Probably these rocks are the result of accumulations of fine ash in an aqueous environment.

A variety of sedimentary rock types, both calcareous and clastic, occur within the Taylor Nunatak section. Limestones include micrites, some with cherty layers, and clastic oosparites. Clastics are represented by volcanic breccias and clean, coarse-grained arenites composed of volcanic rock fragments and crystals of quartz and feldspar winnowed of any fine fraction.

In the area around Lubbock Ridge and Cathedral Peaks, the porphyries are interbedded with argillites and cross-bedded feldspathic quartzites. Most of the quartz is undiagnostic, although some that is embayed indicates local volcanic origin. However, a few grains of rutilated quartz and of perthite indicate a plutonic provenance for some of this rock. Farther to the north, in the vicinity of Mt. Orndorff and Nilsen Peak, the sedimentary rocks include volcaniclastic conglomerates, quartzites, and argillites, as well as several coarsely crystalline calcareous units containing layers of sandy shale and quartz clasts.

## Duncan Mountains and Adjacent Areas

Two sequences of metamorphic rocks, the Duncan and Fairweather Formations, are well exposed in the Duncan Mountains and crop out in a scattering of localities along the ice shelf both to the west and to the east (Fig. 5.13). Both formations were named and mapped by McGregor (1965), who also designated the Henson Marble for a thick, coarsely crystalline unit at Mt. Henson, west of

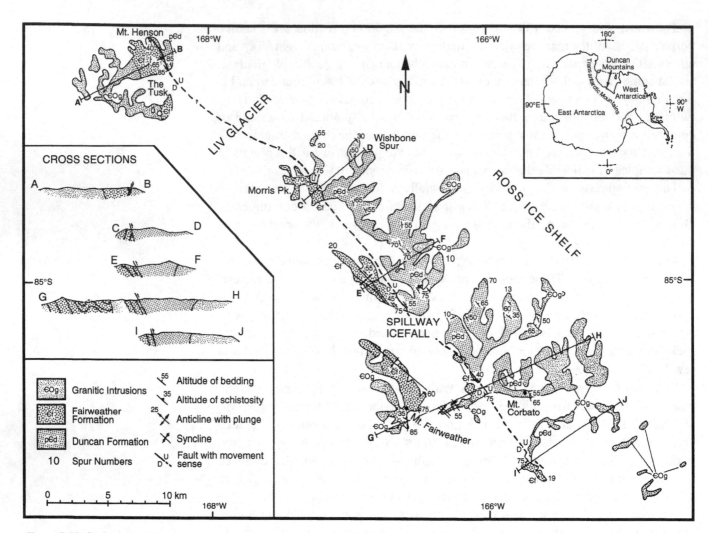

**Figure 5.13.** Geological map and cross sections of the Duncan Mountains and Mt. Henson. After Stump (1981a). Used by permission.

the Duncan Mountains across Liv Glacier. Wade and Cathey (1986) later proposed that the Henson Marble be given member status and be extended to include the marbles in the Taylor Formation to the west of Epidote Peak. Whether these marbles are in fact stratigraphic equivalents is conjectural.

McGregor (1965) interpreted the contact between Duncan and Fairweather Formations in the Duncan Mountains as conformable, while recognizing a fault between Duncan Formation and Henson Marble at Mt. Henson. Stump (1981a) later determined that the contact in the Duncan Mountains is also a fault.

## Duncan Formation

The Duncan Formation is a sequence of dark-colored schists and hornfelses that crop out in the Duncan Mountains and at Mt. Henson. Close to the fault with Fairweather Formation, a strongly developed foliation obliterates all original bedding in Duncan Formation, but in the majority of its outcrops, bedding is well preserved. Nevertheless, these rocks have been transformed by static recrystallization, so that in most samples, hornfels textures are complete. Only in the most coarse-grained samples do vestiges of original quartz grains remain. These are poorly sorted and matrix-supported, and were surely graywacke before metamorphism. Adjacent to granitic plutons the schists have been converted to highly contorted gneisses and migmatites (McGregor, 1965).

Assuming no repetition of bedding by major folds, Stump (1985) estimated the thickness of the Duncan Formation to be a minimum of 4,600 m. Bedding varies from a few centimeters to 2 m in thickness. Laminations are the most common sedimentary structure, but graded beds and dish structures have been reported (McGregor, 1965; Nilsen et al., 1977). Although the rocks are largely pelitic in composition, calc–silicate layers, indicating minor calcite in the original sediment, are scattered throughout the section, and fairly pure quartzites crop out near the ice shelf north of Spillway Icefall. McGregor (1965) reports several thin beds of impure limestone from the "northeastern corner of the Duncan Mountains," but since the mountains are elongate in a northwesterly direction the location of this occurrence is uncertain. Deposition by turbidites in a deep-sea fan setting has been suggested by Stump (1985).

## Fairweather Formation

The Fairweather Formation crops out at several localities in the area of the Duncan Mountains, with its widest development along the ridgeline between the Duncan Mountains and Mt. Fairweather, which McGregor (1965) informally designated the type locality. Stump (1985) measured this section, which is repeated in one portion by a syncline and cut out in another by a syntectonic pluton. [Stump (1985) contains an error in the numbering of segments of the type section. Specifically, No. 1 in Figure 26 corresponds to Section 4 in Table 6, No. 2 corresponds to Section 5, No. 3 to Section 3, No. 4 to Section 2, and No. 5 to Section 1.] Other localities of Fairweather Formation occur to the northwest of the Duncan Mountains at Mt. Henson and at Mt. Roth and Mt. Justman.

Like the Taylor Formation, the Fairweather Formation is composed of a bimodal suite of silicic porphyries and lesser basalts, calcareous and clastic sedimentary rocks, and cherts (Stump, 1985). The gross stratigraphy is also similar, with the basalts occurring low in the section and the clastic sedimentary rocks high. However, the degree of metamorphism of the Fairweather Formation is greater than the Taylor, with no vestiges of original groundmass textures remaining in the porphyries. A foliation produced by cataclastic deformation laces through portions of the Fairweather Formation, and as with the Duncan Formation, static recrystallization has produced hornfels textures that overprint the entire rock.

The silicic porphyries of the Fairweather Formation occur throughout the section, except at the very top. Individual units, which range from a few meters to more than 500 m in thickness, are massive or foliated, lacking any preserved primary layering. A 225-m-thick volcanic breccia occurs in the lower portion of the type section.

The porphyries contain phenocrysts of quartz, plagioclase, and in some samples K-feldspar. In general the quartz has been stretched and polygonized, although in the best-preserved cases embayments remain. Most of the plagioclase is euhedral and twinned, but has been sericitized or saussuritized. Both microcline and orthoclase varieties of K-feldspar are present. In the majority of samples a portion of the phenocrysts have been fragmented. The groundmass of the porphyries is in an advanced state of recrystallization, with fine- to medium-grained hornfels or foliated textures. Groundmass mineralogy includes quartz + plagioclase and/or K-feldspar, with chlorite ± biotite ± muscovite, intergrown around the framework silicates. Fine-grained amphibole occurs in some portions.

Metabasaltic units are fairly abundant low in the Taylor Formation. The mineralogy of these rocks consists of intergrown combinations of hornblende, epidote, actinolite, biotite, chlorite, plagioclase, and rarely quartz. Epidote veining accounts for up to 50% of the rocks in some places. Vesiculated units occur at several horizons, and one agglomerate has been recorded.

Chertlike rocks, both massive and finely laminated, occur in the middle and lower portions of the formation. In addition to microcrystalline quartz these rocks contain finely disseminated feldspar, opaque minerals, and biotite and/or hornblende.

Clastic and calcareous metasedimentary rocks are found principally in the upper portion of the Fairweather Formation. They are well preserved at several localities in the Duncan Mountains. Cross-bedded quartzites, including calcareous varieties, are common. Also plentiful are metaconglomerates and breccias, whose clasts are hornfels that was originally tuff or fine-grained arenite. Fine-grained calcareous hornfels and impure marble also occur. Blue, gray, and tan, finely crystalline marble and white, coarsely crystalline marbles also crop out at several localities.

## Henson Marble

The marbles in the region of the Duncan Mountains were collectively called the Henson Marble by McGregor (1965), who informally designated Mt. Henson as the type locality. Wade and Cathey (1986) suggested that the name be extended to include marbles associated with the Taylor Formation in the Shackleton Glacier area and that it be reduced to member status given the uncertainty of correlation between localities. On the basis of their protractedly in-press manuscript, Stump (1982, 1985) concurred with Wade and Cathey's (1986) original proposal. Stump (1985) measured the section of Henson Marble and Fairweather Formation on Mt. Henson.

At Mt. Henson the Henson Marble is approximately 500 m thick. The middle portion contains a massive, coarsely crystalline white marble, approximately 65 m thick, that passes on either side into blue, gray, and yellow marbles with discernible bedding. The marbles are interbedded with quartzites toward the contact with Fairweather Formation, which is set between the last massive metafelsite and adjacent marble. Henson Marble is thought to overlie Fairweather Formation on Mt. Henson, though no tops indicators have been observed. The other boundary of the Henson Marble on Mt. Henson is a fault with Duncan Formation.

The Fairweather Formation exposed on Mt. Henson contains several hundred meters of massive porphyritic metafelsite adjacent to the Henson Marble, which farther to the west gives way to biotite–muscovite schists and hornblende-bearing gneisses that grade into a pluton of Granite Harbour Intrusives. The marble is picked up again 2 km south of Mt. Henson at the Tusk, a spectacular 600-m horn adjacent to Liv Glacier (see frontispiece).

Another major occurrence of Henson Marble is on the ridge to the northwest of Mt. Fairweather, where it is folded into a syncline that plunges to the northwest. Here again the marble is thought to overlie the volcanic and clastic rocks of the Fairweather Formation, but the relationship of the northwestern ridge to the northeastern ridge where the type section has been measured is uncertain. Presumably a fault has cut out a portion of the section between the summit of Mt. Fairweather and the marble outcrop to the northwest, for the distance is not adequate to contain the thickness of Fairweather Formation measured to the northeast, where no marble occurs.

Additional occurrences of white marble crop out to the east of the Duncan Mountains south of Bigend Saddle (McGregor, 1965) and at Zigzag Bluff (Grindley et al., 1964).

To the west of the Duncan Mountains and Mt. Henson, in the vicinity of Mt. Roth and Mt. Justman, both silicic metavolcanic rocks and marbles crop out. This area has been visited only in reconnaissance fashion and lies approximately midway between exposures of Taylor Formation in the Shackleton Glacier area and Fairweather Formation in the Duncan Mountains (McGregor, 1965; Stump, 1985). Other white marbles crop out farther to the west at Olds Peak, where they are engulfed in granitic rocks, and at Pallid Peak east of Kosco Glacier, where they are surrounded by snow (Stump, 1976a).

Wade and Cathey (1986) suggested that the white marbles to the west of Epidote Peak be correlated with the Henson Marble. In his interpretation of the stratigraphic order of the three main areas of Taylor Formation, Stump (1985) suggested that the marbles to the west of Epidote Peak may correspond to the limestones at Taylor Nunatak, falling in the middle portion of the Taylor Formation. The true relationship of the limestones at Taylor Nunatak to the more metamorphosed occurrences of Henson Marble is not known. If the Henson Marble is the highest unit of the Fairweather Formation, as suggested by McGregor (1965), and if the limestones at Taylor Nunatak do indeed fall in the middle portion of the Taylor Formation, as suggested by Stump (1985), and if the whole of the Taylor and Fairweather Formations is roughly correlative, then the Taylor Formation limestones are probably not correlative with the Henson Marble. However, not enough is known of the interrelationship between the various isolated areas to draw meaningful conclusions.

## Deformation and Metamorphism

Duncan and Fairweather Formations are in contact in the Duncan Mountains along a major high-angle reverse fault parallel to the long axis of the range (Fig. 5.13). Stump (1981a) named this structure the Spillway fault. Throughout its exposure Duncan Formation strikes northwest and dips moderately to steeply to the northeast. In the portions most distal from the fault, bedding is uniformly tilted and paralleled by a mimetic schistosity produced by muscovite and biotite. Effects of the faulting are present in Duncan Formation for 2–3 km away from the fault. At moderate distances from the fault these are characterized by asymmetrical mesoscopic folds that verge to the southwest and plunge approximately 25°NW. Axial-plane cleavage mainly of muscovite cuts the shorter, southwest-dipping limbs. Close to the fault the bedding is completely transposed by cleavage developed entirely of muscovite. Large poikioblasts of biotite spot this portion of the rock, having grown statically after the fault movement and accompanying oriented mineral growth had ceased.

Deformation of the Fairweather Formation is manifested differently than the Duncan Formation, owing to the lithologies of the rocks involved. Several large asymmetrical folds, with the same southwest vergence and 25°NW plunge as the Duncan Formation, occur in the area, but no mesoscopic folds were produced. Pervasive shearing affected considerable portions of the silicic porphyries adjacent to the fault. During the deformation a small granite pluton intruded Fairweather Formation on the ridge east of Mt. Fairweather. Numerous large angular xenoliths of silicic porphyry are aligned roughly parallel to the foliation in the surrounding Fairweather Formation, with long axes plunging steeply.

At Mt. Henson a fault that brings Duncan Formation against Henson Marble (McGregor, 1965) is probably an extension of the Spillway fault in the Duncan Mountains (Stump, 1981a). On the south side of the peak the fault trace is nearly vertical, but on the north side it swings progressively to the west, attaining a 40°SW dip before reaching the end of the outcrop. The sense of motion on the fault in this area is therefore normal, though the displacement of Duncan Formation upward relative to Fairweather Formation is the same as in the Duncan Mountains, where the sense is high-angle reverse. A large drag fold was produced in Henson Marble near the crest of Mt. Henson during this movement (McGregor, 1965, Fig. 4).

Although dynamic recrystallization is apparent in the Duncan Mountains and at Mt. Henson in association with the Spillway fault, most of the rocks carry a hornfels texture related to intrusion of the Queen Maud–Wisconsin Range batholith immediately adjacent to the southwest. Most of the assemblages are typical of a hornblende–hornfels level of metamorphism (McGregor, 1965).

## O'Brien Peak

A sequence of marbles and silicic metasediments crops out in the vicinity of O'Brien Peak at the margin of the Ross Ice Shelf about midway between the Duncan Mountains and the Leverett Glacier area. The locality was visited by Gould's party in 1929, and returned samples were described by Stewart (1934b). Linder et al. (1965) briefly mentioned the metamorphic rocks and collected granitic samples that were dated by Craddock et al. (1964). The O'Brien Peak area was subsequently mapped by Katz and Waterhouse (1970b), who recorded the general lithologies and stratigraphic subdivisions, and described the structures (Fig. 5.14).

O'Brien Peak is the summit of a small westnorthwest-trending range with a series of north-trending spurs. The metasedimentary rocks strike approximately parallel to the range and generally dip steeply to the south except at the southeastern end, where the dip is to the north. The continuity of several lithologically distinct units can be traced from one spur to the next, although the younging

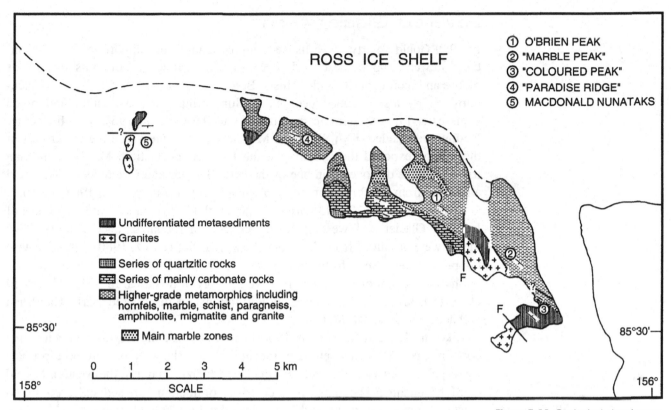

ROSS ICE SHELF

① O'BRIEN PEAK
② "MARBLE PEAK"
③ "COLOURED PEAK"
④ "PARADISE RIDGE"
⑤ MACDONALD NUNATAKS

▦ Undifferentiated metasediments
⊞ Granites

▦ Series of quartzitic rocks
▤ Series of mainly carbonate rocks
▨ Higher-grade metamorphics including
    hornfels, marble, schist, paragneiss,
    amphibolite, migmatite and granite
▨ Main marble zones

85°30'

85°30'

0   1   2   3   4   5 km
SCALE

158°

156°

**Figure 5.14.** Geological sketch map of O'Brien Peak area. After Katz and Waterhouse (1970b). Used by permission.

direction in the sequence is unknown. Within these units tight folding is observed with variably plunging fold hinges.

The sequence is typified by a large proportion of calcareous rocks. The southernmost unit along the main ridgeline is a massive to thick-bedded gray quartzite more than 100 m thick. This is followed to the north by a unit approximately 200 m thick consisting mainly of calcareous rocks with a variety of colors – gray, blue, white, yellow, and brown. Thin-bedded marbles, schistose marbles, calc schists, and quartzose rocks are represented. Northward to the ice shelf the sequence is a complex of high-grade metasedimentary rocks, including mica and hornblende schists, calc–silicates, marbles, amphibolites, and gneisses. The northernmost exposures are considerably migmatized, and granitic and pegmatitic intercalations become prevalent.

At the southeastern end of the range is a distinct sequence separated from the rest of the metasedimentary rocks in the range by a fault-bounded body of granite. Displaying a brittle character, the faulting was postmetamorphism and intrusion, and probably occurred during uplift of the present-day Transantarctic Mountains. Graded bedding in several psammitic units of the metasediments suggests that tops are to the north. The southernmost (lowermost?) portion of the section contains cherts or hornfelses with a number of conglomerates and breccias with white quartzitic clasts. Psammitic and argillaceous rocks pass northward into more calcareous sedimentary layers. These are followed by marble and eventually gneisses, but whether this marble-to-silicate transition correlates with the prominent one to the west is uncertain owing to poor exposure.

## Leverett Glacier Area

In 1929 Gould's party made its easternmost landfall at Supporting Party Mountain. Stewert (1934c) described seven samples collected from this locality as biotite and biotite–hornblende schists. Before traversing up Scott Glacier, Blackburn's party also touched bedrock at Supporting Party Mountain and noted gneisses and quartz veins in the nunataks to the west. In 1964–65 a helicopter landing was made on Mt. Webster. Minshew (1967) measured the lower half of the section exposed there, calling it the Leverett Formation. Middle Cambrian trilobites were discovered in one of the beds (Palmer and Gatehouse, 1972), and a rhyolite from the formation was isotopically dated (Faure et al., 1968, 1979).

During the 1977–78 field season Stump et al. (1978) mapped the area north of Leverett Glacier and west of Reedy Glacier in reconnaissance fashion. Two theses were produced from this work (Lowry, 1980; Heintz, 1980), but otherwise no publications were forthcoming. The descriptions that follow are the first published account of the comprehensive geology of this area. In 1992–93 Rowell et al. (1993a) studied the Leverett Formation and discovered Early Cambrian archaeocyathids at Mt. Mahan.

Like the Duncan Mountains 150 km to the west, the Leverett Glacier area contains two distinct sequences of metamorphic rocks. One is composed primarily of pelitic schists of the Party Formation, which crop out in the western Harold Byrd Mountains. The other is a complex of metamorphosed calcareous and clastic sedimentary rocks and volcanics, the Leverett Formation of Minshew (1967), which are found at Mt. Webster, the eastern Harold Byrd Mountains, and the Bender Mountains. Much of the rest of the area is underlain by a variety of plutonic phases, although a gneiss with extremely coarse porphyroblasts occurs in the vicinity of Phlegar Dome.

## Party Formation

The sequence of metamorphic rocks cropping out in the western Harold Byrd Mountains is here named the Party Formation after Supporting Party Mountain. The name "Byrd Formation" would perhaps be more appropriate, but it is preempted by the Byrd Group of the central Transantarctic Mountains. Knowledge of the formation is not complete enough to formalize it. No type locality is designated, nor have any sections been measured; however, the ridge south of Mt. Nichols appears to contain the broadest and most representative outcrop of the formation.

Lithologically the Party Formation consists of pelitic and calc schists, gneisses, minor quartzite, and rare marble. Bedding is apparent in portions of the formation, but in many places there is uncertainty whether the foliation mimics bedding or is wholly of metamorphic origin. Mesoscopic folding in many parts of the formation further obscures the original character of rocks.

The most common rock type is dark gray pelitic schist with abundant biotite, and often muscovite, hornblende, or garnet. Calc schists are also common, with characteristic tan, brown, and green colorations. Diopside is present in some of these rocks. In the vicinity of "Tongue Hills," in addition to the schists the sequence contains fairly pure quartzites. Around Mt. Graham the rocks are

**Figure 5.15.** Quartzite interbedded with metapelite in the Party Formation south of Mt. Nichols.

predominantly quartzofeldspathic gneisses, with thin intercalations of mafic minerals. One 30- to 40-m-thick white marble bed in a sequence dominated by calc schists crops out on the spur 2 km southsoutheast of Mt. Graham.

Where it can be seen with certainty, bedding varies from a few centimeters to a half meter, with quartzose beds 4 km southsouthwest of Mt. Nicols reaching 2 m. Laminations were observed in some places, generally preserved by the growth of biotite, but otherwise no sedimentary structures are known. In the area 4 km southsouthwest of Mt. Nicols the quartzose and pelitic layers alternate regularly (Fig. 5.15). Calc schists occur at certain places in the sequence, but are absent in others. In some portions they are very prevalent; in others they occur for intervals, with intervals of pelitic schists interspersed.

All of the Party Formation is foliated. Mica is parallel to bedding in some instances. Elsewhere it is oriented oblique to bedding at a very low angle, or more typically at a moderate angle, where it is clearly developed parallel to the axial planes of minor folds. In many places the foliation is pervasive, so that its relationship to original bedding is uncertain. Both bedding and foliation strike in an east–west to northwest–southeast direction throughout the western Harold Byrd Mountains. Except at the area 4 km southsouthwest of Mt. Nicols, where it dips to the northeast at moderate angles (30°–40°), bedding dips to the south to southwest, generally at moderate to steep angles (30°–75°). Foliation, where it is axial-planar to folds in bedding, is steep, deviating not more than 15° from

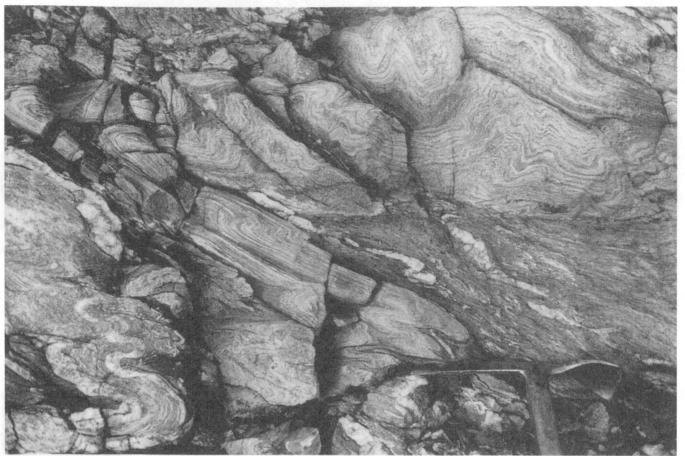

**Figure 5.16.** Complex folding in gneisses of the Party Formation, Supporting Party Mountain.

vertical. The pervasive variety of foliation varies in dip from vertical to low angles to the southwest.

Mesoscopic folds are developed in many portions of the formation. They consistently plunge toward the southeast quadrant between about 20° and 45°, except in the area 4 km southsouthwest of Mt. Nicols, where they plunge westnorthwest at about 25°. The mesofolds appear to occur in domains. In some they have symmetrical limbs; in others they are asymmetrical, with consistent vergence direction either to the northwest or southwest. This appears to indicate that larger-scale folding occurs in the Party Formation, but mapping has not been detailed enough to delineate such structures.

At Supporting Party Mountain and the nunataks west of it, the rocks are pelitic and calc schists containing numerous veinlets of quartz. The rock has been strongly deformed on a fine scale (Fig. 5.16). Both the foliation and the veins have been deformed into numerous tight mesoscopic folds, which in many cases have been sheared out along their axial planes. No axial-plane cleavage from the growth of micas was produced during this folding and shearing. The foliation and shear planes strike northwest–southeast and dip from vertical to low angles to the southwest. The small-scale folds plunge at moderate angles to the southeast.

In the vicinity of Mt. Graham the rocks are gneissic with quartzofeldspathic layers 1–5 cm thick separated by thin (a few millimeters) biotite and hornblende partings. The layering dips between 30° and 45° to the south. The regularity of these layers suggests that they may be relict bedding, but this is not certain. Crosscutting these layers at a moderate angle are shear planes along which movement has been of a thrust nature and north-vergent.

Although not well documented, at least two episodes of deformation appear to have affected the Party Formation. Suggestions of this come from several distinct locations. The first is Supporting Party Mountain, where the foliation is folded throughout. South of Mt. Graham a quartzofeldspathic layer that was isoclinically folded was itself folded by a second deformation that produced kink folds in the surrounding schist. When viewed at a distance, the bedding at the area 4 km southsouthwest of Mt. Nicols appears to have several open folds crossing it at an angle oblique to the direction of the mesoscopic folds measured in outcrop. These large folds were not picked up, however, during the mapping. Beyond these few observations there is no coherent picture of multiple deformation of the Party Formation. Clearly, more detailed study is needed if this is to be sorted out.

Plutons of posttectonic granite intrude the Party Formation at several localities, in all cases with sharp contacts. Porphyroclasts of biotite and of andalusite (up to 3 cm in length) have grown in the schists in the vicinity of the plutons. An unusual series of concordant aplite sills separated by thin partitions of biotite occurs on the wall facing Supporting Party Mountain to the southeast of its summit.

Due to the poor preservation of original sedimentary characteristics, the environment of deposition of the Party Formation cannot be interpreted with assurance. The quartzose layers interbedded with pelites could be from near shore, where currents winnowed away the fines in some beds. Likewise, the marble bed would argue for a shallow marine setting. However, the calcareous fraction in the schists does not rule out the possibility that these beds are of turbidite origin, for calcite may be found in continental slope deposits to considerable depths. The Goldie Formation, for example, has clear indicators of turbiditic origins and contains appreciable calcite.

## Leverett Formation

The Leverett Formation is a highly varied bimodal suite of volcanics (predominantly silicic in composition), clastic and volcaniclastic rocks, marbles, and cherts. Both deformation and metamorphism of these rocks vary considerably throughout the extent of their outcrop. The formation crops out in three separate areas, each of which contains a distinct assemblage of units. The contact with the Party Formation is nowhere exposed.

Deformed trilobite fossils from the Levertt Formation at Mt. Webster could not be identified as to genus, but Palmer and Gatehouse (1972) noted characteristics similar to *Mapania,* a late Middle Cambrian genus found in China and Australia. Rowell et al. (1993a) found Lower Cambrian archaeocyathids in Leverett Formation at Mt. Mahan in the Bender Mountains. These fossil occurrences from two of the three general areas of outcrop indicate that the Leverett Formation spans more than half of the Cambrian series. Rowell et al. (1993a) go so far as to speculate that the third area (eastern Harold Byrd Mountains to Ivory Tower) may be Neoproterozoic. The Leverett Formation as it is herein described may warrant further subdivision, but due to the broad reconnaissance nature of the study by Stump et al. (1978), sufficient data are not at hand to make any formal designations.

At Mt. Webster, where Minshew (1967) first described the Leverett Formation and designated a type section, the lower portion of the sequence is considerably sheared and contains indeterminate clastic or silicic volcanic rocks. Upward the sequence consists of a mixture of recognizable silicic volcanics, volcaniclastics, shales, limestones, and cherts. The uppermost portion is dominated by silicic volcanic rocks, with lesser limestones and clastic sedimentary rocks. Greenstones (chloritized mafic volcanics) comprise an appreciable fraction of the middle and upper portions of the sequence.

The Leverett Formation is deformed into a series of east–west-trending folds connecting the area from the eastern Harold Byrd Mountains to Ivory Tower. These rocks have a more pervasive metamorphic overprint than elsewhere, but bedding is well preserved. Silicic volcanics are a minor constituent in these rocks, except at a few horizons at Cressey Peak and Fadden Peak. In the eastern Harold Byrd Mountains (Mt. Manke and vicinity) quartz and mica schists and impure marbles comprise much of the section. White, coarsely crystalline marbles are conspicuous on Mt. Manke, Cressey Peak, Fadden Peak, and Ivory Tower. Overall, the predominant lithologies are calcareous.

The area of Leverett Formation with the best preservation of its original volcanic and sedimentary characteristics is in the Bender Mountains. A variety of volcaniclastic rocks are well represented, with both fine- and coarse-grained sizes, angular and well-rounded conglomerates, and various clast lithologies including volcanics, cherts, and carbonates. Silicic volcanics show exceptionally well-preserved flow banding and devitrification textures at places. White marble is also appreciable in one portion of the area. Each of these three areas is described in more detail in the sections that follow.

## Mt. Webster

The section of Leverett Formation at Mt. Webster appears to be homoclinal with tops to the north, except in its middle portion, where folding repeats some units. Bedding strikes in an eastnortheasterly direction and dips steeply from about 60°N to overturned 80°S. The lower portion of the formation is considerably sheared, with a pervasive cleavage striking eastsoutheasterly and dipping 80°N to 50°S. This cleavage is less well developed in the upper portions of the formation, but it persists in more shaley layers throughout. The bottom of the formation is truncated by a massive pink postcleavage granite that crops out in the southern portion of the Mt. Webster massif almost to the summit. The upper portion of the formation is covered by snow.

Rocks in the lowermost portion (~250 m) of the Leverett Formation contain megacrysts of quartz, plagioclase, and K-feldspar to about 2 mm set in a fine-grained matrix or groundmass comprising 50–60% of the rock. The highly foliated matrix is composed of variable amounts of quartz, plagioclase, K-feldspar, chlorite, sericite, epidote, and/or amphibole. The quartz megacrysts are polygonized and elongate within the cleavage. The feldspars are also aligned within the cleavage, but show lesser effects of cataclasis.

Perhaps 100 m up into the section, bedding becomes apparent in some portions due to changes in grain size, with intervals varying between 5 cm and 2 m. The rocks are green if chlorite is present, and tan if it is not.

The massive, cleaved lower portions of the Leverett Formation could be either silicic porphyries or clastic sedimentary rocks. If the latter, they probably are reworked volcanics to account for the distribution of megacrysts and matrix material. The bedded portions have probably been water-worked, but conversely they may be tuffaceous rocks layered during eruptive emplacement. The level of deformation in these rocks prevents a clear interpretation. Higher in the section, however, in less cleaved portions massive porphyries do occur that are surely of volcanic origin.

Between approximately 300 and 450 m up into the section, portions of the Leverett Formation take on a brick red coloration, alternating with greens. Some of these rocks are porphyries or volcaniclastics, while others are cryptocrystalline and appear to have been chert or ash lacking phenocrysts. These portions are followed by about 150 m of green porphyritic volcanic or volcaniclastic rocks, overlain by green shaley material with bedding 5–30 cm thick, before an interval to the northern base of the Mt. Webster massif in which the section is scree-covered by rocks of similar appearance.

Farther north in a narrow saddle for about 100 m of ground distance, the section contains folded limestones, volcanic wackes, and mudstones with axial-plane cleavage. Fold plunges vary between 60° and 75° to the eastsoutheast, and cleavage is oriented 100°–110° and nearly vertical. The rocks in this interval are thin-bedded, not more than 15 cm thick, and some of the mudstones are quite fissle. Limestones are gray, weathering brown or tan, and the wackes and mudstones are light to dark green.

Approximately 1,100 m of section continue northward from the narrow saddle across two small summits and connecting ridgelines. The lower 75 m are composed of thin-bedded, alternating limestones and mudstones similar to those in the preceding portion. The trilobite fragments found by Minshew (1967; Palmer and Gatehouse, 1972) presumably came from a limestone in this part of the section.

The upper 1,000 m of the Leverett Formation at Mt. Webster are an interbedded sequence of silicic volcanic porphyries and volcaniclastic sediments (~50%), mafic metavolcanic rocks (approximately 30%), and calcareous units (~20%). Phenocrysts in the porphyries are euhedral quartz and zoned, twinned plagioclase or fragments of the same. K-feldspar also occurs in some units. The groundmass is a fine-grained intergrowth primarily of quartz and feldspar.

In a number of cases massive porphyries grade into shaley bedded units, some of which are calcareous. Another lithology found throughout the section is chertlike rock, which may be either massive or banded and is generally of dark coloration. Conglomerates are also scattered through the section, with clasts of chert or volcanic rock in most cases.

Massive greenstones are interbedded at a number of levels in the upper part of the section. These rocks are altered almost entirely to chlorite. In a few cases ghosts of phenocrysts 1–3 mm in length can be discerned. In rare cases corroded pyroxene grains remain in the rock. Presumably these rocks were originally basaltic volcanics, but their original character is all but lost.

Interbedded throughout the upper part of the section are gray cherty limestones, rarely more than 10 m in thickness, with beds 1–3 cm thick. These rocks commonly contain a fraction of sand grains (quartz or volcanic rock fragments). The cherts are nodular or thinly bedded within the limestones and also occur as thicker, massive beds. White fine-grained marbles also occur at several horizons within the section.

## Eastern Harold Byrd Mountains to Ivory Tower

Extending eastward from the eastern Harold Byrd Mountains to Ivory Tower, Leverett Formation crops out in a series of exposures separated by snowfields. Much of the outcrop occurs in north-trending ridges that display cross sections perpendicular to the east–west structural trend of the formation (Fig. 5.17). The sequence is tightly folded in places, but because of the distances between ridgelines and the reconnaissance nature of the mapping, correlations of individual sections were not made. A futher complication is the possibility of unrecognizable fold closures in areas where massive white marbles crop out. More detailed mapping and section measurement, however, may permit a coherent stratigraphy to be developed for this area.

The westernmost ridgeline of the area, west of Mt. Manke, does contain a section approximately 2,600 m thick that dips uniformly south at about 50° before becoming tightly folded in its southerly exposure. The rocks are metamorphosed to upper greenschist or lower amphibolite facies, as indicated by the presence of kyanite, garnet, epidote, albite, and actinolite. Retrograde chlorite also occurs in some rocks. Bedding within this section is well preserved, as indicated by bedding-plane partings and changes in color and lithology, but it is overprinted in many units by a cleavage oriented at a low angle to the bedding. Sedimentary structures are rare, but cross-bedding in several of the quartzite units indicates that the section is upright, younging to the south. A considerable degree of flattening and stretching has occurred, particularly in the carbonate units, as indicated by conglomerates with marble clasts having long to short axis ratios of approximately 10:1, and thin silicic layers within marbles that occur as widely separated boudins.

The section on the ridgeline west of Mt. Manke is overwhelmingly sedimentary in origin. Several massive units with megacrysts of quartz and plagioclase in a fine-grained groundmass may have originated as silicic volcanics. Also, another unit is a schist composed of actinolite, quartz, and calcite, which contains small vugs with the appearance of vesicles, possibly indicating an intermediate to mafic lava flow.

The predominant lithologies in the section are fine-grained quartzose, pelitic, and calc schists, impure marbles, and quartzites, colored various shades of gray and tan. All gradations exist between these rock types. Bedding is generally fairly thin, varying from 1 to 30 cm and seldom reaching more than a meter in thickness.

The quartz schists contain very little mica, but in some units kyanite is present. The more pelitic schists have abundant biotite, as well as chlorite in some cases.

**Figure 5.17.** Tightly folded metasedimentary rocks of Leverett Formation viewed westward from Mt. Cressey.

If the schists were calcareous, epidote, actinolite, and/or calcite are present. The schistose rocks grade into more coarse-grained varieties in which original clasts of quartz, plagioclase, and rock fragments can be seen. Embayed quartz is readily observed indicating the volcanic provenance of these rocks. Conglomeratic rocks occur at several horizons in the upper part of the section. These contain clasts of marble, schist, and chert.

The marbles in the section are generally fine-grained, with gray, blue, or brown coloration. Typically, a clastic fraction exists in these rocks, up to 25% quartz in some cases. About 800 m up into the section there is a prominent white, nearly pure, coarsely crystalline unit of marble about 100 m in thickness. In addition, several white marble units occur high in the section.

Similar lithologies crop out eastward along the other three ridges of the Mt. Manke group. The degree of deformation also increases to the east, with complex isoclinal folds exposed on Mt. Manke itself (Fig. 5.18).

On the next group of spurs to the east, of which Cressey Peak is the highest point, lithologies of the Leverett Formation vary in their relative proportions. Schistose or argillaceous units predominate in the more southerly outcrops, but these give way northward to a portion of the sequence dominated by silicic metavolcanic rocks and associated volcaniclastic sediments. As is seen elsewhere, embayed quartz, plagioclase, and lesser K-feldspar comprise the pheno-

**Figure 5.18.** Complex folding in Leverett Formation on west face of Mt. Manke.

crysts in the metavolcanics. The northernmost outcrops, in particular the northern end of the long western ridge of the Mt. Cressey group, contain a thick section of white, coarsely crystalline marble.

The rocks around Mt. Cressey strike eastsoutheast and have steep northerly dips. One synformal fold nose was observed in marble about halfway along the middle ridge. A nearly vertical cleavage cuts through much of the exposure. Unfortunately no sedimentary structures were found throughout the sequence, so even if the schistose portion correlates roughly with the rocks around Mt. Manke, the relative stratigraphic position of the metavolcanics and marbles is unknown.

The next outcrop to the east is Fadden Peak. The massif is folded, but the extent to which this has occurred is obscured by abundant skree. The rock types at Mt. Fadden include argillite, quartzite, conglomerate, and pure and impure marble. The argillites are dark green and in part are calcareous. The quartzites can be seen in thin section to contain abundant (30–80%) matrix of quartz and plagioclase, with minor sericite and chlorite, so technically they are wackes. Cross-bedding is commonly developed in this lithology. The conglomerates are matrix-supported, with clasts of quartzite, argillite, marble, chert, and silicic metavolcanics. Calcareous rocks include all gradations from pure fine-grained marbles to calcareous schists. White, coarsely crystalline marble also occurs here.

Ivory Tower, the easternmost outcropping of Leverett Formation in this group of outcrops, is composed almost entirely of massive, coarsely crystalline white marble. Associated with this unit are minor argillaceous and calc–silicate layers, silicic metavolcanics, and conglomerates containing marble, metavolcanic, and chert clasts.

## Bender Mountains

A fault cuts through the Bender Mountains in an eastsoutheasterly direction separating distinct sequences of Leverett Formation. To the north the rocks are primarily sedimentary, are moderately metamorphosed, and contain limited volcanic rocks of intermediate composition. From the standpoint of lithology and metamorphic grade, these rocks bear a resemblance to portions of the sequence in the eastern Harold Byrd–Ivory Tower area.

The rocks to the south of the fault are predominantly silicic volcanics that have been metamorphosed no higher than zeolite grade and in general are dense, brittle rocks. Deformation has caused tilting, fracturing, and shearing in many parts, but owing to the competency of the rocks, some portions remain unaffected, with original volcanic and sedimentary textures exceptionally well preserved.

Lithologies north of the fault include fine-grained quartz, mica, and calc schists, conglomerate, marble, and intermediate volcanics. Clasts in the conglomerates include marble, chert, and fine-grained schist. The rocks are clast-supported with a calc–silicate matrix. Deformation has in general flattened the marble and schist clasts but left the chert intact.

Several thin, light gray marble beds occur in the sequence, and one white, massive, coarsely crystalline marble crops out about midway along the long north-trending ridge on the east side of the Bender Mountains. This marble is intruded by a series of silicic dikes that have caused sulfide and garnet mineralization locally.

An intermediate-composition metavolcanic rock crops out on the ridges to the northeast and northwest of Mt. Mahan. It is characterized by plagioclase and hornblende phenocrysts set in a matrix of chlorite and calcite. This is the only known occurrence of such a volcanic rock within the Liv Group and the only one in which hornblende is a phenocryst phase.

South of the fault, bedding orientations are quite variable. Strike varies between about N45°E and S45°E, with steep to moderate dips to both the north and the south. In the few places where sedimentary structures are preserved, tops are indicated to the south, and the sequence is thought to young in that direction overall. Skree covers much of the ice-free outcrop, so a coherent structural and stratigraphic picture has not been worked out.

Lithologies south of the fault are predominantly silicic volcanic porphyries, cherts, and volcaniclastic rocks. All gradations between these rock types exist. Calc–silicate layers occur in portions, and one thick body of white marble crops out in the vicinity of Mt. Mahan.

The silicic porphyries in the Bender Mountains are similar in appearance to others in the Leverett Formation and the rest of the Liv Group, that is, massive, dense, various dark and light colors. But they appear to be distinct in that the phenocryst population includes quartz and plagioclase, but not K-feldspar. The groundmass of the porphyries, which is too fine-grained to resolve optically, does, however, stain for abundant potassium.

A fair proportion of the rocks in this area are clastic and probably range from strictly tuffaceous volcanics to reworked volcanic sediments. The clasts are generally angular, although some rounded pebble conglomerates exist in particular in association with the marbles. Clast size ranges up to about 2 cm. Some of these rocks contain abundant matrix, others do not. Units are often bedded, but massive accumulations also exist.

Clasts include silicic volcanic rock, chert, limestone, and notably mafic or ultramafic fragments. In their most pristine state the last-named are composed of zoned olivine and twinned pyroxene. Single pyroxene crystals have also been found. More typically these inclusions are altered to amphibole and chlorite. Their presence is another characteristic that distinguishes the rocks in the Bender Mountains from others in the Leverett Formation.

Another abundant rock type is chert. As with cases in the Taylor Formation, it is suspected that much of this material is fine ash deposits, but the groundmass is so fine-grained as not to be resolvable optically. These rocks have the same appearance as the silicic porphyries would have if their phenocrysts were absent. In some cases the cherts contain dispersed clasts of marble.

Flow-banded silicic lavas are found at several localities. In some of them spherulites are abundant and well preserved (Fig. 5.19). The largest known examples are along the ridgecrest east of Peak 1,170±, where they reach a diameter of 6 cm.

Portions of the clastic rocks contain a calcareous fraction, although in general they do not. Several beds of gray marble occur within the clastic portion of the section. In the vicinity of Mt. Mahan a thick white marble crops out. Owing to much lower strength than the surrounding silicic rocks, this unit is highly folded. The marble is also crosscut by a number of diabase dikes on the north face of Mt. Mahan.

The Bender Mountains appear to have been closer to a volcanic eruptive center than either Mt. Webster or the eastern Harold Byrd Mountains–Ivory Tower area. This is indicated by the presence of flow-banded silicic lavas and abundant coarse-grained clastic material.

It is not possible to correlate between the three main areas of Leverett Formation, but some general observations can be stated. The lower portions of the sequences in each area are lacking in limestones or marbles. Thick white marbles are present in each area, with their first appearances substantially high in their respective sections. Clastic or volcaniclastic rocks appear to dominate the lower portions of the sections. Fewer volcanic rocks are found in the eastern Harold Byrd Mountains–Ivory Tower area than in the other two, and the Bender Mountains appear to be closer to a source than Mt. Webster.

A general model for the deposition of the Leverett Formation is one of silicic volcanic islands set in a shallow ocean basin. Eruptions were probably both subaerial and subaqueous, with accumulations of coarse clastic rocks close to source and finer-grained rocks in more distal locations. After an initial period of volcanism and clastic deposition, limestones began accumulating while volcanism continued.

**Figure 5.19.** Spherulites in volcanic rocks of Leverett Formation, Mt. Fiedler.

## Metamorphic Rocks on Watson Escarpment

An assemblage of mica schists and gneisses crops out along Watson Escarpment in the vicinity of Mt. Analogue and Phlegar Dome. No calc schists were observed within the sequence. The orientation of foliation in these rocks is east–west with vertical to steep southerly dips. On Mt. Analogue the rocks are biotite–muscovite schists toward the bottom, giving way to quartzites toward the top. On the low spurs north of Phlegar Dome the rocks are a variety of quartzites, schists, and gneisses. These rocks show lithological variation parallel to the foliation that can be interpreted as original bedding. In a few of the quartzites there is a hint of cross-bedding, suggesting tops to the south. The most striking rock is a gneiss with porphyroblasts of K-feldspar up to 12 cm in length. The megacrysts in portions of the outcrop have an augen form, while elsewhere they are euhedral. On the spurs north of Phlegar Dome the metamorphics are intruded by granitic rocks that carry a strong foliation parallel to that in the country rocks. On the high face of Phlegar Dome a spectacular stoped contact between metamorphics and plutonic rocks is displayed.

The affinity of these metamorphic rocks to either the Party or the La Gorce Formation is uncertain. They certainly are not equivalent to the Leverett Formation, but neither do they appear to have had a calcareous fraction, as is the case with a portion of the Party Formation. Their conversion to metamorphic rock is so complete as to erase essentially all of the original sedimentary character.

## Correlation of Duncan and Party Formations

The Party Formation shares many similarities with the Duncan Formation. Both are pelitic; both contain calc–silicate layers. Quartzose beds and rare marbles are common to both. Even bedding and laminations are the most typical sedimentary characteristics of both. From the lithological and sedimentological standpoint, correlation of the Party and Duncan Formations seems warranted. One can also cite their structural alignment and similar locations on the northern margin of the Queen Maud–Wisconsin Range batholith, although the Party Formation appears to have suffered a more complex deformational history.

Traditionally the Duncan Formation has been included in the Beardmore Group and correlated with the La Gorce and Goldie Formations (McGregor, 1965; Minshew, 1967; McGregor and Wade, 1969; Stump, 1982, 1985). This was based on similarities in lithology, and the interpretation that the Duncan Formation is of turbidite origin.

Wade and Cathey (1986) have suggested a correlation of Duncan Formation with Greenlee Formation, while excluding the Greenlee Formation from correlation with Goldie Formation. One of their arguments is that Greenlee and Duncan Formations are overlain, respectively, by volcanic Taylor and Fairweather Formations. This is based on McGregor's (1965) original mapping of a conformable contact between Duncan and Fairweather Formations, which was later shown to be a fault by Stump (1981a). The other arguments advanced by Wade and Cathey (1986) are that the environments of deposition of the Goldie and Greenlee Formations are "vastly different" and that the former was involved in the Beardmore orogeny, while the latter was not. Implicit in these arguments are the notions that the Duncan Formation was deposited in an environment similar to that of the Greenlee Formation and that it was not involved in the Beardmore orogeny.

Wade and Cathey's (1986) suggestion is not without merit, and the issues raised have important ramifications for the geological development of the Transantarctic Mountains. Regarding deformation, Stump (1981a) concluded that in the Duncan Mountains the Duncan Formation was folded during the Ross orogeny, during a single progressive deformational episode involving the Fairweather Formation and the Spillway fault, and that there was no evidence of Beardmore deformation in those rocks.

Discounting local kinking around the Ackerman fault, only one episode of folding was recorded in the La Gorce Formation. But this was assumed to have been during the Beardmore orogeny, since the folded La Gorce Formation is crosscut by Wyatt Formation, and the Ackerman Formation, conformable with the Wyatt, is itself cut by the Ackerman fault (Ross orogeny) (Stump et al., 1986b). Evidence for two deformations of the Goldie Formation (Beardmore and Ross orogenies) is clear only in the upper Nimrod Glacier area (Stump et al., 1991). Throughout most of the remainder of its outcrop area, the Goldie Formation apparently exhibits only one episode of folding.

Regarding the environment of deposition of these formations, the Greenlee

Formation does appear to be distinct from the Goldie and La Gorce Formations. The latter are fairly clearly deep-sea fan sequences with alternations of gray-wacke and pelite and a variety of diagnostic sedimentary structures. The Greenlee Formation, by contrast, lacks the alternation of fine- and coarse-grained units, in general its bedding thickness is less than the thicker units of Goldie and La Gorce Formations, and it lacks abundant graywacke, with lithologies distributed between fine-grained sandstones and pelites. Regardless of its actual environmental setting, the Greenlee Formation appears not to have been deposited by mass flow processes.

Due to their degree of metamorphism, however, the environment of deposition of the Duncan and Party Formations is not conclusive. McGregor (1965, Fig. 3) pictures features that he interprets as graded beds. Even allowing that some material has been lost by solutioning, as indicated by the ptygmatic fold that crosses the exposure, the thickness of these beds is considerably less than the typical graywacke–pelite packets of the La Gorce and Goldie Formations. Beyond this the only observed sedimentary structures in the Duncan and Party Formations are laminations, which are as abundant in the Greenlee Formation as the Goldie and La Gorce Formations. Furthermore, alternations of metagraywacke and pelite cannot be demonstrated in the Duncan and Party Formations, though again this may be a function of metamorphic grade.

Perhaps the most distinguishing feature of the Duncan and Party Formations is the abundant mica, in particular biotite, throughout the majority of the sequence. This indicates the preponderance of a pelitic fraction in the original sedimentary rock, whether it was graywacke or shale. The Greenlee Formation, if taken to higher metamorphic grades, would probably become schist, but many of the beds would be quartzofeldspathic with minor mica.

I therefore conclude, as in previous summaries (Stump, 1982, 1985), that the Greenlee Formation is distinct from and should not be correlated with the Duncan and Party Formations. However, I now feel that there is less certainty that the Duncan and Party Formations should be included in the Beardmore Group, correlative with the Goldie and La Gorce Formations. In spite of previous interpretations of a turbidite origin for the Duncan Formation (Stump, 1982, 1985), I feel that the evidence is uncertain enough to permit other possibilities. Furthermore, a lesson was learned from northern Victoria Land, where the Robertson Bay Group, which for many years had been correlated with the Beardmore Group on the basis of lithology, was shown to be wholly distinct by the discovery of Tremadocian fossils (Wright et al., 1984).

## Correlation of Liv Group

The Liv Group was established by Stump (1982) to encompass the Taylor, Greenlee, Fairweather, and Leverett Formations. Cropping out intermittently for more than 350 km along the Ross Ice Shelf, these formations display a great range of metamorphic and deformational effects, but have many lithologies in common. Porphyritic silicic volcanic rocks are the most characteristic rock type. The phenocryst assemblage of quartz, plagioclase ± orthoclase, without a ferro-magnesian phase, is distinct from the Wyatt Formation and Thiel Mountains Porphyry, which typically contain phenocrysts of biotite and/or hypersthene. A small component of mafic volcanic rocks is also found in the Taylor, Fairweather, and Leverett Formations, but not in the Wyatt Formation or Thiel Mountains Porphyry. Likewise, limestones and marbles, some of which are quite pure, are characteristic of the Liv Group and lacking in the Wyatt Formation and Thiel Mountains Porphyry.

Control on the age of the Liv Group is poor, though it is better than that on the Duncan and Party Formations. Deformed trilobite fossils from the Leverett Formation at Mt. Webster could not be identified as to genus, but Palmer and Gatehouse (1972) noted characteristics similar to *Mapania,* a late Middle Cambrian genus found in China and Australia. Rowell et al. (1993) found Early Cambrian archaeocyathids in Leverett Formation at Mt. Mahan in the Bender Mountains. Indeterminate trilobites and small tube fossils identified as *Cloudina*? and suggested to be of Lower Cambrian age were found in the Taylor Formation at Taylor Nunatak (Yochelson and Stump, 1977). Rowell and Rees (1989) and Grant (1990) have highlighted the uncertainty of this age assignment.

A Rb/Sr whole-rock isochron from volcanics in the upper part of the Leverett Formation at Mt. Webster gives a date of $493 \pm 9$Ma (IR = 0.7153) (Faure et al., 1979). Although this Early Ordovician date has at times been sighted as consistent stratigraphically with the underlying Middle Cambrian trilobite horizon, Faure et al. (1979) state that because of both the high initial ratio and the degree of deformation and alteration in the field, the number is probably somewhat lower than the actual age of the rock.

## Granite Harbour Intrusives

Much of the 600-km-long area between Ramsey Glacier and the Long Hills is gorged full by the Queen Maud–Wisconsin Range batholith. Reconnaissance mapping has covered much of the outcrop area, but to date no detailed studies have been undertaken. Although contacts have been observed, very little is known about the limits of individual plutons. The only chemical analyses are from a transect of Scott Glacier (Borg, 1983). And we are a long way from a comprehensive understanding of the rocks that comprise this plutonic complex. The composition of the batholith ranges from gabbro to granite, with the most plentiful rock types being granodiorite and quartz monzonite.

McGregor (1965) classified the plutonic rocks in the area that he was first to visit (Axel Heiberg to Shackleton Glacier) as pre- and posttectonic. He recognized differences between primary flow foliation present in some of the posttectonic rocks and secondary metamorphic foliation in the pretectonic rocks, based on both field and microscopic observation. Following the descriptions by Gunn and Warren (1962), McGregor (1965) suggested that much of their weakly foliated Larsen Granodiorite in southern Victoria Land is similar to his posttectonic granodiorite around Strom Glacier and therefore should not be considered syntectonic. Wade and Cathey (1986) described the plutonic rocks in the vicinity of Shackleton Glacier and presented about two dozen modal analyses with compositions ranging from syenogranite to tonalite.

Borg (1983) reported that some of the granitic rocks in the Scott Glacier area contain a cataclastic foliation. Within the upper Scott Glacier area this may be related to the shearing that is seen throughout a considerable portion of the Wyatt Formation and focuses on the Ackerman fault in the La Gorce Mountains (Stump et al., 1986b).

In the other regionally extensive study of the Queen Maud–Wisconsin Range batholith, Murtaugh (1969) observed widespread development of foliation in the plutonic rocks of the Wisconsin Range and Reedy Glacier area, but concluded that it was not possible within the scope of his study to distinguish whether it was due to deformation or to flow during intrusion.

Much of the Queen Maud–Wisconsin Range batholith is composed of granodiorite and quartz monzonite, medium- to coarse-grained, gray or sometimes pale pink, porphyritic or nonporphyritic, mildly foliated or massive, containing biotite and typically hornblende. Granite is also reported from most areas. The contacts between plutons are both sharp and diffuse, with the sharp contacts showing no alteration zone in the intruded rock (McGregor, 1965). Treves (1965) felt that the granodiorite in the upper part of the Ohio Range was an early marginal facies and of the same pluton as the quartz monzonite found beneath.

A hornblende gabbro has been reported by McGregor (1965) from the Gabbro Hills, where it appears to be the oldest rock in the western Queen Maud–Wisconsin Range batholith. It is intruded by diorite porphyry. Borg et al. (1987a, 1990) revisited the locality and found the gabbro to be considerably more restricted than originally mapped by McGregor (1965) and associated with diorite, tonalite, and granodiorite. The present author has observed a mass of gabbro exposed for perhaps 500 m on the ridge north of Mt. Borcik, also enveloped by younger granitic rock.

The most striking phase of the batholith is a coarse-grained rock characterized by porphyroblasts of microcline 6–10 cm in length with some of the biggest up to 20 cm (Murtaugh, 1969). The groundmass minerals are roughly 1 cm in size and fairly equally distributed. The microcline is pink or cream; the groundmass plagioclase is white; biotite is jet black and anastomoses in the foliated phases; quartz is typically a translucent blue. The porphyroblasts are both euhedral and ovoid. Besides those that are strictly microcline, some porphyroblasts are mantled by oligoclase (rapakivi texture), with both types occurring in the same rock. Porphyroblasts grew within the magma, in xenoliths contained therein, and across the contacts of the two.

The rock was observed by Murtaugh (1969) in the Reedy Glacier area and by the author in the Scott Glacier area, where the name "Zanuck Granite" was proposed (Stump et al., 1985). The extent of this rock in the Reedy Glacier area is uncertain, for Murtaugh (1969) lumps it with equigranular plutonics on his map. The best examples of the porphyry that he cites are from the east side of upper Reedy Glacier (Mims Spur and Hatcher Bluffs). The author has observed this rock on the west side of Scott Glacier between Mt. Crockett and Cox Peaks and on the east side in the Gothic Mountains at Mts. Zanuck, Andrews, and Gerdel. Between Mt. Zanuck and Grizzly Peak, the Zanuck Granite is intruded by a medium-grained equigranular granite (Fig. 5.20). Much of the rock is foliated, although portions are not. In less developed cases, biotite defines the fabric, whereas when foliation is strongly developed, the quartz and feldspar are elongate parallel to foliation. Some of the porphyroblasts are granulated at their margins, and groundmass quartz is highly polygonized, indicating that portions

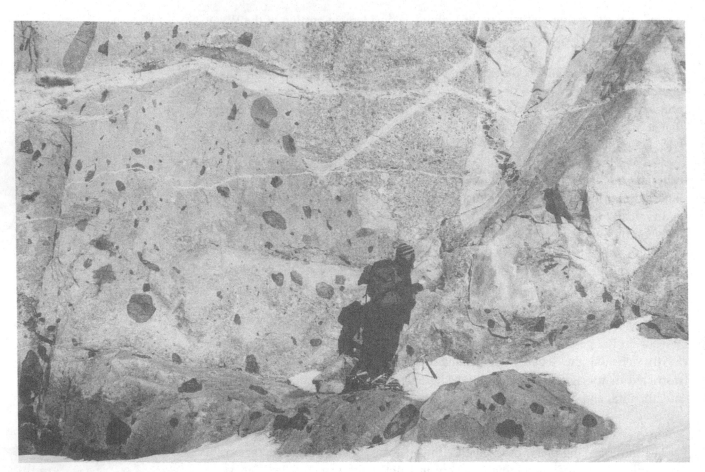

**Figure 5.20.** Contact zone between Zanuck Granite (above figure's head) and crosscutting granodiorite, with several generations of dikes apparent. Locality is at the southeast foot of Grizzly Peak.

of the Zanuck Granite were tectonized during or after emplacement. The massive granite at Grizzly Peak crosscuts foliation in the Zanuck Granite. Zanuck Granite intrudes Wyatt Formation near the summit of Heinous Peak.

Small stocks of two-mica granite and quartz monzonite have been reported from scattered localities on the margin of the batholith adjacent to the ice shelf (Fig. 5.21). In the Duncan Mountains the rock is equigranular with both biotite and muscovite, and it is cut by numerous pegmatites, some of which contain almandine garnet, tourmaline, and/or beryl (McGregor, 1965). At MacDonald Nunataks and O'Brien Peak the rocks are a two-mica granite, cut by some pegmatites, but there the granite is somewhat foliated (Craddock et al., 1964; Linder et al., 1965; Katz and Waterhouse, 1970b). In the Ford Nunataks the rock is an aplitic quartz monzonite with traces of garnet in addition to biotite and muscovite (Murtaugh, 1969).

An unusual tourmaline-bearing granite crops out at two separate localities: at the northern end of Metavolcanic Mountain (Murtaugh, 1969) and at Mt. Mooney (Stump et al., 1985, 1986b). The tourmaline occurs with quartz as ovoid blebs, with the immediate region surrounding the blebs leached of plagioclase and mafic minerals.

A pegmatite with primary andalusite crystals in radiating aggregates up to 0.5 m in length crops out at Szabo Bluff (Burt and Stump, 1983). The presence of

**Figure 5.21.** Leucocratic sills and dikes intruding Duncan Formation in Duncan Mountain to the east of Spillway Icefall.

andalusite rather than muscovite or sillimanite indicates that the pegmatite must have crystallized under extremely restricted conditions at relatively low pressures and moderate temperatures. The andalusite has in part altered to muscovite and corundum, described in detail by Ahn and Buseck (1988) and Ahn et al. (1988).

The age of the Queen Maud–Wisconsin Range batholith is very poorly constrained. In many cases analytical data are not presented with published dates, and in some cases the decay constants that were used are not identified (Treves, 1965; Faure et al., 1968, 1979; Mirsky, 1969; Wade and Cathey, 1986). No U/Pb dates have been obtained. A scattering of K/Ar ages are consistent with cooling ages elsewhere in the Transantarctic Mountains. Minshew (1965) published a K/Ar biotite date of $479 \pm 14$ Ma from a granite at Mt. Wilbur at the head of Scott Glacier. McDougall and Grindley (1965) obtained a K/Ar date on muscovite from a pegmatite at the mouth of Axel Heiberg Glacier of 453 Ma (no error bars). Craddock et al. (1964) presented several Rb/Sr mineral dates with assumed initial ratios (0.707) from O'Brien Peak that ranged from $538 \pm 30$ to $466 \pm 20$ Ma.

As mentioned in the section on Wyatt Formation, a Rb/Sr whole-rock date of $613 \pm 22$ Ma on the rapakivi phase (Zanuck Granite) of the Wisconsin Range batholith by Faure et al. (1968) was subsequently revised to $496 \pm 23$ Ma

(IR = 0.7157) by Faure et al. (1979). Faure et al. (1979) also published a Rb/Sr whole-rock date of $607 \pm 13$ Ma (IR = 0.7115) from a granodiorite on the western side of Nilsen Plateau. Whether any geological significance can be ascribed to these numbers is uncertain.

The only geochemical study of the Queen Maud–Wisconsin Range batholith is that by Borg (1983) on 31 samples collected across the width of the batholith in the Scott and Leverett Glaciers area. For these samples the $SiO_2$ content varies between 58 and 76%. All but four of the samples were identified as I-types, having relatively high $Na_2O/K_2O$ weight ratios (>0.6). Except for four samples from the Berry Peaks, which have a markedly different signature, the samples show a systematic increase in $K_2O$ content across the batholith in a southwesterly direction, as is typical of continental margin batholiths generated by subduction (e.g., Sierra Nevada; Bateman and Dodge, 1970). The samples from the Berry Peaks are distinct in that they have an unusually high $K_2O$ content (4.16–4.92%) for their $SiO_2$ content (59.92–68.30%) and fall well off the trend of the other I-type samples.

Four samples were identified as S-type, including the Zanuck Granite at Mt. Gerdel. That sample and another from Price Bluff contain altered cordierite and possibly altered hypersthene.

The isotopic study by Borg et al. (1990) mentioned in the preceding chapter included two samples from east and west of Ramsey Glacier and four samples from the Gabbro Hills area to the east of Shackleton Glacier. The Ramsey Glacier samples had IRs of 0.7090 and 0.7120 and $\varepsilon_{Nd}$'s of $-1.55$ and $-3.36$, and were included in the Beardmore Glacier block of 1.6-Ga continental crust. With a decidedly different signature, the samples from the Gabbro Hills have IRs of 0.7045–0.7059. Their $\varepsilon_{Nd}$ is $+1.72$ to $+0.44$, $\varepsilon_{Sr}$ is $+7.8$ to $+28.1$, and $\delta^{18}O$ is $+6.00$ to $+7.12$. Borg et al. (1990) interpret these to indicate that the granitic rocks east of Shackleton Glacier were derived mainly from mantle differentiates incorporating small amounts of continental material, probably sedimentary rocks. To round out their crustal model the authors add the Gabbro Hills block to their Beardmore Glacier and Miller Range blocks.

The Gabbro Hills are the northwest extremity of the Queen Maud–Wisconsin Range batholith, which intrudes continuously to the Ohio Range, 500 km to the east, subparallel to the structural trend in the surrounding country rocks. The extent to which sources with the characteristics of the Gabbro Hills block underlie the batholith remains to be demonstrated. Borg (1983) drew a boundary between S- and I-type granitic rocks through the central Scott Glacier area to the north of Albanus Glacier, which may approximate an extension of the boundary between the Beardmore Glacier and Gabbro Hills blocks. Further analyses by Borg and DePaolo (1994) on plutonic rocks from the Horlick Mountains show those rocks to be more like the plutonics in the outer portion of their Beardmore block rather than the Gabbro Hills block, with $\varepsilon_{Nd}$ values of $-0.49$ to $-2.77$ and $\varepsilon_{Sr}$ values of $+12.5$ to $+135.9$ and one low value of $-19.7$.

The several small two-mica plutons mentioned earlier, all of which are located along the northern margin of the main batholith, could be indicative of a continental crustal source outboard of the Gabbro Hills block. Alternatively, these small stocks, located as they are at the boundary between the main batholith and adjacent metasediments, could have been generated by melting of the local sedimentary wedge (Duncan and Party Formations). Klobcar and Holloway (1986) have shown that from a geochemical standpoint the S-type granites in the Wilson terrane of northern Victoria Land could have been formed by melting of Robertson Bay Group metagraywacke.

# 6 Thiel Mountains

## Geological Summary

Between the Horlick and Pensacola Mountains, in that region where the East and West Antarctic Ice Sheets merge to inundate the range, the Thiel Mountains reach above the ice as an isolated set of escarpments 70 km in length. Not only is the area of the Thiel Mountains the smallest of any covered by a chapter in this book, they are the simplest geologically. Only three basement units are exposed: the Thiel Mountains Porphyry and Mt. Walcott Formation, probably corresponding to the Wyatt and Ackerman Formations of the Scott–Reedy Glaciers area, and several small granitic plutons. A unique aspect of the Thiel Mountains is that bedding in the Mt. Walcott Formation is essentially flat-lying. Some very broad warps are displayed in the Thiel Mountains Porphyry (Ford, 1964), but otherwise the Thiel Mountains were not affected by the compressive deformation of the Ross orogeny seen elsewhere in the Transantarctic Mountains.

Ford and his co-workers chose not to name the units that they mapped and described. Formal designations were made by Storey and Macdonald (1987) for the Mt. Walcott Formation, and by Pankhurst et al. (1988) for the Thiel Mountains Porphyry and Reed Ridge Granites.

## Chronology of Exploration

One of the last appreciable areas of bedrock to be discovered on the continent, the Thiel Mountains were sighted and positioned in 1958–59 from a distance of about 50 km by a U.S. geophysical ground traverse looping between Byrd Station and the Horlick Mountains. In 1959–60 an airborne geophysical traverse following the 88°W longitude landed at Smith Knob near the southeastern end of the mountains, where samples were taken from a moraine and subsequently dated (Thiel, 1961; Craddock et al., 1964).

During the 1960–61 and 1961–62 field seasons the U.S. Geological Survey fielded ground parties that mapped the Thiel Mountains in their entirety (Ford and Aaron, 1962; Schmidt and Ford, 1969). No further geological work was done until 1983–84, when a joint U.S.–U.K. airborne party landed during a survey of the nunataks between the Ellsworth and Thiel Mountains (Storey and Dalziel, 1987). Slight revisions to the earlier mapping were presented by Pankhurst et al. (1988) (Fig. 6.2).

**Figure 6.1.** View toward Mt. McKelvey underlain by Thiel Mountains Porphyry, with sedimentary rocks of Mt. Walcott Formation in foreground. Photo by Arthur B. Ford, U.S. Geological Survey.

## Bedrock Geology

## Thiel Mountains Porphyry

Areally the most extensive unit in the Thiel Mountains is the Thiel Mountains Porphyry, the so-called cordierite-bearing hypersthene–quartz–monzonite porphyry of Ford (1964a). Its field appearance in most exposures is dark gray and massive, although faint light and dark layering is discernible at some localities

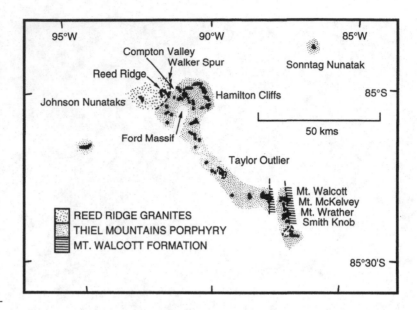

**Figure 6.2.** Generalized geological map of Thiel Mountains. After Pankhurst et al. (1988). Used by permission.

(Figs. 6.1 and 6.3). Aligned, light-colored, lenticular masses of cumulophyric feldspar have been interpreted by Ford (1964a, Fig. 6) as being due to magmatic flowage. These are similar to the features in the Wyatt Formation interpreted by Borg (1980, Fig. 5) to be flattened pumice fragments. A coarse breccia, possibly a flow breccia, is exposed at one locality along Hamilton Cliffs.

The porphyry is crystal rich with phenocrysts of plagioclase, quartz, K-feldspar, hypersthene, and cordierite, typically accounting for 40–70% of the rock (Ford, 1964a). The groundmass is a microcrystalline intergrowth of quartz and alkali feldspar. The most abundant phenocrysts are zoned plagioclase, of intermediate composition ($An_{35-50}$), which are euhedral or fragmented. Quartz crystals are rounded and embayed. Hypersthene is common throughout, generally comprising several percent of the phenocryst population, rarely up to 10%. It is strongly pleochroic and typically rimmed by biotite and/or chlorite. Most crystals are anhedral and some are deeply embayed. Cordierite is a minor (<1%) but pervasive phenocryst in much of the Thiel Mountains Porphyry. Crystals are 1–3 mm in diameter and may be angular to rounded. Most show some alteration to chlorite and sericite, and some are complete pseudomorphs of these minerals. Microprobe studies show the hypersthene to be iron-rich and the cordierite to be a magnesian variety (Ford and Himmelberg, 1976).

That hypersthene and cordierite occur in small xenoliths associated with laboradorite led Ford (1964b) to suggest that the larger crystals are xenocrystic and derived by partial melting of charnockitic crust beneath the Thiel Mountains. Geochemical studies by Vennun and Storey (1987) support this hypothesis.

Interpretations of the nature of emplacement of the Thiel Mountains Porphyry have varied. Although he acknowledged that a volcanic origin was possible, Ford (1964b) at first favored a shallow intrusive origin for the porphyry, citing the lack of well-developed bedding and the uniformity of composition, both laterally and vertically, throughout the entire body. However, Ford and Sumsion (1971) reinterpreted the porphyry as a crystal-rich tuff, emphasizing broken phenocrysts,

**Figure 6.3.** View to east with Reed Ridge Granites cropping out in the high cliffs at the northern end of Reed Ridge. The more distant Walker Ridge is underlain by Thiel Mountains Porphyry. Photo by Arthur B. Ford, U.S. Geological Survey.

eutaxitic structures, and the presence of vague bedding to indicate a volcanic origin. Intrusion cannot be dismissed completely, certainly, for Storey and Macdonald (1987) report two sills within the Mt. Walcott Formation at its type locality. As has been demonstrated for the Wyatt Formation, probably both hypabyssal and extrusive phases of the Thiel Mountains Porphyry exist.

## Mt. Walcott Formation

Flat-lying sedimentary rocks crop out at two localities toward the southeastern end of the Thiel Mountains, Mt. Walcott and Mt. Wrather (Fig. 6.1). In both cases the strata are faulted against the Thiel Mountains Porphyry. Storey and Macdonald (1987) designated Mt. Walcott as the type locality, and presented a 100-m measured section.

The lowest 7 m are a breccia with fragments of porphyry set in a coarse sand matrix. The remainder of the section consists of alternating sandstone, mudstone, and minor conglomerate, with several thin beds of limestone in the lower portion. The sandstones contain a variety of grains indicating a volcanic provenance, including embayed quartz, and lithic fragments of felsite and quartz andesite.

Highly altered alkali and plagioclase feldspar are also present. Limestones are impure, containing interbeds of calcareous siltstone or sandstone on a centimeter scale.

In the lower portion of the section the sandstones alternate with lesser mudstones. The sandstones are planar or cross-bedded, with symmetrical ripple marks on the tops of some units. Higher in the section the mudstones are replaced by pebbly sandstones, some of which contain lenses of conglomerate. The sandstones are structureless or normal-graded, and the pebbly sandstones may be either normal- or reverse-graded. The uppermost portion of the section contains dark silty mudstones with abundant calcareous concretions and thin silty sandstone beds. Story and Macdonald (1987) have interpreted the sequence as indicative of a volcanically active land area inundated by rising sea level and have proposed correlation of the Mt. Walcott Formation with the Ackerman Formation of the La Gorce Mountains.

Stromatolites were reported by Schmidt and Ford (1969) from one of the thin limestone beds. Storey and Macdonald (1987) questioned this interpretation, suggesting instead soft-sediment deformation for the structures. However, they also identified a number of fossil fragments from the limestones, including echinoderm plates, possible sponge spicules, tube-shaped fossils of unknown affinity, and a variety of burrows. Although none are diagnostic enough to permit an age assignment, they nevertheless are the first evidence of a post-Precambrian age for the nongranitic basement rocks in the Thiel Mountains.

## Reed Ridge Granites

At several localities, in particular in the southern portion of the mountains, the Thiel Mountains Porphyry is discordantly cut by small bodies of massive biotite granodiorite and quartz monzonite (Schmidt and Ford, 1969). The rock is white to gray and porphyritic, and contains traces of primary muscovite (Vennum and Storey, 1987). The Reed Ridge Granites, as well as the Thiel Mountains Porphyry, are cut by dikes of coarse-grained pegmatites that at places contain tourmaline, aplite, and muscovite–biotite alaskite.

## Isotopic Age Determinations

Until work by Pankhurst et al. (1988), dating in the Thiel Mountains suffered from a lack of published analytical data and discussions of results. The first dates on the Thiel Mountains Porphyry, and for that matter the first isotopic dates from the interior of Antarctica, were obtained by the lead-$\alpha$ method, published by Ford et al. (1963), Aaron and Ford (1964), and Ford (1964). In fairness, it should be said that the authors did publish their counting data, but the method has subsequently been deemed unreliable. The dates from two samples were 670 $\pm$ 50 and 620 $\pm$ 70 Ma and, from magnetic and nonmagnetic splits of zircons from a third sample, 530 $\pm$ 60 and 630 $\pm$ 60 Ma, respectively. The older ages were cited by Grindley and McDougall (1969) to date the Beardmore orogeny.

Faure et al. (1979) published a Rb/Sr whole-rock isochron date of 646 $\pm$ 77

**Table 6.1.** Rb/Sr Whole-Rock Ages from Thiel Mountains

| Rock | No. of samples | Age (Ma) | IR | MSWD |
|---|---|---|---|---|
| Thiel Mountains Porphyry | 17 | 493±24 | 0.7143 | 10.9 |
| Dacite sills | 5 | 480±18 | 0.7150 | 1.6 |
| Mt. Walcott Formation | 10 | 495±6 | 0.7138 | 5.8 |
| Reed Ridge Granites | 13 | 491±12 | 0.7153 | 3.3 |

*Source:* Pankhurst et al. (1988).

Ma (IR = 0.7069) for the Thiel Mountains Porphyry, but presented neither analytical data nor the number of samples used to generate the isochron.

Dates on the Reed Ridge Granites were obtained by a number of methods. A Rb/Sr whole-rock isochron by Faure et al. (1979) was 531 ± 41 Ma (IR = 0.7115). Craddock et al. (1964) presented a Rb/Sr determination on muscovite with an assumed IR of 0.707 that gave a date of 528 ± 20 Ma. Several K/Ar dates presented by Schmidt and Ford (1969) without analytical data are consistent with the time of cooling of the Granite Harbour Intrusives elsewhere in the Transantarctic Mountains.

Using the Rb/Sr whole-rock method, Pankhurst et al. (1988) dated the Thiel Mountains Porphyry, volcaniclastic rocks of the Mt. Walcott Formation, dacite sills intruding the same, and the Reed Ridge Granites. The dates are summarized in Table 6.1. The degree of fit of the data points to the isochrons (MSWD) establishes the reliability of these dates. Pankhurst et al. (1988) note that if 4 of the 17 points of the Thiel Mountains Porphyry data are eliminated, including the two most altered samples, the enhanced isochron has an MSWD of 2.6 and an age of 485 ± 11 Ma (IR = 0.7146).

The marked similarity between the initial ratios of each sample set, the similarity between their ages, as well as the close similarity between trace- and rare-earth-element compositions of the magmatic rocks (Vennum and Storey, 1987) argue for formation of all rock units during the same igneous episode. A whole-rock isochron combining all 45 data points yields an age of 502 ± 5 Ma (IR = 0.7137, MSWD = 7.9), which Pankhurst et al. (1988) interpret as the closest estimate of the time of magmatism in the Thiel Mountains and, by correlation, the time of formation of the Wyatt and Ackerman Formations in the Horlick Mountains–Scott Glacier area as well.

# 7 Pensacola Mountains

## Geological Summary

In the Pensacola Mountains the relationship of nearly horizontal Beacon Super-group unconformably overlying a deeply eroded basement breaks down, for there the Ross orogeny produced only broad warps in the Cambrian sedimentary rocks and was accompanied by little magmatism. The transition to the Beacon Supergroup is also uncertain, for the assignment of the Neptune Group as lower Beacon or as late Ross is now under question. Furthermore, all of the rocks including the late Paleozoic were affected by a Late Permian or Early Triassic folding event locally called the Weddellian orogeny (Ford, 1972). This deformation produced upright folds in rocks of the Beacon Supergroup and corresponds to folding in the Ellsworth Mountains of West Antarctica and the Cape Mountains of South Africa, broadly named the Gondwanide orogeny by du Toit (1937). Accompanied by neither magmatism nor metamorphism, this deformation in the Pensacola Mountains is unique within the Transantarctic Mountains.

Throughout the Pensacola Mountains, structures of the successive deformations are co-axial and co-planar. In general, the younger phases of deformation have not produced penetrative structures in the older rocks, but the deformational sequence is well established by unconformities. Because the Weddellian orogeny is younger than the time frame covered by this book, this activity will not be discussed further.

The oldest group of rocks in the Pensacola Mountains is the Patuxent Formation, a turbidite sequence in places associated with basaltic and silicic igneous rocks. This is unconformably overlain by the fossil-dated Middle Cambrian Nelson Limestone, followed by the Gambacorta and Wiens Formations, volcanic and clastic sedimentary units, respectively. The succeeding rocks belong to the Neptune Group, an Ordovician (?)–Devonian (?) clastic sequence that unconformably overlies the Cambrian rocks in places and the Patuxent Formation in others.

## Chronology of Exploration

The Pensacola Mountains were first sighted in January 1956, during a reconnaissance flight from McMurdo Station to the Filchner Ice Shelf. The first landfall was made two seasons later at the Dufek Massif by the Ellsworth Station IGY traverse. The geologist N. B. Aughenbaugh (1961) reported preliminary findings

**Figure 7.1.** Tightly folded and cleaved metagraywacke of the Patuxent Formation, Patuxent Range. Photo by Dwight L. Schimdt, U.S. Geological Survey.

on the Jurassic layered basic intrusion that crops out extensively in the Forrestal Range and at the Dufek Massif.

During three successive seasons in the 1960s, parties of the U.S. Geological Survey completed mapping throughout the Pensacola Mountains. In 1962–63 the Patuxent Mountains were mapped by ground parties (Schmidt and Ford, 1963; Schmidt et al., 1964). This activity was followed in 1963–64 by more ground-party work in the Neptune Range (Schmidt, 1964; Schmidt et al., 1965, 1978). A helicopter-supported expedition in 1965–66 completed mapping of the Pensacola Mountains, including reconnaissance sampling in the Argentina Range (Huffman and Schmidt, 1966; Schmidt and Ford, 1966, 1969; Ford et al., 1978a, b), and carried out a geophysical survey of the region (Behrendt et al., 1974).

A joint expedition of the USSR and the German Democratic Republic under-took studies in the Neptune Range during the 1978–79 and 1979–80 field seasons (Frischbutter, 1981, 1982a, b, 1985; Hofmann and Samsonov, 1982; Kaiser et al., 1982; Weber, 1982; Frischbutter and Vogler, 1985). During the 1987–88 season a joint U.S.–U.K. airborne expedition visited scattered localities through-out the Pensacola Mountains and Argentina Range, and examined the southern Neptune Range in detail (Macdonald et al., 1991; Storey et al., 1992). A ground-based U.S. party studied parts of the Neptune and Argentina Ranges in 1989–90 (Rowell et al., 1990, 1992; Evans et al., 1991).

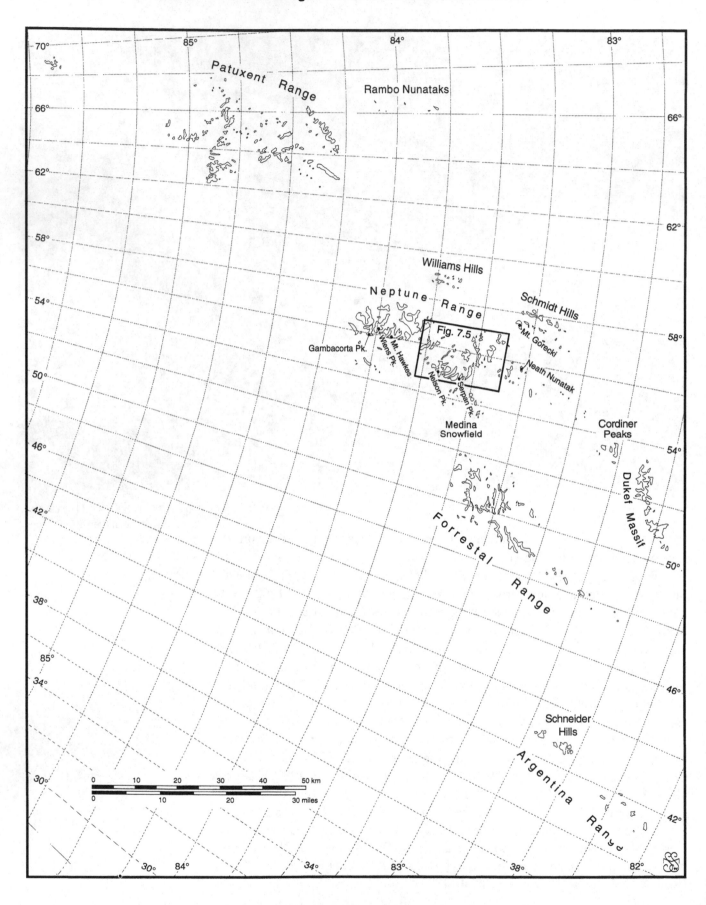

# Bedrock Geology

## Patuxent Formation

The Patuxent Formation is the most widespread unit of the Pensacola Mountains. It comprises more than 90% of the Patuxent Mountains, for which it is named, all of the Rambo Nunataks, Williams Hills, and Schmidt Hills, most of the axial portion of the Neptune Range, and a small area at the southwestern end of the Forrestal Range (Schmidt and Ford, 1969; Ford et al., 1978a, b; Schmidt et al., 1978). The occurrences in the Williams Hills and Schmidt Hills differ from those elsewhere, in that associated with the sedimentary rocks is a copious bimodal suite of magmatic rocks, with both extrusive and intrusive phases. Also, it is perhaps significant that the Williams and Schmidt Hills contain no occurrences of the younger sedimentary rocks (Nelson Limestone, etc.) that are widespread throughout the rest of the Neptune Range. Although the sedimentary character of the Patuxent Formation is similar throughout its exposure, the marked difference in content of basalt and rhyolite, and the presence or absence of Paleozoic rocks within a few kilometers, implies some degree of juxtapositioning of the Williams and Schmidt Hills with the rest of the Neptune Range. Schmidt et al. (1978) have inferred a fault beneath the ice between the areas.

The sedimentary portions of the Patuxent Formation, which are similar throughout their outcrop area, are a medium to dark greenish gray, rhythmically alternating sequence of graywackes and pelites. A thickness in excess of 10,000 m has been estimated (Schmidt et al., 1978), although not more than 900 m of continuous section have been measured on the limbs of folds (Schmidt et al., 1964; Storey et al., 1992). A pronounced cleavage obscures original bedding characteristics in many areas, but sedimentary structures are preserved well enough at some places to indicate the nature of deposition. These include graded bedding, sole markings, cross-bedding, and laminations. Channels filled with coarse-grained sand up to 15 m deep and 100 m wide have been observed at a few places (Schmidt et al., 1964; Storey et al., 1992). The Patuxent Formation most likely originated as a turbidite sequence in a continental slope-rise setting.

In general the graywackes are not coarser than medium-grained. Clasts are composed primarily of quartz, with lesser rock fragments, minor plagioclase and K-feldspar, and rare muscovite (Williams, 1969; Storey et al., 1992). The majority of the quartz is angular to subangular and clear, but some grains are embayed, indicating a volcanic origin, and others are compound, probably derived from quartzite or quartz schist. The rock fragments include chert (mainly volcanically derived), slate, and minor limestone and mafic volcanics. The graywackes typically contain more than 50% matrix composed of quartz, sericite, chlorite, and in some cases calcite, indicating metamorphism under lower greenschist conditions (Schmidt et al., 1964; Williams, 1969; Frischbutter, 1981).

Discontinuous beds and lenses of conglomerate and grit occur sparsely within the sequence. The most abundant clasts are angular to subangular pebbles and granules of sandy and silty limestone, lithologies that are otherwise unknown in the Patuxent Formation (Williams, 1969). These suggest that a shallow-water carbonate shelf existed synchronously with the deeper-water basin in which the Patuxent turbidites were accumulating, but that its location was beyond the area

**Figure 7.2** (*facing page*). Location map of Pensacola Mountains.

of the present-day Pensacola Mountains unless it has subsequently been eroded away.

A light-colored quartzite unit, which may be used as a marker horizon, crops out in the eastern part on the Patuxent Mountains. The bed is primarily well-sorted angular quartz with only minor silicic matrix. Williams (1969) has interpreted this as a graywacke in which the silt and clay fraction has been winnowed away.

The Patuxent Formation shares many characteristics with the Beardmore Group, including abundant matrix, matrix calcite, similar clast lithology, bedding characteristics, sedimentary structures, and the interpretation of deposition as turbidites. It differs by virtue of identified rock fragments and embayed quartz, both of which may be owing to the presence of appreciable volcanism in a portion of the formation. The small amount of volcanic rock present at Cotton Plateau in the central Transantarctic Mountains apparently did not provide a significant clastic input to the Goldie Formation.

**Magmatic Rocks of the Patuxent Formation.** A bimodal suite of volcanic and shallow intrusive rocks crops out extensively within the Patuxent Formation in the Schmidt and Williams Hills, but is practically nonexistent in its other areas of outcrop (Schmidt, 1969; Schmidt and Ford, 1969; Schmidt et al., 1978; Frischbutter and Vogler, 1987; Boyd, 1991; Storey et al., 1992). The little that has been identified elsewhere includes a minor occurrence of diabase in the southern Patuxent Range (Schmidt et al., 1964) and several basaltic dikes along the Washington Escarpment. Schmidt et al. (1978) subdivided and mapped the magmatic rocks within the Patuxent Formation based on chemistry and field appearance. These include the Gorecki Felsite Member, Williams Basalt Member, Pillow Knob Basalt Member, and Diabase of Schmidt Hills.

The Gorecki Felsite is less extensive than the basalts, but does crop out in both the Schmidt and Williams Hills. The member is composed of light to medium gray porphyritic tuffs and volcanic breccias of rhyolite and dacite composition. Schmidt (1969) describes them as quartz keratophyres. Eutaxitic shard structure in some samples suggests an ash-flow origin.

The basalt members all appear to be co-magmatic, but with different emplacement modes. The Williams Basalt Member includes lava flows 2–30 m thick, and Diabase of Schmidt Hills is the designation for sills 2–150 m thick of basaltic composition. Schmidt et al. (1978) acknowledge that the distinction between flows and sills cannot be made at all outcrops. Pillow Knob Basalt Member is applied to all outcrops where pillow structure is present. Igneous rocks make up as much as 30% of the outcrop of Patuxent Formation in the Schmidt Hills, and perhaps 15% in the Williams Hills, where the lavas have a composite thickness of up to 1,000 m.

All of the rocks are dark gray to dark green. From less altered samples of the diabase the original mineralogy appears to have been olivine–augite–labradorite. Alteration to chlorite and sodic plagioclase is extensive, but whether this is due to spilitization at the time of emplacement or to later metamorphism is uncertain (Schmidt, 1969).

Storey et al. (1992) have analyzed samples of Pillow Knob Basalt Member from both Schmidt and Williams Hills and sills from the Schmidt Hills for

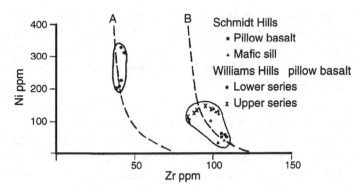

**Figure 7.3.** Nickel versus zirconium discrimination diagram for basalts of Patuxent Formation, showing differences between the "lower series" (A) at Williams Hills and the "upper series" at Williams Hills plus basalts from Schmidt Hills (B). After Storey et al. (1992). Used by permission.

major, trace, and rare earth elements, and report that the rocks are hypersthene-normative olivine tholeiites. They found that the lowermost 30 m of basalts ("lower series") in the Williams Hills are more primitive, and that the upper 270 m of basalts there ("upper series") and the basalts in the Schmidt Hills are more evolved (Fig. 7.3). Storey et al. (1992) further concluded that the "upper series" in the Williams Hills could be related to the sills and flows in the Schmidt Hills by fractional crystallization.

Faure et al. (1979) published Rb/Sr whole-rock isochron dates for the Gorecki Rhyolite and a basalt sill from the Schmidt Hills, but without reporting analytical data, number of samples, or goodness of fits, so one is unable to judge the reliability of the dates. The date for the Gorecki Rhyolite is $792 \pm 38$ Ma (IR = 0.7074), and for the basalt $767 \pm 57$ Ma (IR = 0.7064). These ages are within error of the date on gabbro within Goldie Formation from the Nimrod Glacier area ($762 \pm 24$ Ma; Borg et al., 1990) and suggest a temporal link between the Beardmore Group and the Patuxent Formation.

From the geochemical data, as well as lithological characteristics of the Patuxent Formation, Storey et al. (1992) concluded that the tectonic setting was continental rather than oceanic or accretionary and that the Patuxent Formation and equivalent rocks in the Beardmore Group formed in an intracontinental rift. They rationalize the oceanic character of the basalts in the Goldie Formation (Borg et al., 1990) as due to generation at a later time in the evolution of the rift than that which produced the magmatism in the Patuxent Formation.

## Schneider Hills Limestone

Early Cambrian limestone has been reported from several localities in the vicinity of the Pensacola Mountains. The only known in situ occurrence is in the Schneider Hills of the southern Argentina Range where archeocyathids are found (Konyuschkov and Shulyatin, 1980). Rowell et al. (1992) have suggested the informal name "Schneider Hills limestone" for this sequence. Morainic boulders around Mt. Spann in the northern Argentina Range contain similar archeochathids and Lower Cambrian (Botomian) trilobites, as well as Middle Cambrian trilobite associations similar to the Nelson Limestone cropping out in the Neptune Range (Palmer and Gatehouse, 1972).

A single loose boulder containing archeocyathids has been reported from the Neptune Range (Schmidt et al., 1965). Erratics of Lower Cambrian limestone are reported from the Whichaway Nunataks 150 km east of the Argentina Range (Stephenson, 1966). In addition, a fragmented block of archeocyathid-bearing limestone was dredged from the Weddell Sea at 62°10′S, 41°20′W by the Scottish National Antarctic Expedition of 1902–4. Descriptions of the specimens are given by Gordon (1920).

Along the west-facing bluffs of the Schneider Hills the Lower Cambrian limestones dip steeply to the west. Much of the exposure preserves sedimentary features in detail, but 1- to 10-m-thick shear zones roughly parallel to bedding disrupt the sequence, of which neither top nor bottom is exposed (Rowell et al., 1990). The lower part of the succession contains several hundred meters of bioturbated limestone and calc arenite. Some of the units are bioclastic with archaeocyathid fragments; some contain ooids.

The succession continues with two large boundstone bodies apparently separated by 150–200 m of thin-bedded bioturbated limestone (Rowell et al., 1990). The lower boundstone unit is about 200 m thick; the upper one is thicker. The primary bounding material is cyanobacteria. The units appear platformal in shape and have burrowed, flanking beds in association. Rowell et al. (1990, 1992) have emphasized that these reeflike features are among the largest found on earth for their time period.

## Nelson Limestone

Throughout the southern and east-central Neptune Range the Patuxent Formation is overlain above a marked angular unconformity by Middle Cambrian Nelson Limestone (Figs. 7.4 and 7.5). The top of the limestone is variously overlain by the conformable Gambacorta Formation, basal beds of the unconformable Neptune Group, or is faulted. Four small nunataks at the southern end of the Patuxent Mountains contain similar-appearing limestone, although no fossils have been found there. An informal type locality has been designated 1.5 km south of Nelson Peak, where the discordance with Patuxent Formation is nearly 90° (Schmidt et al., 1965). The Nelson Limestone is dated at middle to late Middle Cambrian by trilobites at several horizons, reaching down to within 10 m of the base of the formation (Palmer and Gatehouse, 1972; Rowell et al., 1992).

The basal part of the Nelson Limestone is composed of clastic sedimentary rocks (conglomerate, sandstone, and shale) varying in thickness from a few centimeters to more than 100 m (Schmidt et al., 1965, 1978; Rowell et al., 1990). The lower part of the clastic unit infills channels and hollows in the Patuxent Formation with up to 90 m of relief. Fine-pebble conglomerate is composed of clasts of quartz, cleaved phyllite, and volcanics derived from the underlying formation. The lower portion of the clastic unit contains planar and trough cross-bedding, interpreted to represent deposition in a fluvial system (Rowell et al., 1990). The upper portion of the unit grades upward into burrowed marine sandstones with brown, green, and red coloration.

The overlying limestones total about 200–400 m in thickness (Fig. 7.6). Cleavage obscures many of the occurrences, but primary structures are preserved at some localities. The sequence contains a variety of limestone lithologies (Schmidt et al., 1965; Rowell et al., 1990, 1992; Evans et al., 1991). In the lower

**Figure 7.4.** Openly folded Nelson Limestone unconformably overlying Patuxent Formation in core of anticline in east-central Neptune Range just west of Nelson Peak. Photo by Dwight L. Schmidt, U.S. Geological Survey.

and upper portions the rocks are primarily thin-bedded gray limestones and limey shales. These units variously contain oolites, oncolites, nodules, and cross-bedded pellets. Some beds are bioturbated and others show desiccation features. The middle portion of the formation is a thick-bedded, massive limestone, typically light gray.

## Gambacorta Formation

The Gambacorta Formation is a silicic volcanic complex that conformably overlies the Nelson Limestone in the southern and east-central parts of the Neptune Range (Schmidt et al., 1965, 1978). The rocks are rhyolites and dacites varying in color from dark brown to reddish brown to light green. Lithologies include ash-flow tuffs, lavas, volcanic breccias, and agglomerates. Thin interbeds of red and green volcanically derived sandstone, similar to the overlying Wiens Formation, are intercalated near the top and bottom of the formation. The Gambacorta Formation attains a maximum thickness of more than 1,500 m near Gambacorta Peak and thins northward to around 160 m in the vicinity of Nelson Peak.

Schmidt et al. (1978) have subdivided and separately mapped the Hawkes Rhyodacite Member of the Gambacorta Formation in the southern Neptune Range. This is a dark green welded tuff of rhyodacitic composition. Phenocrysts

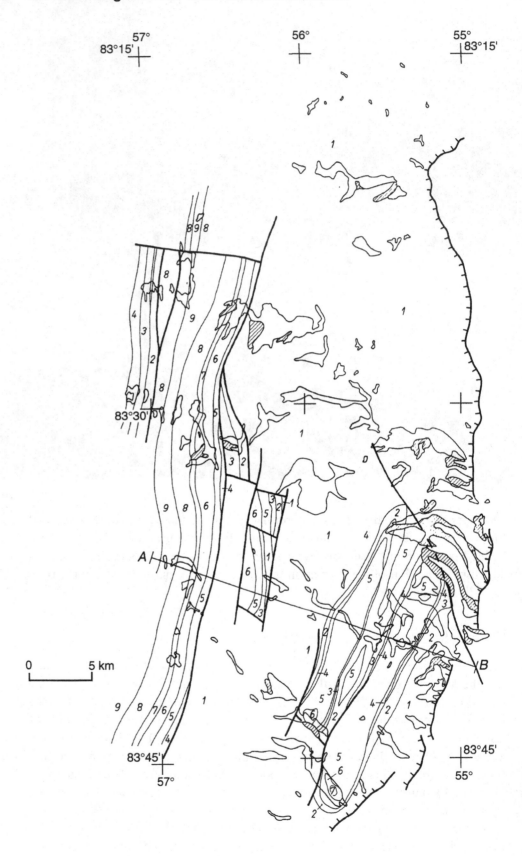

**Figure 7.5.** Geological and tectonic maps of central Neptune Range. After Weber (1982). Used by permission.

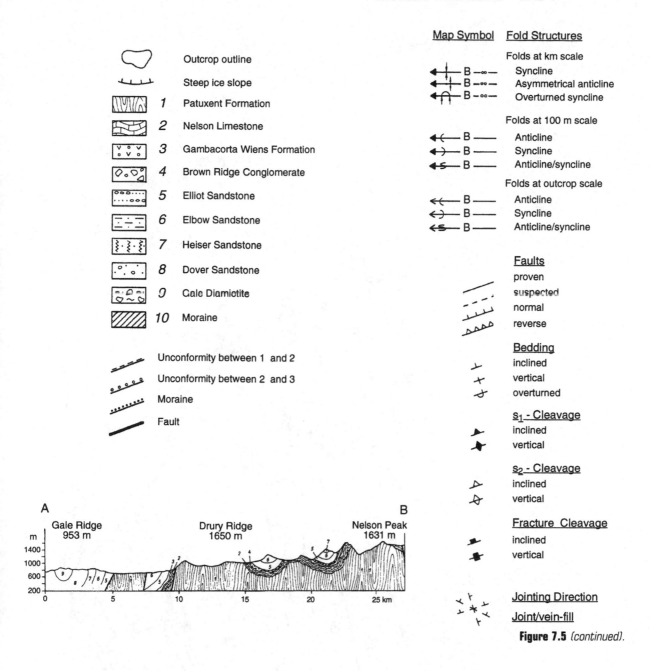

**Map Symbol** **Fold Structures**

Folds at km scale
Syncline
Asymmetrical anticline
Overturned syncline

Folds at 100 m scale
Anticline
Syncline
Anticline/syncline

Folds at outcrop scale
Anticline
Syncline
Anticline/syncline

**Faults**
proven
suspected
normal
reverse

**Bedding**
inclined
vertical
overturned

**s₁ - Cleavage**
inclined
vertical

**s₂ - Cleavage**
inclined
vertical

**Fracture Cleavage**
inclined
vertical

**Jointing Direction**
**Joint/vein-fill**

**Figure 7.5** *(continued)*.

Legend:

Outcrop outline
Steep ice slope
1 Patuxent Formation
2 Nelson Limestone
3 Gambacorta Wiens Formation
4 Brown Ridge Conglomerate
5 Elliot Sandstone
6 Elbow Sandstone
7 Heiser Sandstone
8 Dover Sandstone
9 Gale Diamictite
10 Moraine

Unconformity between 1 and 2
Unconformity between 2 and 3
Moraine
Fault

of broken quartz, plagioclase, and K-feldspar, as well as flattened pumice fragments, are set in a fine-grained groundmass containing relict eutaxitic shard structure. The main outcrops of the member, which attain a thickness of more than 1,000 m, are interpreted as caldera fill based on a circular gravity low (Behrendt et al., 1974) and concentric border faults near Mt. Hawkes.

In the Rb/Sr study mentioned previously, Faure et al. (1979) presented two whole-rock isochron ages, one for rhyolites of the Gambacorta Formation at $553 \pm 35$ Ma (IR $= 0.7057$) and one for the Hawkes Rhyodacite Member at $556 \pm 81$ Ma (IR $= 0.7054$). Based on the timescale of Bowring et al. (1993), the median ages are inconsistent with the stratigraphic relationship of Gambacorta Formation overlying fossil-dated Middle Cambrian rocks.

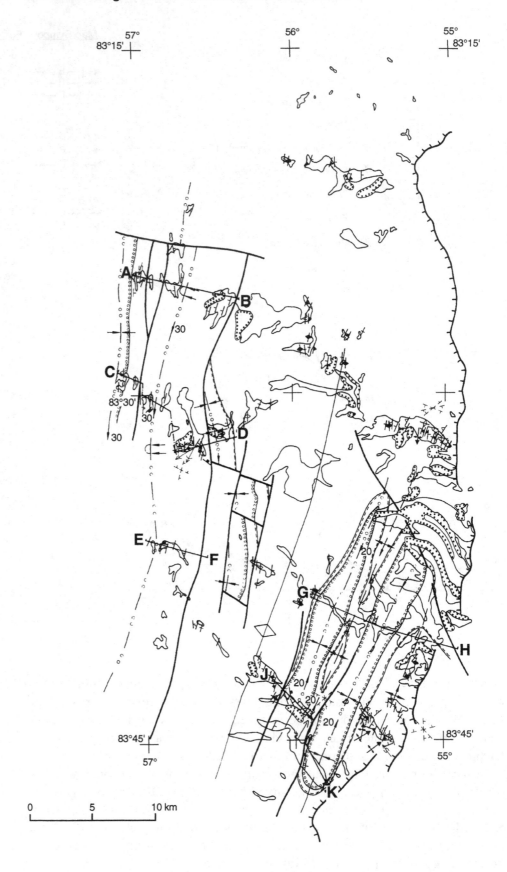

**Figure 7.5** *(continued).*

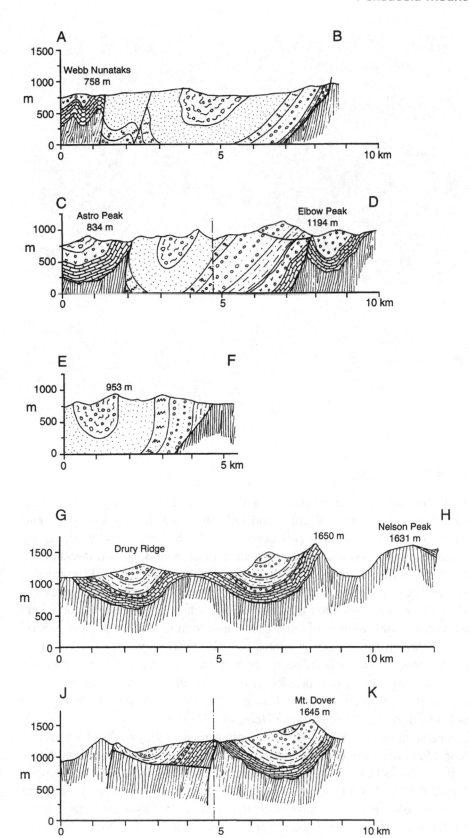

**Figure 7.5** (continued).

**Figure 7.6.** Anticlinal fold in Nelson Limestone along eastern portion of Gale Ridge. Photo by Dwight L. Schmidt, U.S. Geological Survey.

## Wiens Formation

Overlying the Gambacorta Formation is the Wiens Formation, a sedimentary sequence consisting mainly of green and reddish brown thin-bedded shale and sandstone. Several beds of light gray oolitic limestone, 3–10 m meters thick, are contained within the formation. The formation reaches a maximum thickness of about 300 m in the southern Neptune Range. Whereas earlier work reported a conformable contact between the Wiens and Gambacorta Formations, the 1987–88 U.S.–U.K. expedition found that an erosion surface on the Gambacorta Formation with as much as 10 m of relief exists at several localities (B. Storey, pers. comm.). Although the contact is probably a disconformity where it has been observed, cleaved clasts of Gambacorta Formation in the basal Wiens Formation indicate that post-Gambacorta pre-Wiens deformation had occurred somewhere in the area. In the northern Neptune Range in the vicinity of Neith Nunatak, the Wiens Formation rests directly on the Nelson Limestone.

The sandstones of the Wiens Formation are primarily graywackes by virtue of their considerable matrix content, but they differ markedly from the turbidites of the Patuxent Formation, in color, clast compositions, and sedimentary structures (Williams, 1969). Most of the clasts were probably derived from the underlying Gambacorta volcanics. The quartz is clear and in some cases contains embayments; feldspar is more abundant than in the Patuxent Formation, and the K-feldspar lacks microcline grid twinning. Rock fragments are of both felsic and basaltic types. The matrix of the graywackes is primarily quartz, with various amounts of calcite, hematite, sericite, and chlorite. The rocks are generally thinly bedded, with thin cross-bedding common. Mudcracks and raindrop impressions are found in the lower part of the formation, indicating terrestrial deposition, but they change to green sandstones and carbonate units higher in the formation, signaling a marine transgression (D. Macdonald, pers. comm.).

## Serpan Gneiss and Median Granite

A dark to light gray banded gneiss crops out at Serpan Peak in the east-central Neptune Range (Schmidt et al., 1978). Portions are biotite granite in composition; others are hornblende–biotite–sphene granodiorite. The outcrop has a thickness of about 700 m.

This rock is thought to be the border phase of a pluton that is not exposed but indicated by abundant granitic drift in local moraines. The so-called Median Granite is a light gray to light red coarse-grained biotite granite. A negative gravity anomaly indicates that the pluton of this granite may occur in a circular area from Serpan Peak eastward under the Median Snowfield (Behrendt et al., 1974).

## Neptune Group

The Neptune Group, more than 2,000 m thick, crops out throughout the Neptune Range and in the Cordiner Peaks and the southern part of the Forrestal Range. Its basal conglomerates overlie an unconformity that at various places oversteps Wiens, Gambacorta, Nelson, and Patuxent Formations, although the angular relationship is slight except with the Patuxent Formation. Deep karst with well-developed breccia occurs at the top of the Nelson Limestone at several localities (D. Macdonald, pers. comm.). The Neptune Group is unfossiliferous but is conformably overlain by the Dover Sandstone, considered to be Middle Devonian based on micro- and macrofloras (Schopf, 1968). Thus the group is bracketed to be not older than Middle Cambrian or younger than Middle Devonian. The unconformity beneath it may correspond to the unconformity between the Douglas Conglomerate and Shackleton Limestone in the central Transantarctic Mountains or to the Kukri peneplain that separates the Ross orogen and the Beacon Supergroup throughout the rest of the Transantarctic Mountains. In either case the deformation and erosion are much more subtle in the Pensacola Mountains than elsewhere, and the transition to rocks of known Devonian age is conformable rather than an unconformity eroded to the core of the Ross orogen.

Schmidt et al. (1965) divided the Neptune Group into four formations, in ascending order: Brown Ridge Conglomerate, Elliott Sandstone, Elbow Sandstone, and Heiser Sandstone. Schmidt et al. (1978) added the Neith Conglomerate for the basal conglomerate exposed on Neith Nunatak. The 1987–88 U.S.–U.K. expedition (pers. comm.) considered that facies representative of each of the previously named formations exist in the Neptune Group, but that the stratigraphic order is not certain, except for the Heiser Sandstone being on top. In either case the group overall is a fining-upward sequence.

**Brown Ridge and Neith Conglomerates.** The Brown Ridge and Neith Conglomerates are poorly bedded to massive reddish brown conglomerates with moderately rounded clasts up to 1 m in diameter. Interspersed are minor lenses of reddish sandstone. The conglomeratic clasts were derived largely from the underlying Patuxent Formation and Gambacorta Formation, with minor input from Nelson Limestone. Thickness varies from 0 to 1,000 m.

**Elliott Sandstone.** The Elliot Sandstone is coarse-grained reddish sandstone, in places with pebbly horizons. Volcanic rock fragments and volcanically derived quartz are abundant in the lower portion of the formation, but decrease upward. Cross-bedding is a common feature. Thickness varies from around 1,400 m near Gambacorta Peak to zero in the northern Neptune Range. Cropping out near Wiens Peak about 30 m above the base of the formation is a 70-m-thick flow breccia consisting of moderately welded fragments of porphyritic rhyolite.

**Elbow Formation.** The Elbow Formation is characterized by interbedded reddish mudstone and fine-grained sandstone. Cross-bedding is typical, laminations are plentiful, and ripple marks and mudcracks are persistent components. Thickness is about 300 m. Cathcart and Schmidt (1977) describe primary sedimentary phosphorite from the formation.

**Heiser Sandstone.** The Heiser Sandstone is green, and contains alternating sandstone and lesser mudstone. Cross-bedding is common, but the formation is characterized by abundant tubular burrows oriented vertical to bedding. Whereas the underlying formations are interpreted to have been deposited subaerially, the Heiser Sandstone by virtue of its bioturbation is considered to have been shallow or marginal marine.

## Structure

As mentioned in the introduction to this chapter, three main periods of deformation have affected the Pensacola Mountains, the first associated with the Beardmore orogeny (as defined by the angular unconformity between Goldie Formation and Shackleton Limestone in the central Transantarctic Mountains), the second the Ross orogeny, and the third the Permo-Triassic Weddellian orogeny. The earliest deformation occurred under lower-greenschist conditions of metamorphism, but the later two were nonmetamorphic. Whereas each orogenic episode produced penetrative deformation on the immediately preceding rocks, overprinting relationships are generally not seen in the successively older sequences. Unconformities, therefore, have been the best indicators of repeated deformation within the Pensacola Mountains.

The Patuxent Formation was folded before deposition of the Nelson Limestone (Schmidt et al., 1964, 1965; Frischbutter, 1982a). The folds are tight to isoclinal upright structures (Figs. 7.7 and 7.8). Most are symmetrical, but some disharmonic folding does occur locally. The trend of the folds is northeast–southwest in the Neptune Range and northern Patuxent Mountains and north–south in the southern Patuxent Mountains. Hinges plunge gently to both the north and south. Nearly vertical axial-plane cleavage is common. In the eastern and central Neptune Range this cleavage is nearly parallel to bedding, where folding is isoclinal.

**Figure 7.7.** Folded metagraywacke of Patuxent Formation, Patuxent Range. Photo by Arthur B. Ford, U.S. Geological Survey.

Weber (1982) and the 1987–88 U.S.–U.K. expedition (pers. comm.) reported two cleavages that were developed in the Patuxent Formation in the main portion of the Neptune Range before deposition of Nelson Limestone. The first ($S_1$) is a steep slatey cleavage, and the second ($S_2$) a moderately dipping spaced or crenulation cleavage (Weber, 1982) or a pressure solution cleavage (B. Storey, pers. comm.). Two cleavages were not observed in the Schmidt and Williams Hills, or at Rambo Nunataks and the Patuxent Range. However, clasts of Patuxent Formation with both cleavages were found in basal breccia within the Nelson Limestone. The unconformity beneath the Nelson Limestone is markedly angular, with discordances up to 90°.

Deformation during the Ross orogeny produced broad, symmetrical, open folds in the Cambrian formations (Nelson, Gambacorta, and Wiens) with wavelengths of around 5 km (Schmidt et al., 1965) (Fig. 7.6). Locally, disharmonic, asymmetrical, variably plunging folds are developed, in particular around the boundaries between limestones and volcanics. Axial-plane cleavage is variably developed within the limestones and argillites, and a spaced fracture cleavage occurs in some of the volcanics.

Within the Argentina Range the limestone dips steeply west on the west limb of a large anticline. At the southern end of the Patuxent Mountains, the limestone exposed in four small nunataks has a consistent east–west strike and southerly

**Figure 7.8.** Tightly to isoclinally folded Patuxent Formation, Patuxent Range. Photo by Arthur B. Ford, U.S. Geological Survey.

dip, in marked contrast to the trend of the Patuxent Formation to the north and east. Whether this is due to a different direction of folding in the limestones or an intervening fault is not known (Schmidt et al., 1964).

The notion that the sequence of events in the Pensacola Mountains was the same as those in the central Transantarctic Mountains [i.e., desposition of Neoproterozoic turbidites (Goldie and Patuxent Formations) synchronously folded during the Beardmore orogeny before deposition of Early to Middle Cambrian limestones (Shackleton, Nelson, and Schneider Hills Limestones) synchronously folded during the Ross orogeny] has been challenged by Rowell et al. (1992). They point out, and rightly so, that the Nelson Limestone constrains the age of the Patuxent Formation to be only pre–Middle Cambrian and then speculate that the deformation of the Schneider Hills Limestone and the Patuxent may have been early Middle Cambrian, with a later episode of deformation folding the Nelson Limestone.

To this author this seems an unwarranted stretch of the data. Although the reliability of the Rb/Sr data of Faure et al. (1979) is uncertain, the dates for the Gorecki Rhyolite and the sill at Schmidt Hills are consistent and definitely Neoproterozoic. Some of the Patuxent Formation could be Early Cambrian, but that remains to be proved. Further, from regional considerations – that is, the pre–Early Cambrian timing of deformation of the Beardmore Group – the more "parsimonious interpretation" (to use the phrase of Rowell et al., 1992, p. 31) would be that the folding of the Patuxent Formation was pre–Early Cambrian. One could still argue that what has been mapped as Patuxent Formation is in fact a folded Proterozoic sequence overlain by a similar-appearing Early Cambrian sequence, then folded in the earliest Middle Cambrian, but there is no evidence to distinguish two episodes of Patuxent deposition or an unconformity contained within. One bit of evidence that could be used to support a similar age of a part of the Patuxent Formation and the Schneider Hills Limestone is the limestone clasts found in a part of the Patuxent Formation (Williams, 1969). Also, the recognition of two cleavages in the Neptune Range before deposition of the Nelson Limestone could be used to support the idea of pre–Early Cambrian and early Middle Cambrian episodes of deformation in the Neptune Range, but this would preclude an Early Cambrian age for the Patuxent Formation there if it contained a pre–Early Cambrian cleavage.

Macdonald et al. (1991) reported a disconformity in the southern Neptune Range between Wiens and Gambacorta Formations with more than 10 m of relief. While this may simply represent erosion on a volcanic complex with considerable primary relief, cleaved clasts of Gambacorta Formation within the basal Wiens suggest that deformation had affected the Gambacorta before deposition of Wiens.

Because the Ross deformation was less severe than the Beardmore, the unconformity beneath the Neptune Group is not as angular as that beneath the Nelson Limestone. Nevertheless, erosion was considerable in places, and the Neptune Group rests on Wiens Formation, Gambacorta Formation, Nelson Limestone, and Patuxent Formation at various places in Neptune Range.

In the southern Neptune Range the 1987–88 U.S.–U.K. expedition found that deposition in the basal Neptune Group is structurally controlled (Macdonald et al., 1991). A growth fold is indicated where bedding is thicker in the core of a syncline and thins toward the limbs. This is cited as evidence of continued movement in the area, extending the time encompassed by Ross deformation into the episode of deposition of the Neptune Group. This and the movements before the desposition of the Wiens Formation indicate that deformation associated with the Ross orogeny occurred over a protracted period of time in the Pensacola Mountains, whether one accepts the speculations of Rowell et al. (1992) or not.

# 8 Synthesis

## Ross Orogen

The physical and temporal limit on the Ross orogen, expressed for 2,500 km from northern Victoria Land to the Ohio Range, is the Kukri peneplain overlain by the Devonian to Triassic Beacon Supergroup. Before Beacon deposition, the orogen was eroded to its crystalline core with exhumation of 10–20 km in many places. In the Pensacola Mountains, where pre-Beacon exhumation was considerably less, the Cambrian formations are followed by the Neptune Group, which precedes Devonian and younger equivalents of the Beacon Supergroup. There the Ross orogeny is marked by more subtle discordances and a paucity of plutonic rocks, and the entire sequence was deformed in a Permo-Triassic event (Gondwanide or Weddellian orogeny) unrecorded elsewhere in the Transantarctic Mountains.

The goal of this book has been to detail the history of the Ross orogen, and now at last it is time for a summing up. Although the stratigraphy, deformation, and magmatism of each major region are reasonably well established, at least in relative terms, uncertainties remain as to whether specific units are correlative among the regions and as to the precise equivalence of fossil-dated and isotopically dated chronologies. In Antarctica one has always the excuse that what is unknown is hidden beneath the ice. Conversely, one is free to speculate on what lies hidden there as well. Compared with most orogenic belts, the Transantarctic Mountains *are* narrow and do not expose the full width of the Ross orogen, but even if all the ice were to be removed, the questions of correlation and continuity would probably not be fully answered. It is the nature of orogenic belts to change along their lengths, to contain regions with discrete characteristics echoing the irregularities of continental margins at their time of formation (breakup), and the vagaries of interaction as they converge, so that the end result is a collage of blocks or terranes with both shared and unique histories.

Each of the preceding chapters began with a treatment of the history of exploration and geological thought for the region, followed by the details of the geology as presently understood. The one group of references that have been neglected throughout are syntheses, which must be acknowledged for this account to be complete. The geologists of the heroic era were busy filling tomes with descriptions. Their works were exhaustive, but few in number, and the literature had not reached the critical mass that warrants a synthesis.

By the decade following the IGY most of the regions on the continent had been touched by geologists. Because a comprehensive view was for the first time possible, and perhaps because the literature was still small enough to be manageable, a number of papers were published that covered the geology of the continent as a whole. Among these were Adie (1962), Anderson (1965), Ford (1964), Gunn (1963), Hamilton (1964, 1967), Harrington (1965), Klimov et al. (1964), Voronov (1964), and Warren (1965).

The benchmark for Antarctic geology is the American Geographical Society Antarctic Map Folio 12, published in 1969 and 1970. This work brought together in map form at a scale of 1:1,000,000 most of what was known of the continent at that time. The individual sheets pertaining to the Transantarctic Mountains have been referenced in the appropriate chapters and include Gair et al. (1969), northern Victoria Land; Warren (1969), southern Victoria Land; Grindley and Laird (1969), central Transantarctic Mountains; McGregor and Wade (1969), Queen Maud Mountains; Mirsky (1969), Horlick Mountains; and Schmidt and Ford (1969), Pensacola and Thiel Mountains. Craddock (1970a, b, c, d) synthesized the geology, tectonics, paleontology, and isotopic dating for the entire continent at a scale of 1:5,000,000.

As the numbers began to accumulate, compilations of isotopic dates were presented by three groups at the First International Symposium on Antarctic Geology in Cape Town: Angino and Turner (1964), Picciotto and Coppez (1964), and Ravich and Krylov (1964). New Zealand K/Ar dating in Antarctica was listed by Hulston and McCabe (1972). A mammoth compilation of Antarctic geochronology was presented by Stuiver and Brazius (1985).

Several syntheses of Antarctic geology were produced shortly after the publication of the AGS folio: Grikurov (1972), Grikurov et al. (1972), my own first efforts in the field (Stump, 1973), and Elliot (1975), the last of which became the standard reference for the geology of Antarctica as a whole. Malcolm Laird has offered several excellent syntheses of the Ross orogen, with particular emphasis on the lower Paleozoic sedimentary sequences (including Laird, 1981a, b, 1991a, b). Stratigraphic compendia that include equivalent sequences in Australia, New Zealand, and Antarctica are Webby et al. (1981), Brown et al. (1982), and Shergold et al. (1985). Two recent syntheses have been published by the author (Stump, 1992a, b).

Before retracing the history of the Ross orogen, along with its various uncertainties, several generalities can be made to set the framework. The culminating event in the orogen, which unites all of its autochthonous portions, is the intrusion of voluminous, subduction-related plutonic rocks (Granite Harbour Intrusives) at around 510 Ma. The onset of this pulse caught the final stage of deformation of the orogen, but most of the plutonism was posttectonic. The beginnings of tectonic activity (deformation, metamorphism, and plutonism) were close to the end of the Proterozoic and continued into the Late Cambrian or Early Ordovician, but for intervals in the Early and Middle Cambrian carbonate shelf sedimentation was widespread. The active phase of the orogen was preceded by passive margin sedimentation in the Neoproterozoic, which commenced perhaps around 750 Ma. According to the postulates of Bell and Jefferson (1987) and Moores (1991), this passive margin resulted from the breakout of Laurentia from Gondwanaland.

## Precursors

The Ross orogen developed along the margin of the East Antarctic craton. Crystalline rocks of both Archean and Proterozoic age encircle the perimeter of East Antarctica in a scattering of coastal outcrops. The extent to which the cratonic material is found in the Transantarctic Mountains is, however, uncertain. Isotopic signatures on plutonic rocks of the Granite Harbour Intrusives indicate that Paleoproterozoic continental crust $T_{DM} = 2.0–1.7$ Ga underlies most of the Transantarctic Mountains (Borg et al., 1990; Borg and DePaolo, 1991, 1994).

The Nimrod Group of the Miller and Geologists Ranges has traditionally been viewed as cratonic margin to the Ross orogen. These amphibolite-grade metamorphics are composed of mica schists and lesser micaceous quartzite, marble, calc schist, and amphibolite. Both Argosy Formation and the structurally juxtaposed Miller Formation have $T_{DM}$'s in the range 2.7–2.9 Ga. Orthogneiss that appears to intrude Argosy Formation on "Camp Ridge" has produced a Pb/Pb zircon date of 1.7 Ga, interpreted as the time of initial crystallization of this rock (Borg et al., 1990; Goodge et al., 1991a). Other early isotopic studies that produced Paleo- or Mesoproterozoic dates for the Nimrod Group were problematic because of highly scattered Rb/Sr whole-rock data (Gunner and Faure, 1972) or anomalous K/Ar dates shown to be due to excess argon (McDougall and Grindley, 1965; Adams et al., 1982a; Goodge and Dallmeyer, 1992). Although the timing of the metamorphism and accompanying deformation of the Nimrod Group remains to be dated accurately, it seems most likely, following the discussion in Chapter 3, that this was a Paleoproterozoic episode.

Whether the Nimrod Group has any correlatives in the Transantarctic Mountains is uncertain. Borg et al. (1989) have drawn attention to the similarity between Miller Formation and the Horney Formation on the north side of Byrd Glacier, but other than field observation, no other study has been done on those rocks. Various authors, recently including myself (Stump, 1992a), have suggested that the Koettlitz Group of southern Victoria Land may be correlative with the Nimrod Group. The correlation of Koettlitz and Skelton Groups is also an uncertainty, but one that seems fairly likely due to their proximity and the similarity of their lithologies, albeit of different metamorphic grades. If the latter correlation is valid, it would appear to preclude the other, since Rowell et al. (1993b) have reported a $T_{DM}$ of 700–800 Ma for a pillow basalt within the Skelton Group.

## Passive Margin Phase

Assuming the Koettlitz and Skelton Groups are correlative, these rocks belong to the succession deposited in the passive-margin basin established during rifting of the East Antarctic craton. The link is to the pillow basalts and gabbro in the Goldie Formation of the central Transantarctic Mountains, dated by a Sm/Nd mineral isochron at $762 \pm 24$ Ma (Borg et al., 1990).

The sedimentary deposits at this time are characterized by quartzose turbidites lacking rock fragments and representative of a plutonic–metamorphic provenance (e.g., Smit and Stump, 1986). A minor calcareous fraction is also typical of these rocks. Similar sedimentary rocks (Beardmore Group), metamorphosed no higher than lower greenschist grade, occur on the inboard side of the main batholiths in the central Transantarctic Mountains (Goldie Formation), the Queen Maud Mountains (La Gorce Formation), and the Pensacola Mountains (Patuxent Formation). Other possible equivalents of higher metamorphic grade and positioned outboard of the batholiths are the Duncan Formation of the Duncan Mountains and the Party Formation of the Harold Byrd Mountains.

Limited volcanism indicative of rifting is found within the turbidites. This includes a bimodal suite occurring in the Pensacola Mountains, of which the basalts are olivine tholeiites erupted in a continental setting (Storey et al., 1992). The volcanism in the Goldie Formation in the Nimrod Glacier area is also bimodal (see Chapter 3), although the isotopic signature of these pillow basalts is oceanic rather than continental (Borg et al., 1990). Basaltic volcanism occurs in the Goldie Formation in the Ramsey Glacier area as well (Wade and Cathey, 1986).

The contact with basement on which the Neoproterozoic turbidite sequence was deposited is nowhere seen. However, the Goldie Formation conformably overlies the shallow-water Cobham Formation in the Cobham Range, just outboard of the Nimrod Group of the Miller and Geologists Ranges. The lithology of the Cobham Formation includes pelitic schist, quartzite, calc schist, and marble. If the Nimrod Group is in fact representative of the craton along which the Beardmore Group was deposited, then the position of the Cobham Formation is near shore, with the Beardmore basin deepening in an outboard direction. On the basis of isotopic signatures of the Goldie Formation and the Nimrod Group, Borg et al. (1990) suggested that a source for some of the Goldie Formation lay outboard of Antarctica, what they called the "Beardmore microcontinent." This interpretation is consistent with a source from present-day western North America during the rifting of Laurentia and before it had completely drifted away.

The varied lithologies of the Koettlitz Group include schists, calc schists, and marbles; the Skelton Group contains limestone and volcanics in addition to graywacke. If the correlation of these rocks with the Beardmore Group is correct, they contain a considerably larger fraction of shallow-water lithologies than is found south of Byrd Glacier.

Whether the correlation can be carried into northern Victoria Land is also uncertain. Body fossils have been reported from the Priestley Formation, indicating a post-Proterozoic age (Casnedi and Pertusati, 1991); however, crystallization ages of early phases of Granite Harbour Intrusives and Rb Sr dates of metamorphism on Wilson and Lanterman metamorphics indicate that the sedimentation of these rocks must have been Precambrian (Black and Sheraton, 1990; Adams and Hörndorf, 1991). The relationship of the Berg Group, Wilson Group, and Priestley Formation in northern Victoria Land is itself uncertain. The evidence at hand indicates that sedimentation occurred in both the Neoproterozoic and the Cambrian. That metamorphism and intrusion occurred in one portion, before deposition of another, suggests that the sedimentation was in two episodes rather than continuous. This is consistent with the two-stage pattern of sedimentation seen from the central Transantarctic to the Pensacola Mountains.

In summary, rifting at around 750 Ma was followed by near-shore shallow marine deposition and deeper-water turbidite deposition. Although the sequences have been metamorphosed and deformed to various degrees, it appears likely that this episode of sedimentation is recorded throughout the Transantarctic Mountains. A final aside: Neoproterozoic glaciations that have been well documented worldwide (Hambrey and Harland, 1985) may find representation in the Goldie Formation (Stump et al., 1988) and more tentatively in the Hobbs Formation (Skinner, 1992).

## Initiation of Tectonic Activity

The onset of tectonic activity in the Ross orogen is revealed in different ways in different parts of the Transantarctic Mountains. Several dates on plutonic rocks in Victoria Land are older than the norm. These include (1) deformed orthogneiss intruding Wilson metamorphics in the Daniels Range that has given a U/Pb zircon age of 544 ± 4 Ma (Black and Sheraton, 1990), (2) posttectonic quartz syenite intruding Skelton Group adjacent to Cocks Glacier, dated by U/Pb zircon at 550.5 ± 4 Ma (Rowell et al., 1993b), and (3) the Carlyon Granodiorite in the Brown Hills, dated by Rb/Sr whole-rock isochron at 568 ± 54 Ma (Felder and Faure, 1990). The orthogneiss in the Daniels Range was generated at the peak of anatectic conditions in that area and was involved in ongoing deformation. The quartz syenite by Cocks Glacier crosscuts Skelton Group carrying two fold phases, and preceded or accompanied a third phase. The relationship of the Carlyon Granodiorite to metamorphic country rocks is unknown. What this data indicate is that, by uppermost Proterozoic, plutonism associated with deep-seated metamorphism and deformation had commenced throughout Victoria Land.

Two other older plutonic dates have been published, one on the Surgeon Island granodiorite, a Rb/Sr whole-rock isochron of 599 ± 21 Ma (Vetter et al., 1984). Since this is located on the northern margin of the Robertson Bay terrane, it may be related to basement of that terrane and may therefore be allochthonous (Borg and DePaolo, 1991) or, less likely, a far traveled klippe of Wilson terrane (Kleinschmidt et al., 1992b). The other older plutonic date is a Rb/Sr whole-rock isochron of 607 ± 13 Ma on a granodiorite on the west side of Nilsen Plateau. This is the only date of such antiquity south of Byrd Glacier, and whether it has geological significance has been questioned in Chapter 4. Nor have field relations with La Gorce Formation, the local metasedimentary unit, been described.

Goodge et al. (1993c) have documented a series of plutons in the Miller Range of the central Transantarctic Mountains that began around 540 Ma, slightly later than the intrusions just discussed. The older of these plutons carries a fabric from the Endurance shear zone that had begun movement before the initial plutonism.

Throughout the rest of the Transantarctic Mountains beyond Victoria Land, folding of Beardmore Group under lower-greenschist conditions appears to be the earliest tectonic activity. Goldie and Cobham Formations were folded in the middle Early Cambrian before deposition of the Shackleton Limestone. This is marked by an angular unconformity at Cotton Plateau (Stump et al., 1991) and probably at several other localities identified by Laird et al. (1971) throughout the Nimrod Glacier area, although these show evidence of tectonism (Rees et al., 1985).

An angular unconformity also exists in the Pensacola Mountains between folded Patuxent Formation and Middle Cambrian Nelson Limestone. Although the age of the Nelson Limestone indicates that the Patuxent Formation could have been deformed in the Early or earliest Middle Cambrian and could itself be Early Cambrian in age (Rowell et al., 1992), the more parsimonious interpretation from evidence to date is that the Patuxent Formation is time equivalent and correlative with the Goldie Formation, and the deformation is synchronous with the Goldie folding in the Nimrod Glacier area.

By analogy the deformation of the La Gorce Formation, located between the Patuxent and Goldie Formations, was of the same episode. The imprint is simple and, in the La Gorce Formation and most of the Goldie, apparently was not overprinted by later deformation. Multiple cleavage, imparted both pre– and post–Nelson Limestone, is characteristic of the Patuxent Formation, and two generations of folding and cleavage pre– and post–Shackleton Limestone are seen in Goldie Formation at Cotton Plateau, though elsewhere in the Goldie only one episode of folding has been observed.

The deformation of the Beardmore Group has been named the Beardmore orogeny (Grindley and McDougall, 1969). As discussed in Chapter 3, all attempts to date this event isotopically have in the final analysis failed. Nevertheless, the field relations indicate that the folding took place in the Neoproterozoic, as was originally claimed. How far back in the Neoproterozoic is speculative, with approximately 750 Ma being the upper bound. However, since tectonism throughout an orogenic belt is generally driven by plate-scale movement, it is reasonable to link within the same time frame the older plutonism in Victoria Land and the deformation of the Beardmore Group throughout the rest of the Transantarctic Mountains.

## Tectonic Continuity

Within portions of the Transantarctic Mountains the metamorphism and plutonism that had begun in the Neoproterozoic appear to have been more or less continuous features until the final cooling of the orogen in the Ordovician. In northern Victoria Land in the Wilson terrane, peak metamorphism and anatectic conditions were reached around 540–535 Ma. Deformation was active during this period and then waned. A scattering of dates on both plutonic and metamorphic rocks carries down to the interval 480–470 Ma, when widespread posttectonic plutonism occurred and rapidly cooled. In the Lanterman Range cooling ages begin to lock in slightly earlier, at around 500 Ma.

If the Priestley Formation is indeed Cambrian in age, then an episode of sedimentation must be sandwiched into this scenario of ongoing tectonism in northern Victoria Land. Perhaps for a period between around 535 and 480 Ma, compressive deformation and burial ceased locally and a sedimentary basin was established that was later closed, involving the Priestley Formation (and Berg Group?) in deformation and metamorphism. The uncertainty of the association of the various blocks within the Wilson terrane (e.g., Bradshaw et al., 1985) reinforces this possibility.

In southern Victoria Land there is no evidence for Cambrian sedimentation. Deformation and metamorphism had begun before intrusion of the quartz syenite into Skelton Group at 550 Ma and appears to have continued until the early stages of intrusion of the widespread gray granites. Dating is not well constrained, but the main phase of intrusion in southern Victoria Land appears to have been somewhere in the range of 510–490 Ma (see Chapter 2). Cooling to argon retention temperatures had occurred in parts of the region by 500 Ma. Late-stage posttectonic intrusions followed around 470 Ma.

What appear to be lacking in southern Victoria Land are dated events in the period between about 550 and 510 Ma. Presumably during this time the metamorphic rocks that had been buried deeply remained there, and plutonism was not active until later. One exception is the $^{40}$Ar/$^{39}$Ar plateau age of $534 \pm 5$ Ma on hornblende from the Carlyon Granodiorite, indicating cooling to about 550°C by that time (Felder and Faure, 1990).

In the Miller Range in the central Transantarctic Mountains Goodge et al. (1993c) have documented ongoing plutonism from about 540 until 500 Ma. Units intruded during the period 541–521 Ma carry L–S fabrics associated with the Endurance shear zone that juxtaposed Miller Formation over Argosy Formation. The shearing presumably had begun before 541 Ma and waned during the period 541–521 Ma, since younger rocks are incrementally less sheared. The lineations show a top-to-the-southsoutheast sense of shear oblique to the east–west shortening seen in the Goldie Formation and Shackleton Limestone to the east. A 515-Ma pegmatite intruding the shear zone is unsheared. Posttectonic Hope Granite is dated at around 500 Ma. Cooling ages span a period from about 520 to 450 Ma in the Miller Range.

The disparity of geological events in the Ross orogen is most acutely focused in the central Transantarctic Mountains, where the apparently continuous nature of the tectonism in the Miller Range is counterposed to the Early Cambrian shelf sedimentation of the Shackleton Limestone, now located as close as 20 km across Marsh Glacier.

## Cambrian Sedimentation

Following folding and erosion of the Beardmore Group, carbonate shelf sedimentation was established over the older rocks. In the central Transantarctic Mountains this is represented by the Shackleton Limestone, a shallow subtidal to intertidal shelf deposit with well-developed archaeocythan–microbial reefs (Rees et al., 1989b). The formation accumulated for a fairly short period of time, primarily in the Botomian, perhaps spanning parts of the Atdabanian and Toyonian, reaching a thickness of around 2,000 m.

The Botomian-aged Schneider Hills Limestone crops out in the Argentina Range to the northeast of the Pensacola Mountains. The 1,200-km distance between this and the Shackleton Limestone is covered with ice, but it is possible that similar carbonates spanned the distance as a continuous shelf of quiescent deposition.

Outboard of the Shackleton Limestone and Schneider Hills Limestone are limestones associated with volcanic rocks. In the Queen Maud–Horlick Mountains these are represented by the Liv Group. The volcanics are a bimodal suite, primarily felsic in composition. Early Cambrian fossils have been found in the Taylor Formation (Yochelson and Stump, 1977) and the Leverett Formation (Rowell et al., 1993a), and Middle Cambrian fossils are also known from the Leverett Formation (Palmer and Gatehouse, 1972).

In the Pensacola Mountains the Nelson Limestone is Middle Cambrian in age, unconformably overlies the folded and eroded Patuxent Formation, and is associated with volcanics of the Gambacorta Formation (Schmidt et al., 1978). What the record appears to show is that along a substantial length of the Transantarctic Mountains (Byrd Glacier to the Argentina Range) limestone shelf deposition occurred during the Early Cambrian, then ceased. Outboard of this a zone of limestone accompanied by volcanism spans the distance from the Queen Maud Mountains to the Pensacola Mountains. Deposition in the Queen Maud Mountains began in the Early Cambrian, but continued into the Middle Cambrian, whereas in the Pensacola Mountains limestone sedimentation and volcanism did not begin until the Middle Cambrian. All of the formations were subsequently folded, but whether the inboard and outboard deformations were synchronous is uncertain.

Rowell and Rees (1989) have proposed that terranes may account for the differences in Cambrian sedimentary and volcanic associations. In their latest iteration (Rowell et al., 1992) they separate an outboard "Queen Maud terrane" from an inboard "marginal cratonic belt," limiting their discussion to the area from Byrd Glacier to the Pensacola Mountains. The concept is useful for separating the volcanic from nonvolcanic sequences.

The overall tectonic setting is still difficult to rationalize. Goodge et al. (1991a, 1993a, b, c) have developed a model of left-lateral oblique subduction as an explanation of the disparity between orogen-parallel sense of shear on the Endurance shear zone and the orogen-normal fold orientation of the Shackleton Limestone. The rationale is that deeper and shallower levels of the same orogen will respond kinematically in different ways to oblique compression, in essence with a decoupling between the two. In their 1993c paper Goodge et al. emphasize temporal overlap of the deformation of Nimrod Group and Shackleton Limestone, without discussing the overlap of Nimrod deformation with Shackleton deposition or the possible overlap of Nimrod and Beardmore deformation. Using the Bowring et al. (1993) timescale and a figure of around 520 Ma for cessation of movement on the Endurance shear zone, it is possible that Shackleton deposition continued until Endurance movement ceased and that deformation of Shackleton Limestone took up after that.

Regardless, the question remains: What is the tectonic setting that can allow deformation along the cratonic margin, while outboard of it carbonate shelf deposition is occurring, all within a setting of subduction at a plate margin? The bimodal volcanics in the Liv Group imply an extensional regime in the Early and Middle Cambrian, temporally overlapping and continuing beyond deposition of the Shackleton Limestone.

Extending the concept of Goodge et al. (1991a) of a decoupling of structural levels to extension as well as compression, it is possible to view the tectonic evolution of the Transantarctic Mountains as follows. Initiation of subduction in the Neoproterozoic triggered deformation, metamorphism, and limited plutonism (at least in Victora Land) throughout the entire Transantarctic Mountains. These effects continued throughout the Cambrian in Victoria Land (with possible Cambrian deposition of portions of the Priestley Formation). The initial compression of the Beardmore Group and Patuxent Formation in the region from Byrd Glacier to the Pensacola Mountains (Beardmore orogeny) was followed by a period of decoupling and extension at relatively shallow crustal levels. This resulted in erosion followed by Early Cambrian carbonate deposition on an apparently stable shelf, synchronous with carbonate deposition and bimodal volcanism in a more outboard position. This activity overlapped in time the record of compressive deformation on the Endurance shear zone in the more inboard Miller and Geologists Ranges. Deformation of the Shackleton Limestone may have begun soon after cessation of movement on the Endurance shear zone, while carbonate deposition and volcanism continued with the Liv Group and the Nelson Limestone and Gambacorta Formation. Alternatively, deformation of the Shackleton may have been synchronous with the deformation of the Middle Cambrian sequences. In either case the shortening of the Early and Middle Cambrian limestones and volcanics is perpendicular to the orogen, except south of Byrd Glacier, where the trends in the Shackleton Limestone arc toward the Ross Sea.

Subsequent to folding of the Shackleton Limestone, a karst surface was formed upon which were deposited conglomerates and coarse sandstones of the Douglas Conglomerate and Dick Formation (Rees et al., 1985). The upper bound on these rocks is still unconstrained except by the Beacon Supergroup (Devonian). Among the varied lithology of the conglomerates, which include abundant clasts of Shackleton Limestone and Goldie Formation, are granites, but whether these are young "posttectonic" Granite Harbour Intrusives, of the older suite, or from the craton remains to be seen. The Douglas Conglomerate and Dick Formation were themselves caught up in further folding and thrust faulting, with thrusting in the vicinity of Mt. Hamilton apparently toward the south parallel to the orogen. This and the arcing of fold hinges south of Byrd Glacier support a kinematic picture of continuing left-lateral oblique subduction. It is as if the crust were pulled away from a corner south of Byrd Glacier and then came pushing back into the same area, while to the north in Victoria Land the orogen remained unextended while subduction proceeded.

## Granite Harbour Intrusives

As has been mentioned already, plutonism began in a scattering of locations in Victoria Land as early as approximately 550 Ma and appears to have continued at one place or another in the Transantarctic Mountains throughout the Cambrian. The main phase of posttectonic plutonism, which produced voluminous batholiths, appears to have occurred between around 500 and 480 Ma, with cooling to argon retention temperatures within 10–20 m.y. In a few localities cooling lingered until about 450 Ma.

Isotopic studies by Borg and co-workers have indicated a greater involvement of cratonic crustal material in the generation of the Granite Harbour Intrusives, across the orogenic belt toward East Antarctica, resulting from subduction in that direction. Plutonism as well as deformation appears to be less developed in the Thiel and Pensacola Mountains than in the rest of the Transantarctic Mountains. Perhaps at that end of the mountain range subduction was less active.

Following the widespread posttectonic phase of Granite Harbour Intrusives in the Early Ordovician, the Ross orogen cooled and was eroded. Throughout a good portion of the Transantarctic Mountains the exhumation brought plutonic and amphibolite-grade metamorphic rocks to the surface, an exception being in the Pensacola Mountains. By the Early to Middle Devonian, deposition of the Beacon Supergroup, initially marine and fluvial sandstones, had begun above the unconformity (Kukri peneplain) from southern Victoria Land to the Ohio Range (Barrett, 1981; Collinson, 1991). Middle Devonian deposition of similar rocks in the Pensacola Mountains was preceded by more than 2,000 m of the Neptune Group that overstepped the older formations across a low-angled unconformity equivalent to the Kukri peneplain elsewhere in the Transantarctic Mountains. The age of the unfossiliferous Neptune Group is uncertain, but it is likely to predate the Devonian. Furthermore, Macdonald et al. (1991) have suggested that waning effects of Ross deformation controlled the trough of deposition of the lower Neptune Group.

In northern Victoria Land, Beacon sedimentation did not begin until the Permian, when glacial deposits accumulated (Laird and Bradshaw, 1981; Skinner, 1981), as they did throughout the Transantarctic Mountains in the Permo-Carboniferous, and for that matter globally. The Devonian record in northern Victoria Land is one of shallow emplacement of Admiralty Intrusives in the Robertson Bay, Bowers, and possibly the Wilson terranes and of local outpourings of volcanic rocks onto each of the three terranes. The Gallipoli volcanics have been observed to unconformably overlie granite in the Wilson terrane. In contrast to the rest of the Transantarctic Mountains during the Devonian, northern Victoria Land was a source of, rather than a site of, sedimentation. In a number of ways the Bowers and Robertson Bay terranes show more affinity to rocks in Marie Byrd Land, Antarctica, and southeastern Australia than they do to the rest of the Transantarctic Mountains.

The assembly of the Robertson Bay and Bowers terranes was most likely an Early Ordovician event occurring around 500 Ma, with an incremental migration of cleavage development across the Robertson Bay Group down to about 460 Ma (Dallmeyer and Wright, 1992), coinciding with the period of voluminous posttectonic intrusion throughout the Transantarctic Mountains. The assembly of the Bowers and Wilson terranes may also have occurred in the Ordovician, but the evidence is less compelling. In both cases thrusting is indicated by structural data, but strike–slip faulting cannot be ruled out.

# Epilogue

A monk sat beneath a tree far from any path or highway where every day he meditated. Late one afternoon, when the shadows were long, a man approached with urgency in his step. The monk asked him to sit and rest, but he declined, stating that he wished to reach the mountain before nightfall. Sensing that the man was a traveler, the monk asked if he would give him something from his travels, that he might meditate on it in the future. Reaching into a small pocket on the side of his bag, the traveler retrieved a pebble, which he handed to the monk. "This pebble comes from the mountain beyond. It was given to me many years ago by an old man who said that it was given to him as a boy by a foreigner who had traveled to those mountains in his youth and had gotten it high up where the ice never melts. You see the facets here? Each time I return, I hold this pebble and imagine the one who collected it, and the one who gave it to me, and think of the myths that surround the mountain. This pebble is for your meditation. Perhaps this will be my last journey to the mountain. Perhaps I will . . ." But the traveler had become inaudible as he hurried on.

# Bibliography

Aaron, J. M., and Ford, A. B. 1964. Isotopic age determinations in the Thiel Mountains, Antarctica. Geological Society of America Special Paper 76, *Abstracts for 1963*, p. 1.

Adams, C. J. 1986. Age and ancestry of the metamorphic rocks of the Daniels Range, USARP Mountains, Antarctica. In E. Stump, ed., *Geological investigations in northern Victoria Land.* Antarctic Research Series, vol. 46. Washington, DC: American Geophysical Union, pp. 25–38.

Adams, C. J. D., Gabites, J., and Grindley, G. W. 1982a. Orogenic history of the Central Transantarctic Mountains: New K–Ar age data on the Precambrian–Early Paleozoic basement. In C. Craddock, ed., *Antarctic geoscience.* Madison: University of Wisconsin Press, pp. 817–26.

Adams, C. J., Gabites, J. E., Laird, M. G., Wodzicki, A., and Bradshaw, J. D. 1982b. Potassium–argon geochronology of the Precambrian–Cambrian Wilson and Robertson Bay Groups and Bowers Supergroup, North Victoria Land, Antarctica. In C. Craddock, ed., *Antarctic geoscience.* Madison: University of Wisconsin Press, pp. 543–48.

Adams, C. J., and Hörndorf, A. 1991. Age of the metamorphic basement of the Salamander and Lanterman Ranges, northern Victoria Land, Antarctica. In M. R. A. Thomson, J. A. Crame, and J. W. Thomson, eds., *Geological evolution of Antarctica.* Cambridge University Press, pp. 149–53.

Adams, C. J., and Kreuzer, H. 1984. Potassium–argon age studies of slates and phyllites from the Bowers and Robertson Bay terranes, north Victoria Land, Antarctica. *Geologisches Jahrbuch,* B60: 265–88.

Adams, C. J., and Whitla, P. F. 1991. Precambrian ancestry of the Asgard Formation (Skelton Group): Rb–Sr age of basement metamorphic rocks in the Dry Valley region, Antarctica. In M. R. A. Thomson, J. A. Crame, and J. W. Thomson, eds., *Geological evolution of Antarctica.* Cambridge University Press, pp. 129–35.

Adams, C. J., Whitla, P. F., Findlay, R. H., and Field, B. F. 1986. Age of the Black Prince Volcanics in the central Admiralty Mountains and possibly related hypabyssal rocks in the Millen Range, northern Victoria Land, Antarctica. In E. Stump, ed., *Geological investigations in northern Victoria Land.* Antarctic Research Series, vol. 46. Washington, DC: American Geophysical Union, pp. 203–210.

Adamson, R. G. 1971. Granitic rocks of the Campbell–Priestley divide, northern Victoria Land, Antarctica. *New Zealand Journal of Geology and Geophysics,* 14: 486–503.

Adie, R. J. 1962. The geology of Antarctica. In H. Wexler, M. J. Rubin, and J. E. Caskey, eds., *Antarctic research.* Geophysical Monograph, vol. 7. Washington, DC: American Geophysical Union, pp. 26–39.

Ahn, A. D., and Buseck, P. R. 1988. Al-chlorite as a hydration reaction product of andalusite: A new occurrence. *Mineralogical Magazine,* 52: 396–99.

Ahn, J. H., Burt, D. M., and Buseck, P. R. 1988. Alteration of andalusite to sheet silicates in a pegmatite. *American Mineralogist,* 73: 559–67.

Allen, A. D., and Gibson, G. W. 1962. Geological investigations in southern Victoria Land, Antarctica. Part 6: Outline of the geology of the Victoria Valley region. *New Zealand Journal of Geology and Geophysics,* 5: 234–42.

Allibone, A. H. 1987. Koettlitz Group metasediments and orthogneisses from the mid Taylor Valley and Ferrar Glacier areas. *New Zealand Antarctic Record,* 8(1): 48–60.

1992. Low pressure/high temperature metamorphism of Koettlitz Group schists, Taylor Valley and upper Ferrar Glacier area, south Victoria Land, Antarctica. *New Zealand Journal of Geology and Geophysics,* 35: 115–27.

Allibone, A. H., Forsyth, P. J., Turnbull, I. M., Sewell, R. J., and Bradshaw, M. A. 1991. Geology of the Thundergut area, southern Victoria Land, Antarctica, 1:50,000. Miscellaneous Geological Map 21. Lower Hutt: New Zealand Geological Survey, Department of Scientific and Industrial Research.

Allibone, A. H., and Norris, R. J. 1992. Segregation of leucogranite microplutons during syn-anatectic deformation: An example from the Taylor Valley, Antarctica. *Journal of Metamorphic Geology,* 10: 589–600.

Amundsen, R. 1912. *The South Pole: An account of the Norwegian Antarctic Expedition in the "Fram," 1910–1912,* 2 vols., translated by A. G. Chater. London: Murray.

Anderson, J. J. 1965. Bedrock geology of Antarctica: A summary of exploration, 1831–1962. In J. B. Hadley, ed., *Geology and paleontology of the Antarctic.* Antarctic Research Series, vol. 6. Washington, DC: American Geophysical Union, pp. 1–70.

Andrews, P. B., and Laird, M. G. 1976. Sedimentology of a Late Cambrian regressive sequence (Bowers Group), northern Victoria Land, Antarctica. *Sedimentary Geology,* 16: 21–44.

Angino, E. E., and Turner, M. D. 1964. Antarctic orogenic belts as delineated by absolute age dates. In R. J. Adie, ed., *Antarctic geology.* Amsterdam: North-Holland, pp. 551–66.

Angino, E. E., Turner, M. D., and Zeller, E. J. 1962. Reconnaissance geology of lower Taylor Valley, Victoria Land, Antarctica. *Geological Society of America Bulletin,* 73: 1553–62.

Anon. 1963. U.S. helicopters in Antarctica. *Bulletin of the U.S. Antarctic Projects Officer,* 4(5): 7–9.

Armienti, P., Ghezzo, C., Innocenti, F., Manetti, P., Rocchi, S., and Tonarini, S. 1990a. Palaeozoic and Cainozoic intrusives of Wilson terrane: Geochemical and isotopic data. *Memorie della Società Geologica Italiana,* 43: 67–75.

1990b. Isotope geochemistry and petrology of granitoid suites from Granite Harbour Intrusives of the Wilson Terrane, north Victoria Land. *European Journal of Mineralogy,* 2: 103–23.

Aughenbaugh, N. B. 1961. *Preliminary report on the geology of the Dufek Massif.* Glacialogical Report 4. International Geophysical Year World Data Center A, Glacialogy, pp. 155–93.

Babcock, R. S., Plummer, C. C., Sheraton, J. W., and Adams, C. J. 1986. Geology of the Daniels Range, north Victoria Land, Antarctica. In E. Stump, ed., *Geological investigations in northern Victoria Land.* Antarctic Research Series, vol. 46. Washington, DC: American Geophysical Union, pp. 1–24.

Barrett, P. J. 1981. History of the Ross Sea region during the deposition of the Beacon Supergroup 400–180 million years ago. *Journal of the Royal Society of New Zealand,* 11: 447–58.

Barrett, P. J., and Elliot, D. H. 1973. Reconnaissance geological map of the Buckley Island quadrangle, Transantarctic Mountains, Antarctica. Map A-3. Reston, VA: U.S. Geological Survey, United States Antarctic Research Program.

Barrett, P. J., Elliot, D. H., Gunner, J. D., and Lindsay, J. F. 1968. Geology of the Beardmore Glacier area, Transantarctic Mountains. *Antarctic Journal of the United States,* 3: 102–6.

Barrett, P. J., Lindsey, J. F., and Gunner, J. 1970. Reconnaissance geological map of the Mount Rabot quadrangle, Transantarctic Mountains, Antarctica. Map A-1. Reston, VA: U.S. Geological Survey, United States Antarctic Research Program.

Bateman, P., and Dodge, F. W. C. 1970. Variations of major chemical constituents across the central Sierra Nevada batholith. *Geological Society of America Bulletin,* 81: 409–20.

Behrendt, J. C., and Cooper, A. 1991. Evidence of rapid Cenozoic uplift of the shoulder escarpment of the West Antarctic rift system and a speculation on possible climate forcing. *Geology,* 19: 315–19.

Behrendt, J. C., Henderson, J. R., Meister, L. J., and Rambo, W. 1974. *Geophysical investigations of the Pensacola Mountains and adjacent glacierized area of Antarctica.* Professional Paper 844. Reston, VA: U.S. Geological Survey.

Bell, R. T., and Jefferson, C. W. 1987. An hypothesis for an Australian–Canadian connection in the Late Proterozoic and the birth of the Pacific Ocean. In *Proceedings, Pacific Rim Congress '87.* Parkville, Victoria: Australian Institute of Mining and Metallurgy, pp. 39–50.

Biagini, R., Di Vincenzo, G., and Ghezzo, C. 1991a. Petrology and geochemistry of peraluminous granitoids from Priestley and Aviator glacier region, northern Victoria Land, Antarctica. *Memorie della Società Geologica Italiana,* 46: 205–30.

1991b. Mineral chemistry of metaluminous granitoids between David and Campbell Glaciers, Victoria Land (Antarctica). *Memorie della Società Geologica Italiana,* 46: 231–47.

Black, L. P., and Sheraton, J. W. 1990. The influence of Precambrian source components on the

U–Pb zircon age of a Palaeozoic granite from northern Victoria Land, Antarctica. *Precambrian Research,* 46: 275–93.

Blackburn, Q. A. 1937. The Thorne Glacier section of the Queen Maud Mountains. *Geographical Review,* 27: 598–614.

Blank, H. R., Cooper, R. A., Wheeler, R. H., and Willis, I. A. G. 1963. Geology of the Koettlitz–Blue Glacier region, southern Victoria Land, Antarctica. *Transactions of the Royal Society of New Zealand, Geology,* 2: 79–100.

Blattner, P. 1978. Geology and geochemistry of Mt. Dromedary Massif, Koettlitz Glacier area, southern Victoria Land: Preliminary report. *New Zealand Antarctic Record,* 1(2): 16–19.

Borg, S. G. 1980. Petrology and geochemistry of the Wyatt Formation and the Queen Maud batholith, upper Scott Glacier area, Antarctica. M.S. thesis, Arizona State University.

1983. Petrology and geochemistry of the Queen Maud batholith, central Transantarctic Mountains, with implications for the Ross orogeny. In R. L. Oliver, P. R. James, and J. B. Jago, eds., *Antarctic earth science.* Canberra: Australian Academy of Science, pp. 165–69.

1989. Application of isotopic studies to the granite and metamorphic basement rocks of the Transantarctic Mountains, Antarctica. *Memorie della Società Geologica Italiana,* 33: 17–23.

Borg, S. G., and DePaolo, D. J. 1986. Geochemical investigations of lower Paleozoic granites of the Transantarctic Mountains. *Antarctic Journal of the United States,* 21: 41–43.

1990. Crustal basement provinces of the Transantarctic Mountains, Ross Sea sector. *Antarctic Journal of the United States,* 25: 29–31.

1991. A tectonic model of the Antarctic Gondwana margin with implications for southeastern Australia: Isotopic and geochemical evidence. *Tectonophysics,* x: xx–xx.

1994. Laurentia, Australia, and Antarctica as a Late Proterozoic supercontinent: Constraints from isotopic mapping. *Geology,* 22: 307–10.

Borg, S. G., DePaolo, D. J., Daley, E. E., and Sims, K. W. W. 1991. Studies of granitic and metamorphic rocks, Horlick and Whitmore Mountains area. *Antarctic Journal of the United States,* 26: 24–25.

Borg, S. G., DePaolo, D. J., and Smith, B. M. 1988. Geochemistry of Paleozoic granites of the Transantarctic Mountains. *Antarctic Journal of the United States,* 23: 25–29.

1990. Isotopic structure and tectonics of the central Transantarctic Mountains. *Journal of Geophysical Research,* 95: 6647–67.

Borg, S. G., DePaolo, D. J., Wendlandt, E. D., and Drake, T. G. 1989. Studies of granites and metamorphic rocks, Byrd Glacier area. *Antarctic Journal of the United States,* 24: 19–21.

Borg, S. G., Goodge, J. W., Bennett, V. C., DePaolo, D. J., and Smith, B. K. 1987a. Geochemistry of granites and metamorphic rocks: Central Transantarctic Mountains. *Antarctic Journal of the United States,* 22: 21–23.

Borg, S. G., Goodge, J. W., DePaolo, D. J., and Mattinson, J. M. 1986a. Field studies of granites and metamorphic rocks: Central Transantarctic Mountains. *Antarctic Journal of the United States,* 21: 43–45.

Borg, S. G., and Stump, E. 1987. Paleozoic magmatism and associated tectonic problems of northern Victoria Land, Antarctica. In G. D. McKenzie, ed., *Gondwana Six: Structure, tectonics and geophysics.* Geophysical Monograph vol. 40. Washington, DC: American Geophysical Union, pp. 67–75.

Borg, S. G., Stump, E., Chappell, B. W., McCulloch, M. T., Wyborn, D., Armstrong, R. L., and Holloway, J. R. 1987b. Granitoids of northern Victoria Land, Antarctica: Implications of chemical and isotopic variations to regional crustal structure and tectonics. *American Journal of Science,* 287: 127–69.

Borg, S. G., Stump, E., and Holloway, J. R. 1986b. Granitoids of northern Victoria Land, Antarctica: A reconnaissance study of field relations, petrography and geochemistry. In E. Stump, ed., *Geological investigations in northern Victoria Land.* Antarctic Research Series, vol. 46. Washington DC: American Geophysical Union, pp. 115–188.

Borsi, L., Ferrara, G., and Tonarini, S. 1989. Rb–Sr and K–Ar data on Granite Harbour Intrusives from Terra Nova Bay and Priestley Glacier (Victoria Land, Antarctica). *Memorie Società Geologica Italiana,* 33: 161–69.

Bouma, A. H. 1962. *Sedimentology of some flysch deposits.* Amsterdam: Elsevier.

Bowring, S. A., Grotzinger, J. P., Isachsen, C. E., Knoll, A. H., Pelechaty, S. M., and Kolosov, P. 1993. Calibrating rates of Early Cambrian evolution. *Science,* 261: 1293–98.

Boyd, W. W. 1991. Crustal evolution in the Pensacola Mountains: Inferences from chemistry and petrology of the igneous rocks and nodule-bearing lamprophyre dykes. In M. R. A. Thomson, J. A. Crame, and J. W. Thomson, eds., *Geological evolution of Antarctica*. Cambridge University Press, pp. 557–61.

Bradshaw, J. D. 1985. Terrane boundaries and terrane displacement in northern Victoria Land, Antarctica. *Geological Society of Australia Abstracts*, 14: 30–33.

1987. Terrane boundaries and terrane displacement in northern Victoria Land, Antarctica. Some problems and constraints. In E. C. Leitch and E. Scheibner, eds., *Terrane accretion and orogenic belts*. Geodynamics Series, vol. 19. Washington DC: American Geophysical Union, pp. 199–205.

1989. Terrane boundaries in North Victoria Land. *Memorie della Società Geologica Italiana*, 33: 9–15.

Bradshaw, J. D., Begg, J. C., Buggish, W., Brodie, C., Tessensohn, F., and Wright, T. O. 1985a. New data on Paleozoic stratigraphy and structure in north Victoria Land. *New Zealand Antarctic Record*, 6(3): 1–6.

Bradshaw, J. D., and Laird, M. G. 1983. The pre-Beacon geology of northern Victoria Land. In R. L. Oliver, P. R. James, and J. B. Jago, eds., *Antarctic earth science*. Canberra: Australian Academy of Science, pp. 98–101.

Bradshaw, J. D., Laird, M. G., and Wodzicki, A. 1982. Structural style and tectonic history in northern Victoria Land. In C. Craddock, ed., *Antarctic geoscience*. Madison: University of Wisconsin Press, pp. 809–16.

Bradshaw, J. D., Weaver, S. D., and Laird, M. G. 1985b. Suspect terranes in north Victoria Land, Antarctica. In D. C. Howell, D. L. Jones, A. Cox, and A. Nur, *Circum-Pacific Terrane Conference, Proceedings*. Stanford, CA: Stanford University Press, pp. 36–39.

Brown, A. V., Coper, R. A., Corbett, K. D., Daily, B., Green, G. R., Grindley, G. W., Jago, J., Laird, M. G., Vandenberg, A. H. M., Vidal, G., Webby, B. D., and Wilkinson, H. E. 1982. *Late Proterozoic to Devonian sequences of southeastern Australia, Antarctica and New Zealand and their correlation*. Special Publication no. 9. Sydney: Geological Society of Australia.

Buggisch, W., and Repetski, J. E. 1987. Uppermost Cambrian (?) and Tremadocian conodonts from Handler Ridge, Robertson Bay terrane, north Victoria Land, Antarctica. *Geologisches Jahrbuch*, B66: 145–85.

Buggish, W., and Kleinschmidt, G. 1991. Recovery and recrystallization of quartz and "crystallinity" of illite in the Bowers and Robertson Bay terranes, northern Victoria Land, Antarctica. In M. R. A. Thomson, J. A. Crame, and J. W. Thomson, eds., *Geological evolution of Antarctica*. Cambridge University Press, pp. 155–59.

Burgener, J. D. 1975. Petrography of the Queen Maud Batholith, central Transantarctic Mountains. M.S. thesis, University of Wisconsin.

Burgess, C. J., and Lammerink, W. 1979. Geology of the Shackleton Limestone (Cambrian) in the Byrd Glacier area. *New Zealand Antarctic Record*, 2(1): 12–16.

Burrett, C. F., and Findlay, R. H. 1984. Cambrian and Ordovician conodonts from the Robertson Bay Group, Antarctica, and their tectonic significance. *Nature*, 307: 723–26.

Burt, D. M., and Stump, E. 1983. Mineralogical investigation of andalusite-rich pegmatites from Szabo Bluff, Scott Glacier area. *Antarctic Journal of the United States*, 18: 49–52.

Caironi, V. 1990. Preliminary results on the characterization of the Granite Harbour Intrusives (Victoria Land, Antarctica) through zircon typology data. *Memorie della Società Geologica Italiana*, 43: 77–86.

Capponi, G., Messiga, B., Piccardo, G. B., Scambelluri, M., Traverso, G., and Vannucci, R. 1990. Metamorphic assemblages in layered amphibolites and micaschists from the Dessent Formation (Mountaineer Range, Antarctica). *Memorie della Società Geologica Italiana*, 43: 87–95.

Carmignani, L., Ghezzo, C., Gosso, G., Lombardo, B., Meccheri, M., Montrasio, A., Pertusati, P. C., and Salvini, F., 1988. Geological map of the area between David and Mariner Glaciers, Victoria Land, Antarctica. Programma Nazionale di Ricerche in Antartide, CNR-ENEA.

1989. Geology of the Wilson terrane in the area between David and Mariner Glaciers, Victoria Land (Antarctica). *Memorie della Società Geologica Italiana*, 33: 77–97.

Casnedi, R., and Pertusati, P. C. 1991. Geology of the Priestley Formation (northern Victoria Land, Antarctica) and its significance in the context of Early Palaeozoic continental reconstruction. *Memorie della Società Geologica Italiana*, 46: 145–56.

Castelli, D., Lombardo, B., Oggiano, G., Rossetti, P., and Talarico, F. 1991. Granulite facies rocks of the Wilson terrane (northern Victoria Land): Campbell Glacier. *Memorie della Società Geologica Italiana,* 46: 197–203.

Cathcart, J. B., and Schmidt, D. L. 1977. *Middle Paleozoic sedimentary phosphate in the Pensacola Mountains, Antarctica.* Professional Paper 456-E. Reston, VA: U.S. Geological Survey, pp. 1–17.

Chapman, F. 1916. Report on a probable calcareous alga from the Cambrian limestone breccia found in Antarctica at 85° S. *Reports of the Scientific Investigations, British Antarctic Expedition, 1907–09, Geology,* 2(4): 81–84.

Chappell, B. W., and White, A. J. R. 1974. Two contrasting granite types. *Pacific Geology,* 8: 173–74.

Collinson, J. W. 1991. The palaeo-Pacific margin as seen from East Antarctica. In M. R. A. Thomson, J. A. Crame, and J. W. Thomson, eds., *Geological evolution of Antarctica.* Cambridge University Press, pp. 199–204.

Compston, W., Williams, W. I., Kirschvink, J. L., Zichao, Z., and Guogan, M. 1992. Zircon U–Pb ages for the Early Cambrian time-scale. *Journal of the Geological Society, London,* 149: 171–84.

Coney, P. J., Jones, D. J., and Monger, J. W. H. 1980. Cordilleran suspect terranes. *Nature,* 288: 329–33.

Cook, Y. 1991. Deformation in the St. Johns Range, southern Victoria Land. *New Zealand Antarctic Record,* 11(1): 19–28.

Cooper, A. K., and Davey, F. J., eds. 1987. *The Antarctic continental margin: Geology and geophysics of the western Ross Sea.* Earth Science Series, vol. 5B. Houston: Circum-Pacific Council for Energy and Mineral Resources.

Cooper, J. A., Jenkins, R. J. F., Compston, W., and Williams, I. S. 1992. Ion-probe zircon dating of a mid-Early Cambrian tuff in South Australia. *Journal of the Geological Society, London,* 149: 185–92.

Cooper, R. A., Begg, J. G., and Bradshaw, J. D. 1990. Cambrian trilobites from Reilly Ridge, northern Victoria Land, Antarctica, and their stratigraphic implications. *New Zealand Journal of Geology and Geophysics,* 33: 55–66.

Cooper, R. A., Jago, J. B., MacKinnon, D. I., Shergold, J. H., and Vidal, G. 1982. Late Precambrian and Cambrian fossils from northern Victoria Land and their stratigraphic implications. In C. Craddock, ed., *Antarctic geoscience.* Madison: University of Wisconsin Press, pp. 629–33.

Cooper, R. A., Jago, J. B., MacKinnon, D. I., Simes, J. E., and Braddock, P. E. 1976. Cambrian fossils from the Bowers Group, northern Victoria Land, Antarctica (preliminary note). *New Zealand Journal of Geology and Geophysics,* 19: 283–88.

Cooper, R. A., Jago, J. B., Rowell, A. J., and Braddock, P. 1983. Age and correlation of the Cambro-Ordovician Bowers Supergroup, northern Victoria Land. In R. L. Oliver, P. R. James, and J. B. Jago, eds., *Antarctic earth science.* Canberra: Australian Academy of Science, pp. 128–31.

Cox, S. C. 1987. Origin of Olympus Granite-Gneiss. *New Zealand Antarctic Record,* 8(1): 42–47.

1992. Garnet–biotite geothermometry of Koettlitz Group metasediments, Wright Valley, south Victoria Land, Antarctica. *New Zealand Journal of Geology and Geophysics,* 35: 29–40.

Cox, S. C., and Allibone, A. H. 1991. Petrogenesis of orthogneisses in the Dry Valleys region, South Victoria Land. *Antarctic Science,* 3: 405–17.

Craddock, C., ed. 1969. *Geologic maps of Antarctica.* New York: American Geographical Society, folio 12.

1970a. Fossil map of Antarctica. American Geographical Map Folio Series, folio 12, plate XVIII.

1970b. Radiometric map of Antarctica. American Geographical Map Folio Series, folio 12, plate XIX.

1970c. Geologic map of Antarctica. American Geographical Map Folio Series, folio 12, plate XX.

1970d. Tectonic map of Antarctica. American Geographical Map Folio Series, folio 12, plate XXI.

Craddock, C., Gast, P. W., Hansen, G. N., and Linder, H. 1964. Rubidium–strontium ages from Antarctica. *Geological Society of America Bulletin,* 75: 237–40.

Craw, D., and Findlay, R. H. 1984. Hydrothermal alteration of Lower Ordovician granitoids and Devonian Beacon Sandstone at Taylor Glacier, McMurdo Sound, Antarctica. *New Zealand Journal of Geology and Geophysics,* 27: 465–75.

Crawford, A., Green, D. H., and Findlay, R. H. 1984. A preliminary petrographical-geochemical survey of dyke rocks from North Victoria Land, Antarctica. *Geologisches Jahrbuch,* B60: 153–65.

Crittenden, M. D., Jr., Coney, P. J., and Davis, G. H., eds. 1980. *Cordilleran metamorphic core complexes.* Memoir 153. Boulder, CO: Geological Society of America.

Crowder, D. F. 1968. *Geology of a part of northern Victoria Land, Antarctica.* Professional Paper 600-D. Reston, VA: United States Geological Survey, pp. 95–107.

Dahl, P. S., and Palmer, D. F. 1981. Field study of orbicular rocks in Taylor Valley, southern Victoria Land. *Antarctic Journal of the United States,* 16: 47–49.

   1983. The petrology and origin of orbicular tonolite from western Taylor Valley, southern Victoria Land, Antarctica. In R. L. Oliver, P. R. James, and J. B. Jago, eds., *Antarctic earth science.* Canberra: Australian Academy of Science, pp. 156–59.

Dallmeyer, R. D., and Wright, T. O. 1992. Diachronous cleavage development in the Robertson Bay terrane, northern Victoria Land, Antarctica: Tectonic implications. *Tectonics,* 11: 437–48.

Damaske, D., Kothe, J., and Dürnbaum, H.-J. 1989. The GANOVEX IV Expedition: Planning, implementation and logistics. *Geologisches Jahrbuch,* E38: 15–40.

Davey, F. J. 1981. Geophysical studies in the Ross Sea region. *Journal of the Royal Society of New Zealand,* 11: 465–79.

David, T. W. E., and Priestley, R. E. 1914. Glacialogy, physiography, stratigraphy, and tectonic geology of South Victoria Land, with short notes on paleontology by T. Griffith Taylor. *British Antarctic Expedition, 1907–9, Reports on the Scientific Investigations, Geology,* vol. 1. London: W. Heinemann.

David, T. W. E., Smeeth, W. F., and Schofield, J. A. 1896. Notes on Antarctic rocks collected by Mr C. E. Borchgrevink. *Proceedings of the Royal Society of New South Wales,* 29: 461–92.

Debenham, F. 1921. The sandstone, etc., of the McMurdo Sound, Terra Nova Bay, and Beardmore Glaicer regions. *British Antarctic ("Terra Nova") Expedition, 1910, Geology,* 1(4): 103–19.

Debrenne, F., and Kruse, P. D. 1986. Shackleton limestone archaeocyaths. *Alcheringa,* 10: 235–78.

   1989. Cambrian Antarctic archaeocyaths. In J. A. Crame, ed., *Origins and evolution of the Antarctic biota.* Special Publication no. 47. London: Geological Society, pp. 15–28.

Delisle, G., and Fromm, K. 1984. Paleomagnetic investigation of Ferrar Supergroup rocks, north Victoria Land, Antarctica. *Geologisches Jahrbuch,* B60: 41–55.

Deutsch, S., and Grögler, N. 1966. Isotopic age of Olympus Granite-Gneiss (Victoria Land, Antarctica). *Earth and Planetary Science Letters,* 1: 82–84.

Deutsch, S., and Webb, P. N. 1964. Sr/Rb dating on basement rocks from Victoria Land: Evidence for a 1,000 million-year-old event. In R. J. Adie, ed., *Antarctic geology.* Amsterdam: North-Holland, pp. 557–62.

Doumani, G. A., and Minshew, V. H. 1965. General geology of the Mount Weaver area, Queen Maud Mountains, Antarctica. In J. B. Hadley, ed., *Geology and paleontology of the Antarctic.* Antarctic Research Series, vol. 6. Washington DC: American Geophysical Union, pp. 127–39.

Dow, J. A. S., and Neall, V. E. 1972. A summary of the geology of the lower Rennick Glacier, Northern Victoria Land, Antarctica. In R. J. Adie, ed., *Antarctic geology and geophysics.* Oslo: Universitetsforlaget, pp. 339–44.

   1974. Geology of the lower Rennick Glacier, northern Victoria Land, Antarctica. *New Zealand Journal of Geology and Geophysics,* 17: 659–714.

Dürbaum, H.-J., Damaske, D., and Roland, N. W. 1989. Scientific program of the GANOVEX IV expedition. *Geologisches Jahrbuch,* E38: 7–14.

du Toit, A. L. 1937. *Our wandering continents.* Edinburgh: Oliver & Boyd.

Eastin, R. 1970. Geochronology of the basement rocks of the central Transantarctic Mountains, Antarctica. Ph.D. dissertation, Ohio State University.

Eastin, R., and Faure, G. 1972. *Geochronology of the basement rocks of the Pensacola Mountains, Antarctica.* Geological Society of America Abstracts with Programs, vol. 4, p. 496.

Edgerton, D. G. 1987. Kinematic orientations of Precambrian Beardmore and Cambro-Ordovician Ross orogenies, Goldie Formation, Nimrod Glacier area, Antarctica. M.S. thesis, Arizona State University.

Ellery, S. G. 1988. Geological observations from the lower Wright Valley, southern Victoria Land, Antarctica. *New Zealand Antarctic Record,* 8(3): 35–48.

Elliot, D. H. 1970. Beardmore Glacier investigations, 1969–70: Narrative and geological report. *Antarctic Journal of the United States,* 5: 83–85.

   1975. Tectonics of Antarctica: A review. *American Journal of Science,* 275A: 45–106.

Elliot, D. H., Barrett, P. J., and Mayewski, P. A. 1974. Reconnaissance geological map of the Plunket Point quadrangle, Transantarctic Mountains, Antarctica. Map A-4. Washington DC: U.S. Geological Survey, United States Antarctic Research Program.

Elliot, D. H., and Coats, D. A. 1971. Geological investigations in the Queen Maud Mountains. *Antarctic Journal of the United States,* 6: 114–18.

Elliot, D. H., and Foland, K. A. 1986. Potassium–argon age determinations of the Kirkpatrick Basalt, Mesa Range. In E. Stump, ed., *Geological investigations in northern Victoria Land.* Antarctic Research Series, vol. 46. Washington, DC: American Geophysical Union, pp. 279–88.

Engel, S. 1984. Petrogenesis of contact schists in the Morozumi Range, north Victoria Land. *Geologisches Jahrbuch,* B60: 167–86.

— 1987. Contact metamorphism by the layered gabbro at Spatulate Ridge and Apostrophe Island, north Victoria Land, Antarctica. *Geologisches Jahrbuch,* B66: 275–301.

Evans, K. R., and Rowell, A. J. 1990. Small shelly fossils from Antarctica: An Early Cambrian faunal connection with Australia. *Journal of Paleontology,* 64: 692–700.

Evans, K. R., Rowell, A. J., and Rees, M. N. 1991. Sea-level fluctuations and the evolution of a Middle Cambrian carbonate ramp in the Neptune Range. *Antarctic Journal of the United States,* 26: 47–49.

Faure, G. 1986. *Principles of isotopic geology,* 2d ed. New York: Wiley.

Faure, G., Eastin, R., Ray, P. T., McLelland, D., and Shultz, C. H. 1979. Geochronology of igneous and metamorphic rocks, central Transantarctic Mountains. In B. Laskar and C. S. Raja Rao, eds., *Fourth International Gondwana Symposium: Papers.* Calcutta: Hindustan Publishing, pp. 805–13.

Faure, G., and Felder, R. P. 1984. Lithium-bearing pegmatite and bismuth–antimony–lead–copper-bearing veinlets on Mt. Madison, Byrd Glacier area. *Antarctic Journal of the United States,* 19: 13–14.

Faure, G., and Gair, H. S. 1970. Age determinations of rocks from northern Victoria Land, Antarctica. *New Zealand Journal of Geology and Geophysics,* 13: 1024–26.

Faure, G., and Jones, L. M. 1974. Isotopic composition of strontium and geologic history of the basement rocks of Wright Valley, southern Victoria Land, Antarctica. *New Zealand Journal of Geology and Geophysics,* 17: 611–27.

Faure, G., Murtaugh, J. G., and Montigny, R. J. E. 1968. The geology and geochronology of the basement complex of the central Transantarctic Mountains. *Canadian Journal of Earth Sciences,* 5: 555–60.

Felder, R. P., and Faure, G. 1979. Investigation of an anomalous date for the Lonely Ridge granodiorite, Nilsen Plateau, Transantarctic Mountains. *Antarctic Journal of the United States,* 14: 24.

— 1980. Rubidium–strontium age determination of part of the basement complex of the Brown Hills, central Transantarctic Mountains. *Antarctic Journal of the United States,* 15: 16–17.

— 1990. Age and petrogenesis of the granitic basement rocks, Brown Hills, Transantarctic Mountains. *Zentralblatt für Geologie und Paläontologie,* 1: 45–62.

Fenn, G., and Henjes-Kunst, F. 1992. Field relations of Granite Harbour Intrusives and associated dikes from the USARP Mountains, north Victoria Land, and Prince Albert Mountains, central Victoria Land, Antarctica. *Polarforschung,* 60: 110–12.

Ferrar, H. T. 1907. Report on the field geology of the region explored during the "Discovery" Antarctic Expedition, 1901–4. In *British Antarctic Expedition, 1901–1904, Natural History,* vol. 1: *Geology.* London: Museum of Natural History, pp. 1–100.

Field, B. D., and Findlay, R. H. 1982. Preliminary report on the sedimentology of the Robertson Bay Group, north Victoria Land. *New Zealand Antarctic Record,* 4(2): 20–23.

— 1983. Sedimentology of the Robertson Bay Group, northern Victoria Land. In R. L. Oliver, P. R. James, and J. B. Jago, eds., *Antarctic earth science.* Canberra: Australian Academy of Science, pp. 102–6.

Findlay, R. H. 1978. Provisional report on the geology of the region between the Renegar and Blue Glaciers, Antarctica. *New Zealand Antarctic Record,* 1(1): 39–44.

— 1985. The Granite Harbour Intrusive Complex in McMurdo Sound: Progress and problems. *New Zealand Antarctic Record,* 6(3): 10–22.

— 1986. Structural geology of the Robertson Bay and Millen terranes, northern Victoria Land, Antarctica. In E. Stump, ed., *Geological investigations in northern Victoria Land.* Antarctic Research Series, vol. 46. Washington, DC: American Geophysical Union, pp. 91–114.

1987. A review of the problems important for interpretation of the Cambro-Ordovician paleoge-ography of northern Victoria Land (Antarctica), Tasmania, and New Zealand. In G. D. McKenzie, ed., *Gondwana Six: Structure, tectonics and geophysics.* Geophysical Monograph, vol. 40. Washington, DC: American Geophysical Union, pp. 49–66.

1990a. Silurian and Devonian events in the Tasman orogenic zone, New Zealand and Marie Byrd Land and their comparison with northern Victoria Land. *Memorie della Società Geologica Italiana,* 43: 9–32.

1990b. Is the Larsen granodiorite of the Transantarctic Mountains Vendian? A discussion. *New Zealand Antarctic Record,* 10(3): 32–38.

Findlay, R. H. 1992. The age of cleavage development in the Ross orogen, northern Victoria Land, Antarctica: Evidence from $^{40}Ar/^{39}Ar$ whole-rock slate ages – Discussion. *Journal of Structural Geology,* 14: 887–90.

Findlay, R. H., Brown, A. V., and McClenaghan, M. P. 1991. Confirmation of the correlation between Lower Palaeozoic rocks in western Tasmania and northern Victoria Land, Antarctica and a revised tectonic interpretation. *Memorie della Società Geologica Italiana,* 46: 117–33.

Findlay, R. H., and Field, B. D. 1982a. Black Prince volcanics, Admiralty Mountains, Antarctica: Reconnaissance observations. *New Zealand Antarctic Record,* 4(2): 11–14.

1982b. Preliminary report on the structural geology of the Robertson Bay Group, north Victoria Land, Antarctica. *New Zealand Antarctic Record,* 4(2): 15–19.

1983. Tectonic significance of deformations affecting the Robertson Bay Group and associated rocks, northern Victoria Land, Antarctica. In R. L. Oliver, P. R. James, and J. B. Jago, eds., *Antarctic earth science.* Canberra: Australian Academy of Science, pp. 107–12.

Findlay, R. H., and Jordan, H. 1984. The volcanic rocks of Mt. Black Prince and Lawrence Peaks, north Victoria Land, Antarctica. *Geologisches Jahrbuch,* B60: 143–51.

Findlay, R. H., Skinner, D. N. B., and Craw, D. 1984. Lithostratigraphy and structure of the Koettlitz Group, McMurdo Sound, Antarctica. *New Zealand Journal of Geology and Geophysics,* 27: 513–36.

Fitzgerald, P. G. 1992. The Transantarctic Mountains of southern Victoria Land: The application of apatite fission track analysis to a rift shoulder uplift. *Tectonics,* 11: 634–62.

Fitzgerald, P. G., and Gleadow, A. J. W. 1988. Fission track geochronology, tectonics and structure of the Transantarctic Mountains in northern Victoria Land, Antarctica. *Isotope Geoscience,* 73: 169–98.

Fitzgerald, P. G., Sandiford, M., Barrett, P. J., and Gleadow, A. J. W. 1986. Asymmetric extension associated with uplift and subsidence of the Transantarctic Mountains and Ross embayment. *Earth and Planetary Science Letters,* 81: 67–78.

Flory, R. F., Murphy, D. J., Smithson, S. B., and Houston, R. S. 1971. Geologic studies of basement rocks in southern Victoria Land. *Antarctic Journal of the United States,* 6: 119–20.

Flöttmann, T., Gibson, G. M., and Kleinschmidt, G. 1993a. Structural continuity of the Ross and Delamerian orogens of Antarctica and Australia along the margin of the paleo-Pacific. *Geology,* 21: 319–22.

Flöttmann, T., and Kleinschmidt, G. 1991a. Opposite thrust systems in northern Victoria Land, Antarctica. Imprints of Gondwana's Paleozoic accretion. *Geology,* 19: 45–47.

1991b. Kinematics of major structures of North Victoria and Oates Lands, Antarctica. *Memorie della Società Geologica Italiana,* 46: 273–82.

1993. The structure of Oates Land and implications for the structural style of northern Victoria Land, Antarctica. *Geologisches Jahrbuch,* E47: 419–36.

Flöttmann, T., Kleinschmidt, G., and Funk, T. 1993b. Thrust patterns of the Ross/Delamerian orogens in northern Victoria Land (Antarctica) and southeastern Australia and their implications for Gondwana reconstructions. In R. F. Findlay, R. Unrug, M. R. Banks, and J. J. Veevers, eds., *Gondwana Eight.* Rotterdam: Balkema, pp. 131–39.

Ford, A. B. 1964a. Cordierite-bearing, hypersthene–quartz monzonite porphyry and its regional importance. In R. J. Adie, ed., *Antarctic geology.* Amsterdam: North-Holland, 429–41.

Ford, A. B. 1964b. Review of Antarctic geology. *Transactions of the American Geophysical Union,* 45: 363–81.

1972. The Weddell orogeny: Latest Permian to early Mesozoic deformation at the Weddell Sea margin of the Transantarctic Mountains. In R. J. Adie, ed., *Antarctic geology and geophysics.* Oslo: Universitetsforlaget, pp. 419–25.

Ford, A. B., and Aaron, J. M. 1962. Bedrock geology of the Thiel Mountains, Antarctica. *Science,* 137: 751–52.

Ford, A. B., and Himmelberg, G. R. 1976. Cordierite and orthopyroxene megacrysts in late Precambrian volcanic rocks of the Thiel Mountains. *Antarctic Journal of the United States,* 11: 260–63.

Ford, A. B., Hubbard, H. A., and Stern, T. W. 1963. Lead-alpha ages of zircon in quartz monzonite porphyry, Thiel Mountains, Antarctica: A preliminary report. Professional Paper 450-E. Reston, VA: United States Geological Survey, pp. 105–7.

Ford, A. B., Schmidt, D. L., and Boyd, W. W., Jr. 1978a. Geologic map of the Davis Valley quadrangle and part of the Cordiner Peaks quadrangle, Pensacola Mountains, Antarctica. United States Antarctic Research Program Map A-10, 1:250,000. Reston, VA: U.S. Geological Survey.

Ford, A. B., Schmidt, D. L., Boyd, W. W., Jr., and Nelson, W. H. 1978b. Geologic map of the Saratoga Table quadrangle, Pensacola Mountains, Antarctica. Map A-9, 1:250,000. Reston, VA: U.S. Geological Survey, United States Antarctic Research Program.

Ford, A. B., and Sumsion, R. S. 1971. Late Precambrian silicic pyroclastic volcanism in the Thiel Mountains. *Antarctic Journal of the United States,* 6: 185–86.

Frezzotti, M. L., and Talarico, F. 1990. Preliminary investigations on fluid inclusions in orthopyroxene–garnet–cordierite granulites from Boomerang Glacier (Deep Freeze Range, North Victoria Land, Antarctica): Evidence of mixed $CO_2$–$CH_4$ fluids. *Memorie della Società Geologica Italiana,* 43: 59–66.

Frischbutter, A. 1981. Gliederung, bau und entwicklung des Transantarktischen Gelurges in bereich der Neptune Range (Antarktis). part 1: Stratigraphie und regional einbindung. *Zeitschrift für Geologische Wissenschaften,* 9: 817–33.

1982a. Gliederung, bau und entwicklung des Transantarktischen Gelurges in bereich der Neptune Range (Antarktis). Part 2: Deformation, metamorphose und gesantentwickling. *Zeitschrift für Geologische Wissenschaften,* 10: 165–80.

1982b. Lithostratigraphische korrelation proterozoisch-paläozoischer strukturzonen zwischen Australien und Antarktis als teilen Gondwanas. *Zeitschrift für Geologische Wissenschaften,* 10: 421–33.

1985. Rasterelktronenmikroskopische untersuchungen zum problem der schieferungsentwicklung in gefalteten schiefern und grauwacken (Präkambrium) der Neptune Range (Antarktis). *Zeitschrift für Geologische Wissenschaften,* 13: 427–42.

Frischbutter, A., and Vogler, P. 1985. Contributions to the geochemistry of magmatic rocks in the upper Precambrian lower Paleozoic profile of the Neptune Range, Transantarctic Mountains, Antarctica. *Zeitschrift für Geologische Wissenschaften,* 13: 345–57.

Gair, H. S. 1964. Geology of upper Rennick, Campbell and Aviator Glaciers, northern Victoria Land. In R. J. Adie, ed., *Antarctic geology.* Amsterdam: North-Holland, 188–98.

1967. The geology from the upper Rennick Glacier to the coast, northern Victoria Land, Antarctica. *New Zealand Journal of Geology and Geophysics,* 10: 309–44.

Gair, H. S., Sturm, A., Carryer, S. J., and Grindley, G. W. 1969. The geology of northern Victoria Land. Antarctic Map Folio Series, Folio 12, XII. Washington, DC: American Geographical Society.

GANOVEX Team. 1987. Geological map of north Victoria Land, Antarctica, 1:500 000 – explanatory notes. *Geologisches Jahrbuch,* B66: 7–79.

GANOVEX-ITALIANTARTIDE. 1991. Preliminary geological-structural map of Wilson, Bowers, and Robertson Bay terranes in the area between Aviator and Tucker Glaciers (northern Victoria Land – Antarctica). *Memorie della Società Geologica Italiana,* 46: 267–72.

Geikie, A. 1897–99. Notes on some specimens of rocks from Antarctic regions with petrographical notes by J. J. H. Teall, F. R. S. (Rocks collected by Capt. Robertson of the "Active" at Dundee Island, 1893, and by Mr. Borchgrevink in 1895 at Cape Adare.) *Proceedings of the Royal Society of Edinburgh,* 22: 66–70.

Ghent, E. D. 1970. Chemistry and mineralogy of the Mt. Falconer pluton and associated rocks, lower Taylor Valley, south Victoria Land, Antarctica. *Transactions of the Royal Society of New Zealand, Earth Sciences,* 8: 117–32.

Ghent, E. D., and Henderson, R. A. 1968. Geology of the Mt. Falconer pluton, lower Taylor Valley, south Victoria Land, Antarctica. *New Zealand Journal of Geology and Geophysics,* 11: 851–80.

Ghezzo, C., Baldelli, C., Biagini, R., Carmignani, L., Di Vincenzo, G., Gosso, G., Lelli, A., Lombardo, B., Montrasio, A., Pertusati, P. C., and Salvini, F. 1989. Granitoids from the David Glacier–Aviator Glacier segment of the Transantarctic Mountains, Victoria Land, Antarctica. *Memorie Società Geologica Italiana,* 33: 143–59.

Gibson, G. M. 1984. Deformed conglomerate in the eastern Lanterman Range, north Victoria Land, Antarctica. *Geologisches Jahrbuch,* B60: 117–41.

1985. Lanterman fault: Boundary between allochthonous terranes, northern Victoria Land, Antarctica. *Geological Society of Australia Abstracts,* 14: 79–82.

1987. Metamorphism and deformation in the Bowers Supergroup: Implications for terrane accretion in northern Victoria Land. In E. C. Leitch and E. Scheibner, eds., *Terrane accretion and orogenic belts.* Geodynamics Series, vol. 19. Washington, DC: American Geophysical Union, pp. 207–19.

Gibson, G. M., Tessensohn, F., and Crawford, A. 1984. Bowers Supergroup rocks west of the Mariner Glacier and possible greenschist facies equivalents. *Geologisches Jahrbuch,* B60: 289–318.

Gibson, G. M., and Wright, T. O. 1985. The importance of thrust faulting in the tectonic development of northern Victoria Land, Antarctica. *Nature,* 315: 480–83.

Gleadow, A. J. W., and Fitzgerald, P. G. 1987. Uplift history and structure of the Transantarctic Mountains: New evidence from fission track dating of basement apatites in the Dry Valleys area, southern Victoria Land. *Earth and Planetary Science Letters,* 82: 1–14.

Goldich, S. S., Nier, A. O., and Washburn, A. L. 1958. A$^{40}$/K$^{40}$ age of gneiss from McMurdo Sound, Antarctica. *Transactions of the American Geophysical Union,* 39: 956–58.

Goodge, J. W., Borg, S. G., Smith, B. K., and Bennett, V. C. 1991a. Tectonic significance of Proterozoic ductile shortening and translation along the Antarctic margin of Gondwana. *Earth and Planetary Science Letters,* 102: 58–70.

Goodge, J. W., and Dallmeyer, R. D. 1992. $^{40}$Ar/$^{39}$Ar mineral age constraints on the Paleozoic tectonothermal evolution of high-grade basement rocks within the Ross orogen, central Transantarctic Mountains. *Journal of Geology,* 100: 91–106.

Goodge, J. W., Hansen, V. L., and Peacock, S. M. 1993a. Multiple petrotectonic events in high-grade metamorphic rocks of the Nimrod Group, central Transantarctic Mountains, Antarctica. In Y. Yoshida, K. Kaminuma, and K. Shiraishi, eds., *Recent progress in Antarctic earth science.* Tokyo: Terra Scientific, pp. 203–9.

Goodge, J. W., Hansen, V. L., Peacock, S. M., and Smith, B. K. 1990. Metamorphic rocks in the Geologists and Miller Ranges, Nimrod Glacier area, central Transantarctic Mountains. *Antarctic Journal of the United States,* 25: 35–36.

Goodge, J. W., Hansen, V. L., Peacock, S. M., Smith, B. K., and Walker, N. W. 1993b. Kinematic evolution of the Miller Range shear zone, central Transantarctic Mountains, Antarctica, and implications for Neoproterozoic to Early Paleozoic tectonics of the East Antarctic margin of Gondwana. *Tectonics,* 12: 1460–78.

Goodge, J. W., Hansen, V. L., and Walker, N. W. 1991b. Geologic relations of the upper Nimrod Glacier region, central Transantarctic Mountains: Evidence of multiple orogenic history. *Antarctic Journal of the United States,* 26: 4–6.

Goodge, J. W., Walker, N. W., and Hansen, V. L. 1993c. Neoproterozoic–Cambrian basement-involved orogenesis within the Antarctic margin of Gondwana. *Geology,* 21: 37–40.

Gordon, W. T. 1920. Scottish National Antarctic Expedition, 1902–04: Cambrian organic remains from a dredging in the Weddell Sea. *Transactions of the Royal Society of Edinburgh,* 52(4): 681–714.

Gould, L. M. 1931. Some geographical results of the Byrd Antarctic Expedition. *Geographical Review,* 21: 177–200.

1935. Structure of the Queen Maud Mountains. *Geological Society of America Bulletin,* 46: 937–84.

Graham, I. J., and Palmer, K. 1987. New precise Rb–Sr mineral and whole-rock dates for I-type granitoids from Granite Harbour, South Victoria Land, Antarctica. *New Zealand Antarctic Record,* 8(1): 72–80.

Grant, S. W. F. 1990. Shell structure and distribution of *Cloudina,* a potential index fossil for the terminal Proterozoic. *American Journal of Science,* 290-A: 261–94.

Grew, E. S., Kleinschmidt, G., and Schubert, W. 1984. Contrasting metamorphic belts in North Victoria Land, Antarctica. *Geologisches Jahrbuch,* B60: 253–63.

Grew, E. S., and Sandiford, M. 1982. Field studies of the Wilson and Rennick Groups, Rennick Glacier area, northern Victoria Land. *Antarctic Journal of the United States,* 17: 7–8.

1984. A staurolite–talc assemblage in tourmaline–phlogopite–chlorite schist from northern Victoria Land, Antarctica, and its petrogenetic significance. *Contributions to Mineralogy and Petrology,* 87: 337–50.

1985. Staurolite in a garnet–hornblende–biotite schist from the Lanterman Range, northern Victoria Land, Antarctica. *Neue Jahrbuecher fuer Mineralogie, Monatshefte,* 9: 396–410.

Grikurov, G. E. 1972. Tectonics of the Antarctandes. In R. J. Adie ed., *Antarctic geology and geophysics.* Oslo: Universitetsforlaget, pp. 163–67.

Grikurov, G. E., Ravich, M. G., and Soloviev. 1972. Tectonics of Antarctica (review). In R. J. Adie ed., *Antarctic geology and geophysics.* Oslo: Universitetsforlaget, pp. 457–68.

Grindley, G. W. 1963. The geology of the Queen Alexandra Range, Beardmore Glacier, Ross Dependency, Antarctica, with notes on the correlation of Gondwana sequences. *New Zealand Journal of Geology and Geophysics,* 6(3): 307–47.

1967. The geomorphology of the Miller Range, central Transantarctic Mountains, with notes on the glacial history and neotectonics of East Antarctica. *New Zealand Journal of Geology and Geophysics,* 10: 557–98.

1972. Polyphase deformation of the Precambrian Nimrod Group, central Transantarctic Mountains. In R. J. Adie, ed., *Antarctic geology and geophysics.* Oslo: Universitetsforlaget, pp. 313–18.

1981. Precambrian rocks of the Ross Sea region. *Transactions of the Royal Society of New Zealand,* 11: 411–23.

Grindley, G. W., and Laird, M. G. 1969. Geology of the Shackleton Coast. Antarctic Map Folio Series, Folio 12, XIV. New York: American Geographical Society.

Grindley, G. W., and McDougall, I. 1969. Age and correlation of the Nimrod Group and other Precambrian Rock Units in the Central Transantarctic Mountains, Antarctica. *New Zealand Journal of Geology and Geophysics,* 12: 391–411.

Grindley, G. W., McGregor, V. R., and Walcott, R. I. 1964. Outline of the geology of the Nimrod–Beardmore–Axel Heiberg Glaciers region, Ross Dependency. In R. J. Adie, ed., *Antarctic geology.* Amsterdam: North-Holland, pp. 206–19.

Grindley, G. W., and Oliver, R. L. 1983. Post-Ross orogeny cratonization of northern Victoria Land. In R. L. Oliver, R. L. James, and J. B. Jago, eds., *Antarctic earth science.* Canberra: Australian Academy of Earth Science, pp. 133–39.

Grindley, G. W., and Warren, G. 1964. Stratigraphic nomenclature and correlation in the western Ross Sea region. In R. J. Adie, ed., *Antarctic geology.* Amsterdam: North-Holland, pp. 314–33.

Gunn, B. M. 1963. Geological structure and stratigraphic correlation in Antarctica. *New Zealand Journal of Geology and Geophysics,* 6: 423–33.

Gunn, B. M., and Walcott, R. I. 1962. The geology of the Mt. Markham Region, Ross Dependency, Antarctica. *New Zealand Journal of Geology and Geophysics,* 5(3): 407–26.

Gunn, B. M., and Warren, G. 1962. *Geology of Victoria Land between Mawson and Mullock Glaciers, Antarctica.* Bulletin 71. New Zealand Geological Survey.

Gunner, J. D. 1969. *Petrography of metamorphic rocks from the Miller Range, Antarctica.* Institute of Polar Studies Report no. 32. Columbus: Ohio State University.

1971. Ida Granite: A new formation of the Granite Harbour Intrusives, Beardmore Glacier region. *Antarctic Journal of the United States,* 6: 194–96.

1976. *Isotopic and geochemical studies of the pre-Devonian basement complex, Beardmore Glacier region, Antarctica.* Institute of Polar Studies Report 41. Columbus: Ohio State University.

1982. Basement geology of the Beardmore Glacier region. In M. D. Turner and J. F. Splettstoesser, eds., *Geology of the central Transantarctic Mountains.* Antarctic Research Series, vol. 26. Washington, DC: American Geophysical Union, pp. 1–9.

Gunner, J. D., and Faure, G. 1972. Rubidium–strontium geochronology of the Nimrod Group, central Transantarctic Mountains. In R. J. Adie, ed., *Antarctic geology and geophysics.* Oslo: Universitetsforlaget, pp. 305–11.

Gunner, J. D., and Mattinson, J. M. 1975. Rb–Sr and U–Pb isotopic ages of granites in the central Transantarctic Mountains. *Geology Magazine,* 112: 25–31.

Haller, J. 1971. *Geology of the East Greenland Caledonides.* New York: Wiley-Interscience.

Hambrey, M. J., and Harland, W. B. 1985. The Late Proterozoic glacial era. *Palaeogeography, Palaeoclimatology, Palaeoecology,* 57: 255–72.

Hamilton, W. B. 1961. *Petrochemistry of probable Paleozoic granitic rocks from the Ross Sea region, Antarctica.* Professional Paper 424-C. Reston, VA: U.S. Geological Survey, pp. C209–C212.

Hamilton, W. B. 1964. Tectonic map of Antarctica: A progress report. In R. J. Adie, ed., *Antarctic geology.* Amsterdam: North-Holland, pp. 676–79.

Hamilton, W. B. 1967. Tectonics of Antarctica. *Tectonophysics,* 4: 555–68.

Hamilton, W. B., and Hayes, P. T. 1960. Geology of Taylor Glacier: Taylor Dry Valley region, South Victoria Land, Antarctica. United States Geological Survey 400-B, pp. B376–B378.

Harland, W. B., Cox, A. V., Llewellyn, P. G., Pickton, C. A. G., Smith, A. G., and Walters, R. 1982. *A geologic time scale.* Cambridge University Press.

Harrington, H. J. 1958. Nomenclature of rock units in the Ross Sea region, Antarctica. *Nature,* 182: 290.

Harrington, H. J. 1965. Geology and morphology of Antarctica. In J. van Mieghem and P. van Oye, eds., *Biogeography and ecology in Antarctica.* Monographiae Biologicae, 15: 1–71.

Harrington, H. J., Wood, B. L., McKellar, I. C., and Lensen, G. J. 1964. The geology of Cape Hallet–Tucker Glacier district. In R. J. Adie, ed., *Antarctic geology.* Amsterdam: North-Holland, pp. 220–28.

1967. *Topography and geology of the Cape Hallett district, Victoria Land, Antarctica.* Bulletin 80. Lower Hutt: New Zealand Geological Survey.

Haskell, T. R., Kennett, J. P., and Prebble, W. M. 1964. Basement and sedimentary geology of the Darwin Glacier area. In R. J. Adie, ed., *Antarctic geology.* Amsterdam: North-Holland, pp. 348–51.

1965a. Geology of the Brown Hills and Darwin Mountains, southern Victoria Land, Antarctica. *Transactions of the Royal Society of New Zealand, Geology,* 2: 231–48.

Haskell, T. R., Kennett, J. P., Prebble, W. M., Smith, G., and Willis, I. A. G. 1965b. The geology of the middle and lower Taylor Valley of South Victoria Land, Antarctica. *Transactions of the Royal Society of New Zealand, Geology,* 2: 169–86.

Heintz, G. M. 1980. Structural geology of the Leverett Glacier area, Antarctica. M.S. thesis, Arizona State University.

Henjes-Kunst, F. 1992. A sampling scheme for geochronological investigations in Devonian–Carboniferous Admiralty Intrusives and their country rocks in north Victoria Land, Antarctica. *Polarforshung,* 60: 113–16.

Herbert, W. W. 1962. The Axel Heiberg Glacier. *New Zealand Journal of Geology and Geophysics,* 5: 681–706.

Hill, D. 1964a. Archaeocyatha from loose material at Plunket Point at the head of Beardmore Glacier (with an appendix by R. L. Oliver). In R. J. Adie, ed., *Antarctic geology.* Amsterdam: North-Holland, pp. 609–22.

1964b. Archaeocyatha from the Shackleton Limestone of the Ross system, Nimrod Glacier area, Antarctica. *Transactions of the Royal Society of New Zealand,* 2: 137–46.

1965. Archaeocyatha from Antarctica and a review of the phylum. Trans-Antarctic Expedition Science Report no. 10, Geology (3).

Howell, D. G., ed. 1985. *Tectonostratigraphic terranes of the circum-Pacific region.* Earth Science Series, vol. 1. Houston: Circum-Pacific Council for Energy and Mineral Resources.

Hofmann, J., and Samsonov, V. V. 1982. Tektonische untersuchungen in der Patuxent Formation der Schmidt Hills (Pensacola Mts., Antarktika). *Freiberger Forschungsheft,* 371: 97–117.

Huffman, J. W., and Schmidt, D. L. 1966. Pensacola Mountains project. *Antarctic Journal of the United States,* 1: 123–24.

Hulston, J. R., and McCabe, W. J. 1972. New Zealand potassium–argon age list: 1. *New Zealand Journal of Geology and Geophysics,* 15: 406–32.

Iltchenko, L. N. 1972. Late Precambrian Acritarcha of Antarctica. In R. J. Adie, ed., *Antarctic geology and geophysics.* Oslo: Universitetsforlaget, pp. 599–602.

Jago, J. B. 1981. Late Precambrian–Early Paleozoic relationships between Tasmania and northern Victoria Land. In M. M. Cresswell and P. Vella, eds., *Gondwana Five.* Rotterdam: Balkema, pp. 199–204.

Jell, J. S. 1981. Silurian tabulate corals from the Cloudmaker Moraine, Beardmore Glacier, Antarctica. *Alcheringa,* 5: 311–16.

Jones, L. M., and Faure, G. 1967. Age of the Vanda porphyry dikes in Wright Valley, southern Victoria Land, Antarctica. *Earth and Planetary Science Letters*, 3: 321–24.

1969. Age of the basement complex of Wright Valley, Antarctica. *Antarctic Journal of the United States*, 4: 204–5.

Jordan, H. 1981. Metasediments and volcanics in the Frolov Ridge, Bowers Mountains, Antarctica. *Geologisches Jahrbuch*, B41: 139–54.

Jordan, H., Findlay, R., Mortimer, G., Schmidt-Thomé, M., Crawford, A., and Müller, P. 1984. Geology of the northern Bowers Mountains, North Victoria Land, Antarctica. *Geologisches Jahrbuch*, B60: 57–81.

Kaiser, G., Werner, K., and Weber, W. 1982. K/Ar – Attersdatierungen an magmatiten der Pensacola Mountains (Antarktika). *Zeitschrift für Geologische Wissenschaften*, 10: 527–30.

Kamenev, E. N., and Kumeev, S. S. 1982. Subdivision of granitoids of southern Victoria Land by means of feldspar structural diffractometry. In C. Craddock, ed., *Antarctic geoscience*. Madison: University of Wisconsin Press, pp. 703–8.

Katz, H. R., and Waterhouse, B. C. 1970a. Geological reconnaissance of the Scott Glacier area, south-eastern Queen Maud Range, Antarctica. *New Zealand Journal of Geology and Geophysics*, 13(4): 1030–37.

1970b. The geologic situation at O'Brien Peak, Queen Maud range, Antarctica. *New Zealand Journal of Geology and Geophysics*, 13(4): 1038–47.

Keiller, I. G. 1988. Comments and observations on dyke intrusives of the Wright Valley. *New Zealand Antarctic Record*, 8(3): 25–34.

Klee, S., Baumann, A., and Thiedig, F. 1992. Age relations of high-grade metamorphic rocks in the Terra Nova Bay area, north Victoria Land, Antarctica: A preliminary report. *Polarforschung*, 60: 101–6.

Kleinschmidt, G. 1981. Regional metamorphism in the Robertson Bay Group area and in the southern Daniels Range, North Victoria Land, Antarctica: A preliminary comparison. *Geologisches Jahrbuch*, B41: 201–28.

1983. Trends in regional metamorphism and deformation in northern Victoria Land, Antarctica. In R. L. Oliver, P. R. James, and J. B. Jago, eds., *Antarctic earth science*. Canberra: Australian Academy of Science, pp. 119–22.

1992a. The southern continuation of the Wilson thrust. *Polarforshung*, 60: 124–27.

1992b. Structural observations in the Robertson Bay terrane, and their implications. *Polarforschung*, 60: 128–32.

Kleinschmidt, G., Buggish, W., and Flöttmann, T. 1992a. Compressional causes for the early Paleozoic Ross orogen – Evidence from Victoria Land and the Shackleton Range. In Y. Yoshida, K. Kaminuma, and K. Shiraishi, eds., *Recent progress in Antarctic earth science*. Tokyo: Terra Scientific, pp. 227–33.

Kleinschmidt, G., and Matzer, S. 1992. Structural field observations in the basement between Fry and Reeves Glaciers, Victoria Land, Antarctica. *Polarforschung*, 60: 107–9.

Kleinschmidt, G., Matzer, S., Henjes-Kunst, F., and Fenn, G. 1992b. New field data from Surgeon Island, north Victoria Land, Antarctica. *Polarforschung*, 60: 133–34.

Kleinschmidt, G., Mazzoli, C., and Sassi, F. P. 1991. The pressure character of the low-grade metapelites from Robertson Bay terrane and Bowers terrane, northern Victoria Land (Antarctica). *Memorie della Società Geologica Italiana*, 46: 283–89.

Kleinschmidt, G., Roland, N. W., and Schubert, W. 1984. The metamorphic basement complex in the Mountaineer Range, North Victoria Land, Antarctica. *Geologisches Jahrbuch*, B60: 213–51.

Kleinschmidt, G., Schubert, W., Olesch, M., and Rettmann, E. S. 1987. Ultramafic rocks of the Lanterman Range in North Victoria Land, Antarctica: Petrology, geochemistry, and geodynamic implications. *Geologisches Jahrbuch*, B66: 231–73.

Kleinschmidt, G., and Skinner, D. N. B. 1981. Deformation styles in the basement rocks of North Victoria Land, Antarctica. *Geologisches Jahrbuch*, B41: 155–99.

Kleinschmidt, G., and Tessensohn, F. 1987. Early Paleozoic westward subduction at the Pacific margin of Antarctica. In G. D. McKenzie ed., *Gondwana Six: Structure, tectonics and geophysics*. Geophysical Monograph, vol. 40. Washington, DC: American Geophysical Union, pp. 89–105.

Klimov, L. V., Ravich, M. G., and Soloviev, D. S. 1964. Geology of the Antarctic platform. In R. J. Adie, ed., *Antarctic geology*. Amsterdam: North-Holland, pp. 681–91.

Klimov, L. V., and Soloviev, D. S. 1958a. Some features of the geological structure of Wilkes Land, King George V Land and Oates Coast (East Antarctic). *Doklady of the Academy of Sciences of the USSR, Geological Sciences Section,* 123(1–6): 889–92.

1958b. Preliminary report on geological observations in East Antarctica [in Russian]. *Information Bulletin of Soviet Antarctic Expedition,* 1: 27–30. [In English: *Soviet Antarctic Expedition Information Bulletin,* vol. 1. Amsterdam: Elsevier, 1964, pp. 15–18.]

1960. Correlation of the geological formations of the Ross Sea region and Oates Coast [in Russian]. *Information Bulletin of Soviet Antarctic Expedition,* 16: 7–10. [In English: *Soviet Antarctic Expedition Information Bulletin,* vol. 2, Amsterdam: Elsevier, 1964, pp. 171–74.]

Klobcar, C. L., and Holloway, J. R. 1986. Petrography and geochemistry of igneous and metamorphic rocks from the Emlen Peaks and Robertson Bay–Everett Range area, northern Victoria Land, Antarctica. In E. Stump, ed., *Geological investigations in northern Victoria Land.* Antarctic Research Series, vol. 46. Washington, DC: American Geophysical Union, pp. 189–202.

Konyuschkov, K. N., and Shulyatin, O. G. 1980. Ob arkheotsiatakh Antarkidy i ikh sopostavlenii s arkheotsintami Sibiri [On the archaeocyathids of Antarctica and their comparison with archaeocyathids of Siberia]. In I. T. Zhuravleva, ed., *Kembrii Altae-Sayanskoi Skladchatoi Oblasti [Cambrian of the Altay-Sayan folded region].* Akademiya Nauk SSSR, Sibirskoe Otdelenie, Instituta Geologii i Geofiziki, pp. 143–50.

Kothe, J. 1984. The expedition and its logistics. *Geologisches Jahrbuch,* B60: 9–30.

Kothe, J., Tessensohn, F., Thonhauser, W., and Wendebourg, R. 1981. The expedition and its logistics. *Geologisches Jahrbuch,* B41: 3–30.

Kreuzer, H., Delisle, G., Fromm, K., Höhndorf, A., Lenz, H., Müller, P., and Vetter, U. 1987. Radiometric and paleomagnetic results from northern Victoria Land, Antarctica. In G. D. McKenzie, ed., *Gondwana Six: Structure, tectonics, and geophysics.* Geophysical Monograph, vol. 40. Washington, DC: American Geophysical Union, pp. 31–47.

Kreuzer, H., Höhndorf, A., Lenz, H., Vetter, U., Tessensohn, F., Müller, P., Jordon, H., Haue, W., and Besang, C. 1981. K/Ar and Rb/Sr dating of igneous rocks from North Victoria Land, Antarctica. *Geologisches Jahrbuch,* 41: 267–73.

Kyle, P. R., Elliot, D. H., and Sutter, J. F. 1981. Jurassic Ferrar Group tholeiites from the Transantarctic Mountains, Antarctica, and their relationship to the initial fragmentation of Gondwana. In M. M. Cresswell and P. Vella, eds., *Gondwana Five.* Rotterdam: Balkema, pp. 283–87.

Laird, M. G. 1963. Geomorphology and stratigraphy of the Nimrod Glacier, Beaumont Bay region, southern Victoria Land, Antarctica. *New Zealand Journal of Geology and Geophysics,* 6: 465–84.

1964. Petrology of rocks from the Nimrod Glacier–Starshot Glacier region, Ross Dependency. In R. J. Adie, ed., *Antarctic geology.* Amsterdam: North-Holland, pp. 463–72.

1981a. Lower Palaeozoic rocks of Antarctica. In C. H. Holland, ed., *Lower Palaeozoic of the Middle East, eastern and southern Africa, and Antarctica.* London: Wiley, pp. 257–314.

1981b. Lower Palaeozoic rocks of the Ross Sea area and their significance in the Gondwana context. *Journal of the Royal Society of New Zealand,* 11: 425–38.

1989. Evolution of the Cambrian–Early Ordovician Bowers basin, northern Victoria Land, and its relationship with the adjacent Wilson and Robertson Bay terranes. *Memorie della Società Geologica Italiana,* 33: 25–34.

1991a. The late Proterozoic–middle Palaeozoic rocks of Antarctica. In R. J. Tingey, ed., *The geology of Antarctica.* Oxford: Clarendon Press, pp. 74–119.

1991b. Lower-mid-Palaeozoic sedimentation and tectonic patterns on the palaeo-Pacific margin of Antarctica. In M. R. A. Thomson, J. A. Crame, and J. W. Thomson, eds., *Geological evolution of Antarctica.* Cambridge University Press, pp. 177–83.

Laird, M. G., Andrews, P. B., and Kyle, P. R. 1974. Geology of northern Evans Névé, Victoria Land, Antarctica. *New Zealand Journal of Geology and Geophysics,* 17: 587–601.

Laird, M. G., Andrews, P. B., Kyle, P., and Jennings, P. 1972. Late Cambrian fossils and the age of the Ross orogeny, Antarctica. *Nature,* 238: 34–36.

Laird, M. G., and Bradshaw, J. D. 1981. Permian tillites of North Victoria Land, Antarctica. In M. J. Hambrey and W. B. Harland, eds., *Earth's pre-Pleistocene glacial record.* Cambridge University Press, pp. 237–40.

1982. New data on the basement geology of north Victoria Land, Antarctica. *New Zealand Antarctic Record,* 4(2): 3–10.

1983. New data on the lower Paleozoic Bowers Supergroup, northern Victoria Land. In R. L. Oliver, P. R. James, and J. B. Jago, eds., *Antarctic earth science.* Canberra: Australian Academy of Science, pp. 123–26.

Laird, M. G., Bradshaw, J. D., and Wodzicki, A. 1976. Re-examination of the Bowers Group (Cambrian), northern Victoria Land, Antarctica (preliminary note). *New Zealand Journal of Geology and Geophysics,* 19: 275–82.

1982. Stratigraphy of the late Precambrian and early Paleozoic Bowers Supergroup, northern Victoria Land, Antarctica. In D. Craddock, ed., *Antarctic geoscience.* Madison: University of Wisconsin Press, pp. 535–42.

Laird, M. G., Cooper, R. A., and Jago, J. B. 1977. New data on the lower Paleozoic sequence of northern Victoria Land, Antarctica, and its significance for Australian–Antarctic relations in the Palaeozoic. *Nature,* 265: 107–10.

Laird, M. G., Mansergh, G. D., and Chappell, J. M. A. 1971. Geology of the central Nimrod Glacier area, Antarctica. *New Zealand Journal of Geology and Geophysics,* 14: 427–68.

Laird, M. G., and Waterhouse, J. B. 1962. Archeocyathine limestones of Antarctica. *Nature,* 194: 861.

Le Couteur, P. C., and Leitch, E. C. 1964. Preliminary report on the geology of an area south-west of the Upper Tucker Glacier, northern Victoria Land. In R. J. Adie, ed., *Antarctic geology.* Amsterdam: North-Holland, pp. 229–36.

Linder, H., de Hills, S. M., and Thiel, E. C. 1965. Basement complex in the Queen Maud Mountains, central Transantarctic Mountains. In J. B. Hadley, ed., *Antarctic geology and paleontology.* Antarctic Research Series, vol. 6. Washington, DC: American Geophysical Union, pp. 141–44.

Lindsay, J. F., Gunner, J., and Barrett, P. J. 1973. Reconnaissance geologic map of the Mount Elizabeth and Mount Kathleen quadrangles, Transantarctic Mountains, Antarctica. Geological Map A-2. Reston, VA: U.S. Geological Survey, United States Antarctic Research Program.

Lombardo, B., Cappelli, B., Carmignani, L., Gosso, G., Memmi, I., Montrasio, A., Palmeri, R., Pannuti, F., Pertusati, P. C., Ricci, C. A., Salvini, F., and Talarico, F. 1989. The metamorphic rocks of the Wilson terrane between David and Mariner Glaciers, North Victoria Land, Antarctica. *Memorie della Società Geologica Italiana,* 33: 99–130.

Lombardo, B., Pertusati, P. C., and Ricci, C. A. 1991. Some geological and petrological problems in the Wilson terrane of Borschgrevink Coast (northern Victoria Land, Antarctica). *Memorie della Società Geologica Italiana,* 46: 135–43.

Long, W. E. 1959. *Preliminary report of the geology of the central range of the Horlick Mountains, Antarctica.* Report 825-2. Columbus: Ohio State University Research Foundation, part 7.

1964. The stratigraphy of the Horlick Mountains. In R. J. Adie, ed., *Antarctic geology.* Amsterdam: North-Holland, pp. 352–63.

1965a. Stratigraphy of the Ohio Range, Antarctica. In J. B. Hadley, ed., *Geology and paleontology of the Antarctic.* Antarctic Research Series, vol. 6. Washington, DC: American Geophysical Union, pp. 71–116.

1965b. Stratigraphy of the Thorvald Nilsen Mountains, Queen Maud Range, central Antarctica. Geological Society of America, Special Paper 82, *Abstracts for 1964,* p. 124.

Lopatin, B. G. 1972. Basement complex of the McMurdo "oasis," South Victoria Land. In R. J. Adie, ed., *Antarctic geology and geophysics.* Oslo: Universitetsforlaget, pp. 287–92.

Lowry, P. H. 1980. The stratigraphy and petrography of the Cambrian Leverett Formation, Antarctica. M.S. thesis, Arizona State University.

Macdonald, D. I. M., Storey, B. C., Dalziel, I. W. D., Grunow, A. M., and Isbell, J. L. 1991. Early Palaeozoic sedimentation and tectonics in the Pensacola Mountains. In *Abstracts, Sixth International Symposium on Antarctic Earth Sciences.* Japan: National Institute of Polar Research, pp. 381–82.

Manzoni, M., and Nanni, T. 1977. Paleomagnetism of Ordovician lamprophyres from Taylor Valley, Victoria Land, Antarctica. Pageoph., 115: 961–77.

Matzer, S. 1992. Geological and structural field observations in the Carryer Glacier area, Bowers Mountains, northern Victoria Land, Antarctica. *Polarforschung,* 60: 121–23.

Mawson, D. 1916. Petrology of rock collections from the mainland of South Victoria Land. *British Antarctic Expedition, 1907–09, Report on the Scientific Investigations, Geology,* 2(13): 201–34.

McDougall, I., and Ghent, E. D. 1970. Potassium–argon dates on minerals from the Mt. Falconer area, lower Taylor Valley, southern Victoria Land, Antarctica. *New Zealand Journal of Geology and Geophysics,* 13: 1026–29.

McDougall, I., and Grindley, G. W. 1965. Potassium–argon dates on micas from the Nimrod–Beardmore–Axel Heiberg region, Ross Dependency, Antarctica. *New Zealand Journal of Geology and Geophysics,* 8: 304–13.

McElroy, C. T., and Rose, G. 1987. Geology of the Beacon Heights area, southern Victoria Land, Antarctica, 1:50 000. Miscellaneous Series Map 21. Lower Hutt: New Zealand Geological Survey, Department of Scientific and Industrial Research.

McGregor, V. R. 1965. Geology of the area between the Axel Heiberg and Shackleton glaciers, Queen Maud Range, Antarctica. *New Zealand Journal of Geology and Geophysics,* 8: 314–43.

McGregor, V. R., and Wade, F. A. 1969. Geology of the western Queen Maud Mountains. American Geographical Society Map Folio Series, folio 12, plate XV.

McKelvey, B. C., and Webb, P. N. 1959. Geological investigations in South Victoria Land, Antarctica. Part 1: Geology of Victoria Dry Valley. *New Zealand Journal of Geology and Geophysics,* 2: 120–36.

1961. Geological reconnaissance in Victoria Land, Antarctica. *Nature,* 189: 545–47.

1962. Geological investigations in southern Victoria Land, Antarctica. Part 3: Geology of Wright Valley. *New Zealand Journal of Geology and Geophysics,* 5: 143–62.

McLeod, I. R. 1964. Geological observations in Oates Land. In R. J. Adie, ed., *Antarctic geology.* Amsterdam: North-Holland, pp. 482–86.

McLeod, I. R., and Gregory, C. M. 1967. *Geological investigations along the Antarctic coast between longitudes 108°E and 166°E.* Geology and Geophysics Report 78. Canberra: Australian Bureau of Mineral Resources.

Minshew, V. H. 1965. Potassium–argon age from a granite at Mount Wilber, Queen Maud Range, Antarctica. *Science,* 150: 741–43.

1966. Stratigraphy of the Wisconsin Range, Horlick Mountains, Antarctica. *Science,* 152: 637–38.

1967. Geology of the Scott Glacier and Wisconsin Range areas, central Transantarctic Mountains, Antarctica. Ph.D. dissertation, Ohio State University.

Mirsky, A. 1969. Geology of the Ohio Range–Liv Glacier area. American Geographical Society Map Folio Series, folio 12, plate XVI.

Moores, E. M. 1991. Southwest U.S.–East Antarctic (SWEAT) connection: A hypothesis. *Geology,* 19: 425–28.

Mortimer, G. 1981. Provisional report on the geology of the basement complex between Miers and Salmon Valley, McMurdo Sound, Antarctica. *New Zealand Antarctic Record,* 3(2): 1–8.

Mortimer, G., Schmidt-Thomé, M., and Tessensohn, F. 1984. Stratigraphic problems in the upper part of the Bowers Supergroup, North Victoria Land, Antarctica. *Geologisches Jahrbuch,* B60: 83–103.

Murphy, D. J., 1972. Petrology and deformational history of the basement complex, Wright Valley, Antarctica with special reference to the origin of the augen gneisses. Ph.D. dissertation, University of Wyoming.

Murphy, D. J., Flory, R. F., Houston, R. S., and Smithson, S. B. 1970. Geological studies of basement rocks in South Victoria Land. *Antarctic Journal of the United States,* 5: 102.

Murtaugh, J. G. 1969. Geology of the Wisconsin Range batholith, Transantarctic Mountains. *New Zealand Journal of Geology and Geophysics,* 12: 526–50.

Nathan, S. 1971a. Potassium–argon dates from the area between Priestley and Mariner Glaciers, northern Victoria Land, Antarctica. *New Zealand Journal of Geology and Geophysics,* 14: 504–11.

1971b. Geology and petrology of the Campbell–Aviator divide, northern Victoria Land, Antarctica. Part 2: Paleozoic and Precambrian rocks. *New Zealand Journal of Geology and Geophysics,* 14: 564–96.

Nathan, S., and Schulte, F. J. 1968. Geology and petrology of the Campbell–Aviator divide, north Victoria Land, Antarctica. *New Zealand Journal of Geology and Geophysics,* 11: 940–75.

Nelson, W. H., Schmidt, D. L., and Schopf, J. M. 1968. *Structure and stratigraphy of the Pensacola Mountains, Antarctica.* Special Paper 115. Boulder, CO: Geological Society of America.

Nilsen, T. H., Bartow, J. A., Stump, E., and Link, M. H. 1977. Unusual occurrences of dish structure in the stratigraphic record. *Journal of Sedimentary Petrology,* 46: 1299–1304.

Odin, G. S., Gale, N. H., Auvray, B., Bielski, M., Doré, F., Lancelot, J.-R., and Pasteels, P. 1983. Numerical dating of Precambrian–Cambrian boundary. *Nature,* 301: 21–23.

Olesch, M., and Schubert, W. 1989. Aplite and pegmatite mineralogy of the Granite Harbour Intrusives, North Victoria Land, Antarctica. *Geologisches Jahrbuch,* E38: 257–58.

Oliver, R. L. 1964. Some basement rock relations in Antarctica. In R. J. Adie, ed., *Antarctic geology.* Amsterdam: North-Holland, pp. 259–64.

1972. The geology of an area near the mouth of the Beardmore Glacier, Ross Dependency, Antarctica. In R. J. Adie, ed., *Antarctic geology and geophysics.* Oslo: Universitetsforlaget, pp. 379–85.

Palmer, A. R. 1970. Early and Middle Cambrian trilobites from Antarctica. *Antarctic Journal of the United States,* 5: 162.

Palmer, A. R., and Gatehouse, C. G. 1972. *Early and Middle Cambrian trilobites from Antarctica.* Professional Paper 456-D. Reston: United States Geological Survey.

Palmer, D. F., Bradley, J., and Prebble, W. M. 1967. Orbicular granite from Taylor Valley, South Victoria Land, Antarctica. *Geological Society of America Bulletin,* 78: 1423–28.

Palmer, K. 1987. *XRF analysis of granitoids and associated rocks from South Victoria Land, Antarctica.* Research School of Earth Sciences, Geology Board of Studies, Publication no. 3. Victoria: University of Wellington.

Palmeri, R., and Talarico, F. 1990. Contrasting petrological features between garnet–cordierite–biotite paragneisses from Black Ridge and from Cape Sastrugi, Deep Freeze Range (North Victoria Land, Antarctica). *Memorie della Società Geologica Italiana,* 43: 33–48.

Palmeri, R., Talarico, F., Meccheri, M., Oggiano, G., Pertusati, P. C., Rastelli, N., and Ricci, C. A. 1991. Progressive deformation and low pressure/high temperature metamorphism in the Deep Freeze Range, Wilson terrane, northern Victoria Land, Antarctica. *Memorie della Società Geologica Italiana,* 46: 179–95.

Pankhurst, R. J., Storey, B. C., Millar, I. L., Macdonald, D. I. M., and Vennum, W. R. 1988. Cambro-Ordovician magmatism in the Thiel Mountains, Transantarctic Mountains, and implications for the Beardmore Orogeny. *Geology,* 16: 246–9.

Panttaja, S. K., and Rees, M. N. 1991. Provenance, age, and tectonic setting of lower Paleozoic siliciclastics in the central Transantarctic Mountains. *Geological Society of America, Abstracts with Programs,* 23(5): A364.

Pearce, J. 1982. Trace element characteristics of lavas from destructive plate boundaries. In R. S. Thorpe, ed., *Andesites.* New York: Wiley, pp. 525–48.

Pearn, W. C., Angino, E. E., and Stewart, D. 1963. New isotope age measurement from McMurdo Sound area, Antarctica. *Nature,* 199: 685.

Pettijohn, F. J., Potter, P. E., and Siever, R. 1972. *Sand and sandstone.* New York: Springer.

Picciotto, E., and Coppez, A. 1964. Bibliography of absolute age determinations in Antarctica (addendum). In R. J. Adie, ed., *Antarctic geology.* Amsterdam: North-Holland, pp. 561–69.

Pitcher, W. S. 1983. Granite type and tectonic environment. In K. Hsu, ed., *Mountain building processes.* San Diego, CA: Academic Press, pp. 19–40.

1984. Phanerozoic plutonism in the Peruvian Andes. In R. S. Harmon and B. A. Barreiro, eds., *Andean magmatism: Chemical and isotopic constraints.* Nantwich: Shiva, pp. 152–167.

Plumb, K. A. 1991. New Precambrian time scale. *Episodes,* 14: 139–40.

Plummer, C. C., Babcock, R. S., Sheraton, J. W., Adams, C. J. D., and Oliver, R. L. 1983. Geology of the Daniels Range, northern Victoria Land, Antarctica: A preliminary report. In R. L. Oliver, P. R. James, and J. B. Jago, eds., *Antarctic earth science.* Canberra: Australian Academy of Science, pp. 113–17.

Priestley, R. E., and David, T. W. E. 1912. Geological notes of the British Antarctic Expedition, 1907–09. *11th International Geological Congress, Stockholm, Compte rendu,* 2: 767–811.

Prior, G. T. 1899. Petrographical notes on the rock-specimens collected in Antarctic regions during the voyage of HMS *Erebus* and *Terror* under Sir James Clark Ross in 1839–43. *Mineralogical Magazine,* 12: 69–91.

1902. *Report on rock specimens collected by the Southern Cross Antarctic Expedition.* Report of the Southern Cross Collections. London: British Museum, pp. 321–32.

1907. Report on the rock specimens collected during the *Discovery* Antarctic Expedition, 1901–4. *National Antarctic Expedition, 1901–4, Natural History,* 1: 101–60.

Rastall, R. H., and Priestley, R. E. 1921. The slate–greywacke formation of Robertson Bay. *British Antarctic ("Terra Nova") Expedition, 1910, Natural History Report, Geology,* 1(4b): 121–29.

Ravich, M. G., Klimov, L. V., and Solov'ev, D. S. 1965. The Pre-Cambrian of East Antarctica. *Transactions of the Scientific Research Institute of the Geology of the Arctic of the State Geological Committee of the USSR.* Moscow.

Ravich, M. G., and Krylov, A. J. 1964. Absolute ages of rocks from East Antarctica. In R. J. Adie, ed., *Antarctic geology:* Amsterdam: North-Holland, pp. 579–89.

Rees, M. N., Duebendorfer, E. M., and Rowell, A. J. 1989a. The Skelton Group, southern Victoria Land. *Antarctic Journal of the United States,* 24: 21–24.

Rees, M. N., Girty, G. H., Panttaja, S. K., and Braddock, P. 1987. Multiple phases of early Paleozoic deformation in the central Transantarctic Mountains. *Antarctic Journal of the United States,* 22: 33–35.

Rees, M. N., Pratt, B. R., and Rowell, A. J. 1989b. Early Cambrian reefs, reef complexes, and associated facies of the Shackleton Limestone, Transantarctic Mountains. *Sedimentology,* 36: 341–61.

Rees, M. N., and Rowell, A. J. 1991. The pre-Devonian Palaeozoic clastics of the central Transantarctic Mountains: Stratigraphy and depositional setting. In M. R. A. Thomson, J. A. Crame, and J. W. Thomson, eds., *Geological evolution of Antarctica.* Cambridge University Press, pp. 187–92.

Rees, M. N., Rowell, A. J., and Cole, E. D. 1988. Aspects of the late Proterozoic and Paleozoic geology of the Churchill Mountains, southern Victoria Land. *Antarctic Journal of the United States,* 23: 23–25.

Rees, M. N., Rowell, A. J., and Pratt, B. R. 1985. The Byrd Group of the Holyoke Range, central Transantarctic Mountains. *Antarctic Journal of the United States,* 20: 3–5.

Ribecai, C. 1991. Occurrence of the organic-walled microfossils from the Priestley Formation (Wilson terrane, northern Victoria Land, Antarctica) and the problem of their age: A preliminary report. *Memorie della Società Geologica Italiana,* 46: 157–61.

Ricker, J. 1964. Outline of the geology between Mawson and Priestley Glaciers, Victoria Land. In R. J. Adie, ed., *Antarctic geology.* Amsterdam: North-Holland, pp. 265–75.

Riddols, B. W., and Hancox, G. T. 1968. Geology of the upper Mariner Glacier region, North Victoria Land, Antarctica. *New Zealand Journal of Geology and Geophysics,* 11: 881–99.

Rocchi, S. 1991. Zircon typology of Cambro-Ordovician granitoids from northern Victoria Land, Antarctica. *Memorie della Società Geologica Italiana,* 46: 249–55.

Rodgers, J. 1970. *The tectonics of the Appalachians.* New York: Wiley-Interscience.

Roland, N. W. 1991. The boundary of the East Antarctic craton on the Pacific margin. In M. R. A. Thomson, J. A. Crame, and J. W. Thomson, eds., *Geological evolution of Antarctica.* Cambridge University Press, pp. 161–65.

Roland, N. W., Gibson, G. M., Kleinschmidt, G., and Schubert, W. 1984. Metamorphism and structural relations of the Lanterman metamorphics, North Victoria Land, Antarctica. *Geologisches Jahrbuch,* 60: 319–61.

Roland, N. W., Olesch, M., and Schubert, W. 1989. Geology and petrology of the western border of the Transantarctic Mountains between Outback Nunataks and Reeves Glacier, North Victoria Land, Antarctica. *Geologisches Jahrbuch,* E38: 119–41.

Roland, N. W., and Tessensohn, F. 1987. Rennick faulting – an early phase of Ross Sea rifting. *Geologisches Jahrbuch,* B66: 203–29.

Ross, J. C. 1847. *A voyage of discovery and research in the southern and Antarctic regions during the years 1839–43,* 2 vols. London: John Murray.

Rowell, A. J., Cooper, R. A., Jago, J. B., and Braddock, P. 1983. Paleontology of the lower Paleozoic of northern Victoria Land: Brachiopods with Australian and New Zealand affinities in the Spurs Formation. *Antarctic Journal of the United States,* 18: 18–20.

Rowell, A. J., Evans, K. R., and McKenna, L. W., III. 1993a. Cambrian and possible Proterozoic strata in the Transantarctic Mountains, north of Leverett Glacier. *Antarctic Journal of the United States,* 28: 35–37.

Rowell, A. J., Evans, K. R., and Rees, M. N. 1988a. Fauna of the Shackleton Limestone. *Antarctic Journal of the United States,* 23: 13–14.

Rowell, A. J., and Rees, M. N. 1989. Early Paleozoic history of the upper Beardmore Glacier area: Implications for a major Antarctic structural boundary within the Transantarctic Mountains. *Antarctic Science,* 1: 249–60.

1991. Setting and significance of the Shackleton Limestone, central Transantarctic Mountains. In M. R. A. Thomson, J. A. Crame, and J. W. Thomson, eds., *Geological evolution of Antarctica.* Cambridge University Press, pp. 171–75.

Rowell, A. J., Rees, M. N., and Braddock, P. 1986. Pre-Devonian rocks of the central Transantarctic Mountains. *Antarctic Journal of the United States,* 21: 48–50.

Rowell, A. J., Rees, M. N., Cooper, R. A., and Pratt, B. R. 1988b. Early Paleozoic history of the central Transantarctic Mountains: Evidence from the Holyoke Range, Antarctica. *New Zealand Journal of Geology and Geophysics,* 31: 397–404.

Rowell, A. J., Rees, M. N., Duebendorfer, E. M., Wallin, E. T., Van Schmus, W. R., and Smith, E. I. 1993b. An active Neoproterozoic margin: Evidence from the Skelton Glacier area, Transantarctic Mountains. *Journal of the Geological Society, London,* 150: 677–82.

Rowell, A. J., Rees, M. N., and Evans, K. R. 1990. Depositional setting of the Lower and Middle Cambrian in the Pensacola Mountains. *Antarctic Journal of the United States,* 25: 40–42.

    1992. Evidence of major Middle Cambrian deformation in the Ross orogen, Antarctica. *Geology,* 20: 31–34.

Sandiford, M. 1985. Structural evolution of the Lanterman Metamorphic Complex, northern Victoria Land, Antarctica. *New Zealand Journal of Geology and Geophysics,* 28: 443–58.

Schetelig, J. 1915. Report on the rock specimens collected on Roald Amundsen's South Pole Expedition. *Skrifter Videnskapsselskapets, Mat.-Naturv.,* 1(4): 1–32.

Schmidt, D. L. 1964. Geology of the Pensacola Mountains. *Bulletin of the U.S. Antarctic Projects Officer,* 5(10): 98–101.

    1969. Precambrian and lower Paleozoic igneous rocks, Pensacola Mountains, Antarctica. *Antarctic Journal of the United States,* 4: 203–4.

Schmidt, D. L., Dover, J. H., Ford, A. B., and Brown, R. D. 1964. Geology of the Patuxent Mountains. In R. J. Adie, ed., *Antarctic geology.* Amsterdam: North-Holland, pp. 276–83.

Schmidt, D. L., and Ford, A. B. 1963. U.S. Geological Survey in the Patuxent Mountains. *Bulletin of the U.S. Antarctic Projects Officer,* 4(8): 20–24.

    1966. Geology of the northern Pensacola Mountains and adjacent areas. *Antarctic Journal of the United States,* 1: 125.

    1969. Geology of the Pensacola and Thiel Mountains. American Geographical Society Map Folio Series, folio 12, V.

Schmidt, D. L., Williams, P. L., and Nelson, W. H. 1978. Geologic map of the Schmidt Hills quadrangle and part of the Gambacorta Peak quadrangle, Pensacola Mountains, Antarctica. Map A-8. Reston, VA: U.S. Geological Survey, United States. Antarctic Research Program.

Schmidt, D. L., Williams, P. L., Nelson, W. H., and Ege, J. R. 1965. *Upper Precambrian and Paleozoic stratigraphy and structure of the Neptune Range, Antarctica.* Professional Paper 525-D. Reston, VA: United States Geological Survey. pp. D112–D119.

Schmidt, M. T., Dahl, P. S., and Friberg, L. M. 1984. Petrologic study of metasediments in Taylor Valley, southern Victoria Land. *Antarctic Journal of the United States,* 9: 16–18.

Schopf, J. M. 1968. Studies in Antarctic paleobotany. *Antarctic Journal of the United States,* 3: 176–77.

Schmerer, K., and Burmester, R. 1986. Paleomagnetic results from the Cambro-Ordovician Bowers Supergroup, northern Victoria Land, Antarctica. In E. Stump, ed., *Geological investigations in northern Victoria Land.* Antarctic Research Series, vol. 46. Washington, DC: American Geophysical Union, pp. 69–90.

Schmidt-Thomé, M., and Wolfart, R. 1984. Tremadocian faunas (trilobites, brachiopods) from Reilly Ridge, North Victoria Land, Antarctica. *Newsletter of Stratigraphy, Berlin, Stuttgart,* 13: 88–93.

Schubert, W. 1987. Petrography of the eastern Thompson Spur, Daniels Range, north Victoria Land, Antarctica. *Geologisches Jahrbuch,* B66: 131–43.

Schubert, W., and Olesch, M. 1989. The petrological evolution of the crystalline basement of Terra Nova Bay, north Victoria Land, Antarctica. *Geologisches Jahrbuch,* E38: 277–98.

Schubert, W., Olesch, M., and Schmidt, K. 1984. Paragneiss–orthogneiss relationships in the Kavrayskiy Hills, North Victoria Land. *Geologisches Jahrbuch,* B60: 187–211.

Schüssler, U., Skinner, D. B. N., and Roland, N. W. 1993. Subduction-related mafic to intermediate plutonism. *Geologische Jahrbuch,* E47: 389–418.

Seward, A. C. 1914. Antarctic fossil plants. *British Antarctic ("Terra Nova") Expedition 1910, Natural History Report, Geology,* 1(1): 49 p.

Sheraton, J. W., Babcock, R. S., Black, L. P., Wyborn, D., and Plummer, C. C. 1987. Petrogenesis of granitic rocks of the Daniels Range, northern Victoria Land, Antarctica. *Precambrian Research,* 37: 267–86.

Sheraton, J. W., Thomson, J. W., and Collerson, K. D. 1987. Mafic dyke swarms of Antarctica. In

H. C. Hills and W. F. Fahig, eds., *Mafic dyke swarms*. Special Paper 34. Toronto: Geological Association of Canada, pp. 419–33.

Shergold, J. H., and Cooper, R. A. 1985. Late Cambrian trilobites from the Mariner Group, northern Victoria Land, Antarctica. *BMR Journal of Australian Geology & Geophysics,* 9: 91–106.

Shergold, J. H., Cooper, R. A., Druce, E. C., and Webby, B. D. 1982. Synopsis of selected sections at the Cambrian–Ordovician boundary in Australia, New Zealand, and Antarctica. In M. G. Basset and W. T. Dean, eds., *The Cambrian–Ordovician boundary*. Cardiff: National Museum of Wales Press, pp. 211–27.

Shergold, J. H., Cooper, R. A., MacKinnon, D. I., and Yochelson, E. L. 1976. Late Cambrian Brachiopoda, Mollusca, and Trilobita from northern Victoria Land. *Palaeontology,* 19: 247–91.

Shergold, J. H., Jago, J. B., Cooper, R. A., and Laurie, J. 1985. *The Cambrian system in Australia, Antarctica and New Zealand*. Publication no. 19. Ottawa: International Union of Geological Sciences.

Skeats, E. W. 1916. Report on the petrology of some limestones from the Antarctic. *Reports of the Scientific Investigations, British Antarctic Expedition, 1907–09, Geology,* 2(12): 189–200.

Skinner, D. N. B. 1964. A summary of the geology of the region between Byrd and Starshot Glaciers, South Victoria Land. In R. J. Adie, ed., *Antarctic geology*. Amsterdam: North-Holland, pp. 284–92.

1965. Petrographic criteria of the rock units between the Byrd and Starshot Glaciers, South Victoria Land, Antarctica. *New Zealand Journal of Geology and Geophysics,* 8: 292–303.

1972. A differentiation source for the mafic and ultramafic rocks ("enstatite peridotites"), and some porphyritic granites of Terra Nova Bay. In R. J. Adie, ed., *Antarctic geology and geophysics*. Oslo: Universitetsforlaget, pp. 299–303.

1981. Possible Permian glaciation in north Victoria Land. *Geologisches Jarhbuch,* B41: 261–66.

1982. Stratigraphy and structure of lower grade metasediments of Skelton Group, McMurdo Sound: Does Teall Greywacke really exist? In C. Craddock, ed., *Antarctic geoscience*. Madison: University of Wisconsin Press, pp. 555–63.

1983a. The geology of Terra Nova Bay. In R. L. Oliver, P. R. James, and J. B. Jago, eds., *Antarctic earth science*. Canberra: Australian Academy of Science, pp. 150–55.

1983b. The granites and two orogenies of southern Victoria Land. In R. L. Oliver, P. R. James, and J. B. Jago, eds., *Antarctic earth science*. Canberra: Australian Academy of Science, pp. 160–63.

1983c. *Terra Nova Bay Geological Expedition III, NZARP 1982–83*. Report G72. Lower Hutt: New Zealand Geological Survey, pp. 1–40.

1983d. Terra Nova Bay revisited: Metasediments, mafites, and granites. *New Zealand Antarctic Record,* 4(3): 16–17.

1989. Terra Nova Bay and the Deep Freeze Range: The southern allochthonous border of North Victoria Land, Antarctica. *Memorie della Società Geologica Italiana,* 33: 41–58.

1991a. The Priestley Formation, Terra Nova Bay, and its regional significance. In M. R. A. Thomson, J. A. Crame, and J. W. Thomson, eds., *Geological evolution of Antarctica*. Cambridge University Press, pp. 137–41.

1991b. Metamorphic basement contact relations in the southern Wilson terrane, Terra Nova Bay, Antarctica: The Boomerang thrust. *Memorie della Società Geologica Italiana,* 46: 163–78.

1992. Metasedimentary rocks of western Wilson terrane (Victoria Land – Oates Land) and Gondwana connections to Australia. In Y. Yoshida, K. Kaminuma, and K. Shiraishi, eds., *Recent progress in Antarctic earth science*. Tokyo: Terra Scientific, pp. 219–26.

1993. *Report on participation in Ganovex VII expedition, Terra Nova Bay 1992–93*. Science Report 93/12. Lower Hutt: Institute of Geological and Nuclear Sciences.

Skinner, D. N. B., and Ricker, J. 1968. The geology of the region between the Mawson and Priestley Glaciers, North Victoria Land, Antarctica. *New Zealand Journal of Geology and Geophysics,* 11: 1009–40.

Smillie, R. W. 1987. Petrological evolution of basement granitoids, southern Victoria Land. *New Zealand Antarctic Record,* 8(1): 61–71.

1992. Suite subdivision and petrological evolution of granitoids from Taylor Valley and Ferrar Glacier region, South Victoria Land. *Antarctic Science,* 4: 71–87.

Smit, J. H. 1981. Sedimentology, metamorphism, and structure of the La Gorce Formation, La Gorce Mountains, upper Scott Glacier, Antarctica. M.S. thesis, Arizona State University.

Smithson, S. B., Fikkan, P. R., and Toogood, D. J. 1969. Geological and geophysical studies in the ice-free valley area of Victoria Land. *Antarctic Journal of the United States,* 4: 131–32.

——— 1970. Early geologic events in the ice-free valleys, Antarctica. *Geological Society of America Bulletin,* 81: 207–10.

Smithson, S. B., Murphy, D. J., and Houston, R. S. 1971b. Development of an augen gneiss terrain. *Contributions to Mineralogy and Petrology,* 33: 184–90.

Soloviev, D. S. 1960. The lower Palaeozoic metamorphic slates of the Oates Coast. *Naucho-issledovatelsky Institut Geologii Antarkitki,* 113: 147–58.

Starik, I. E., Krylov, A. J., Ravich, M. G., Krylov, A. Y., and Silin, J. I. 1959. On the absolute ages of rocks of the East Antarctic platform [in Russian]. *Doklady Akademiya Nauk SSSR,* 126(1): 144–46.

——— 1961. The absolute ages of East Antarctic rocks. *Annals of the New York Academy of Science,* 91: 576–82.

Steiger, R. H., and Jäger, E. 1977. Subcommission on geochronology: Convention on the use of decay constants in geo- and cosmochronology. *Earth and Planetary Science Letters,* 36: 359–62.

Stepheson, P. J. 1966. Theron Mountains, Shackleton Range and Whichaway Nunataks, with a section on paleomagnetism of the dolerite intrusions by D. J. Blundell. Trans-Antarctic Expedition 1955–58, Scientific Reports no. 8. London: Trans-Antarctic Expedition Committee.

Stern, T. A., and ten Brink, U. S. 1989. Flexural uplift of the Transantarctic Mountains. *Journal of Geophysical Research,* 94: 10,315–10,330.

Smit, J. H., and Stump, E. 1986. Sedimentology of the La Gorce Formation, La Gorce Mountains, Antarctica. *Journal of Sedimentary Petrology,* 56: 663–68.

Smith, W. C. 1924. The plutonic and hypabyssal rocks of South Victoria Land. *British Antarctic ("Terra Nova") Expedition, 1910, Natural History Report, Geology,* 1(6): 167–227.

——— 1964. The petrography of the igneous and metamorphic rocks of South Victoria Land: Based on the collections made on the British Antarctic (*Terra Nova*) Expedition, 1910–13. In R. J. Adie, ed., *Antarctic geology.* Amsterdam: North-Holland, pp. 419–28.

Smith, W. C., and Debenham, F. 1921. The metamorphic rocks of the McMurdo Sound region. *British Antarctic ("Terra Nova") Expedition, 1910, Natural History Report, Geology,* 1 (5a): 133–44.

Smith, W. C., and Priestley, R. E. 1921. The metamorphic rocks of the Terra Nova Bay region. *British Antarctic ("Terra Nova") Expedition, 1910, Natural History Report, Geology,* 1(5b): 145–65.

Smithson, S. B., Fikkan, P. R., and Houston, R. S. 1971a. Amphibolitization of calc–silicate metasedimentary rocks. *Contributions to Mineralogy and Petrology,* 31: 228–37.

Smithson, S. B., Fikkan, P. R., Murphy, D. R., and Houston, R. S. 1972. Development of augen gneiss in the ice-free valley area, South Victoria Land. In R. J. Adie, ed., *Antarctic geology and geophysics.* Oslo: Universitetsforlaget, pp. 293–98.

Stewert, D. 1934a. The petrography of some Antarctic rocks. *American Mineralogist,* 19: 150–60.

——— 1934b. The petrography of some rocks from South Victoria Land. *Proceedings of the American Philosophical Society,* 74: 307–10.

——— 1934c. A contribution to Antarctic petrography. *Journal of Geology,* 42: 546–50.

Storey, B. C., Alabaster, T., Macdonald, D. I. M., Millar, I. L., Pankhurst, R. J., and Dalziel, I. W. D. 1992. Upper Proterozoic rift-related rocks in the Pensacola Mountains, Antarctica: Precursors to supercontinent breakup? *Tectonics,* 11: 1392–1405.

Storey, B. C., and Dalziel, I. W. D. 1987. Outline of the structural and tectonic history of the Ellsworth Mountains–Thiel Mountains Ridge. In G. D. McKenzie, ed., *Gondwana Six: Structure, tectonics, and geophysics.* Geophysical Monograph 40. Washington, DC: American Geophysical Union, pp. 117–28.

Storey, B. C., and Macdonald, D. I. M. 1987. Sedimentary rocks of the Ellsworth–Thiel Mountains ridge and their regional equivalents. *British Antarctic Survey Bulletin,* 76: 21–49.

Stuckless, J. S. 1975. Geochronology of core samples recovered from DVDP, Lake Vida, Antarctica. *Dry Valley Drilling Project Bulletin,* vol. 6. Wellington: Victoria University.

Stuckless, J. S., and Erickson, R. L. 1975. Rubidium–strontium ages of basement rocks recovered from DVDP hole 6, southern Victoria Land. *Antarctic Journal of the United States,* 10: 302–7.

Stuiver, M., and Brazius, T. F. 1985. Compilation of isotopic dates from Antarctica. *Radiocarbon,* 27: 117–304.

Stump, E. 1973. Earth evolution in the Transantarctic Mountains and West Antractica. In D. H. Tarling and S. K. Runcorn, eds., *Implications of continental drift to the earth sciences,* vol. 2. New York: Academic Press, pp. 909–24.

——— 1974. Volcanic rocks of the Early Cambrian Taylor formation, central Transantarctic Mountains. *Antarctic Journal of the United States,* 9: 228–29.

——— 1976a. On the late Precambrian–early Paleozoic metavolcanic and metasedimentary rocks of the Queen Maud Mountains, Antarctica, and a comparison with rocks of similar age from southern Africa. Institute of Polar Studies Report 62. Columbus: Ohio State University.

——— 1976b. Accretionary lapilli and lithophysal spehrulites from the Taylor Formation, Queen Maud Mountains, Antarctica. *Antarctic Journal of the United States,* 11: 246–48.

——— 1980. Two episodes of deformation at Mt. Madison, Antarctica. *Antarctic Journal of the United States,* 15: 13–14.

——— 1981a. Structural relationships in the Duncan Mountains, central Transantarctic Mountains, Antarctica. *New Zealand Journal of Geology and Geophysics,* 24: 87–93.

——— 1981b. Observations on the Ross orogen, Antarctica. In M. M. Cresswell and P. Vella, eds., *Gondwana Five.* Rotterdam: Balkema, pp. 205–8.

——— 1982. The Ross Supergroup in the Queen Maud Mountains. In C. Craddock, ed., *Antarctic geoscience.* Madison: University of Wisconsin Press, pp. 565–69.

——— 1983. Type locality of the Ackerman Formation, La Gorce Mountains, Antarctica. In R. L. Oliver, R. L. James, and J. B. Jago, eds., *Antarctic earth sciences.* Canberra: Australian Academy of Earth Science, pp. 170–74.

——— 1985. Stratigraphy of the Ross Supergroup, central Transantarctic Mountains. In M. D. Turner and J. F. Splettstoesser, eds., *Geology of the central Transantarctic Mountains.* Antarctic Research Series, vol. 36. Washington, DC: American Geophysical Union, pp. 225–74.

——— 1987. Construction of the Pacific margin of Gondwana during the Pannotios cycle. In G. D. McKenzie, ed., *Gondwana Six: Structure, tectonics and geophysics.* Geophysical Monograph 40. Washington, DC: American Geophysical Union, pp. 77–87.

——— Comp. 1989. Reconnaissance geologic map of the Welcome Mountain quadrangle, Transantarctic Mountains, Antarctica. Reston, VA: Map A-14. U.S. Geological Survey, United States Antarctic Research Program.

——— 1992a. Pre-Beacon tectonic development of the Transantarctic Mountains. In Y. Yoshida, K. Kaminuma, and K. Shiraishi, eds., *Recent progress in Antarctic earth science.* Tokyo: Terra Scientific, pp. 235–40.

——— 1992b. The Ross orogen of the Transantarctic Mountains in light of the Laurentia–Gondwana split. *GSA Today,* 2: 25–31.

Stump, E., Borg, S. G., and Armstrong, R. L. In press. Rb–Sr geochronology of granitic rocks, northern Victoria Land, Antarctica. *Antarctic Science.*

Stump, E., Edgerton, D. G., and Korsch, R. J. 1986a. Structural geological investigations in the Nimrod Glacier area. *Antarctic Journal of the United States,* 21: 35–36.

Stump, E., and Fitzgerald, P. G. 1988. Field collecting for fission-track analysis of uplift history of the Scott Glacier area, Transantarctic Mountains. *Antarctic Journal of the United States,* 23: 12.

——— 1992. Episodic uplift of the Transantarctic Mountains. *Geology,* 20: 161–64.

Stump, E., Holloway, J. R., Borg, S. G., and Lapham, K. E. 1983a. Geological investigations of early to middle Paleozoic magmatic rocks, northern Victoria Land. *Antarctic Journal of the United States,* 17: 17–18.

Stump, E., Korsch, R. J., and Edgerton, D. C. 1991. The myth of the Beardmore and Nimrod orogenies. In M. R. A. Thomson, J. A. Crame, and J. W. Thomson, eds., *Geological evolution of Antarctica.* Cambridge University Press, pp. 143–47.

Stump, E., Laird, M. G., Bradshaw, J. D., Holloway, J. R., Borg, S. G., and Lapham, K. E. 1983b. Bowers graben and associated tectonic features cross northern Victoria Land, Antarctica. *Nature,* 304: 334–36.

Stump, E., Lowry, P. H., Heintz-Stocker, G. M., and Colbert, P. V. 1978. Geological investigations in the Leverett Glacier area. *Antarctic Journal of the United States,* 13: 3–4.

Stump, E., and Maccracken, H. 1983. Reconnaissance geological investigation of basement units, McMurdo Sound region. *Antarctic Journal of the United States,* 18: 46–47.

Stump, E., Miller, J. M. G., Korsch, R. J., and Edgerton, D. G. 1988. Diamictite from Nimrod Glacier area, Antarctica: Possible Proterozoic glaciation on the seventh continent. *Geology,* 16: 225–28.

Stump, E., Self, S., Smit, J. H., Colbert, P. V., and Stump, T. M. 1981. Geological investigations in the La Gorce Mountains and central Scott Glacier area. *Antarctic Journal of the United States,* 16: 55–57.

Stump, E., Sheridan, M. F., Borg, S. G., Lowry, P. H., and Colbert, P. V. 1979. Geological investigations in the Scott and Byrd Glacier areas. *Antarctic Journal of the United States,* 14: 39–40.

Stump, E., Smit, J. H., and Self, S. 1985. Reconnaissance geologic map of the Mt. Blackburn Quadrangle, Antarctica. Map A-11. Reston, VA: U.S. Geological Survey, United States Antarctic Research Program.

1986b. Timing of events during the late Proterozoic Beardmore Orogeny, Antarctica: Geological evidence from the La Gorce Mountains. *Geological Society of America Bulletin,* 97: 953–65.

Stump, E., Splettstoesser, J. F., and Colbert, P. V. 1983c. Northern Victoria Land project, 1981–82. *Antarctic Journal of the United States* 17: 3–4.

Stump, E., White, A. J. R., and Borg, S. G. 1986c. Reconstruction of Australia and Antarctica: Evidence from granites and recent mapping. *Earth and Planetary Science Letters,* 79: 348–60.

Sturm, A. G., and Carryer, S. 1970. Geology of the region between Matusevich and Tucker Glaciers, northern Victoria Land, Antarctica. *New Zealand Journal of Geology and Geophysics,* 13: 408–35.

Sublett, A. G., Hauck, W. C., Huston, R. S., and Smithson, S. B. 1972. Basement geology of Mount Insell area, Victoria Valley, Antarctica. *Antarctic Journal of the United States,* 7: 107–8.

Talarico, F. 1990. Reaction textures and metamorphic evolution of aluminous garnet granulites from the Wilson terrane (Deep Freeze Range, north Victoria Land, Antarctica). *Memorie della Società Geologica Italiana,* 43: 49–58.

Talarico, F., Franceschelli, M., Lombardo, B., Palmeri, R., Pertusati, P. C., Rastelli, N., and Ricci, C. A. 1992. Metamorphic facies of the Ross orogeny in the southern Wilson terrane of northern Victoria Land, Antarctica. In Y. Yoshida, K. Kaminuma, and K. Shiraishi, eds., *Recent progress in Antarctic earth science.* Tokyo: Terra Scientific, pp. 211–18.

Talarico, F., Memmi, I., Lombardo, B., and Ricci, C. A. 1989. Thermo-barometry of granulite rocks from the Deep Freeze Range, north Victoria Land, Antarctica. *Memorie della Società Geologica Italiana,* 33: 131–41.

Tessensohn, F. 1984. Geological and tectonic history of the Bowers structural zone, North Victoria Land, Antarctica. *Geologisches Jahrbuch,* 60: 371–96.

1989. GANOVEX: An earth science programme of the Federal Republic of Germany in the Antarctic. *Memorie della Società Geologica Italiana,* 33: 59–67.

1992. The Antarctic expedition GANOVEX VI: Introduction to field results. *Polarforschung,* 60: 63–66.

1993. Crustal investigations in the Transantarctic Mountains: The concept for the GANOVEX V expedition. *Geologische Jahrbuch,* E47: 39–48.

Tessensohn, F., Duphorn, K., Jordan, H., Kleinschmidt, G., Skinner, D. N. B., Vetter, U., Wright, T. O., and Wyborn, D. 1981. Geological comparison of basement units in North Victoria Land. *Geologisches Jahrbuch,* B41: 31–88.

Tessensohn, F., Kleinschmidt, G., Henjes-Kunst, F., and Fenn, G. 1992a. Gradational east–west increase in metamorphism in the basement rocks of the Helliwell Hills, Wilson terrane, North Victoria Land, Antarctica. *Polarforschung,* 60: 117–20.

Tessensohn, F., Wörner, G. Kleinschmidt, G., Henjes-Kunst, F., Fenn, G., and Matzer, S. 1992b. The Johnnie Walker Formation: A pre–Granite Harbour sub-volcanic unit in the Wilson terrane, lower Mawson Glacier, Victoria Land, Antarctica. *Polarforschung,* 60: 91–95.

Thiedig, F., and Stackebrandt, W. 1992. Agmatites, basaltic intrusions and younger deformations in the "Berliner Mauer," Cape Sibbald, Aviator Glacier tougue, North Victoria Land, Antarctica. *Polarforschung,* 60: 99–100.

Thiel, E. C. 1961. Antarctica, one continent or two? *Polar Record,* 10: 335–48.

Tingey, R. J. 1991. Mesozoic tholeiitic igneous rocks in Antarctica: The Ferrar (Super) Group and related rocks. In R. J. Tingey, ed., *The geology of Antarctica.* Oxford: Clarendon Press, pp. 153–74.

Treves, S. B. 1965. Igneous and metamorphic rocks of the Ohio Range, Horlick Mountains, Antarctica. In J. B. Hadley, ed., *Geology and paleontology of the Antarctic.* Antarctic Research Series, vol. 6. Washington, DC: American Geophysical Union, pp. 117–25.

Ulitzka, S. 1987. Petrology and geochemistry of the migmatites from Thompson Spur, Daniels Range, North Victoria Land, Antarctica. *Geologisches Jahrbuch,* B66: 81–130.

Vennum, W. R., and Storey, B. C. 1987. Petrology, geochemistry, and tectonic setting of granitic rocks from the Ellsworth–Whitmore Mountains crustal block and Thiel Mountains, West Antarctica. In G. D. McKenzie, ed., *Gondwana Six: Structure, tectonics and geophysics.* Geophysical Monograph 40. Washington, DC: American Geophysical Union, pp. 139–50.

Vetter, U., Lenz, H., Kreuzer, H., and Besang, C. 1984. Pre-Ross granites at the Pacific margin of the Robertson Bay terrane, North Victoria Land, Antarctica. *Geologisches Jahrbuch,* B60: 363–69.

Vetter, U., Roland, N. W., Kreuzer, H., Höhndorf, A., Lenz, H., and Besang, C. 1983. Geochemistry, petrography, and geochronology of the Cambro-Ordovician and Devonian-Carboniferous granitoids of northern Victoria Land, Antarctica. In R. L. Oliver, P. R. James, and J. B. Jago, eds., *Antarctic earth sciences.* Canberra: Australian Academy of Science, pp. 140–43.

Vetter, U., and Tessensohn, F. 1987. S- and I-type granitoids of North Victoria Land, Antarctica, and their inferred geotectonic setting. *Geologische Rundschau,* 76: 233–43.

Viti, G., Lombardo, B., and Guiliani, O. 1991. Ordovician uplift pattern in the Wilson terrane of the Borschgrevink Coast, Victoria Land (Antarctica). *Memorie della Società Geologica Italiana,* 46: 257–65.

Vocke, R. D., Jr., and Hanson, G. N. 1981. U–Pb zircon ages and petrogenetic implications for two basement units from Victoria Valley, Antarctica. In L. D. McGinnis, ed., *Dry Valley Drilling Project.* Antarctic Research Series, vol. 33. Washington, DC: American Geophysical Union, pp. 248–55.

Vocke, R. D., Jr., Hanson, G. N., and Stuckless, J. S. 1987. Ages for the Vida granite and Olympus granite gneiss, Victoria Valley, southern Victoria Land. *Antarctic Journal of the United States,* 13: 15–17.

Voronov, P. S. 1964. Tectonics and neotectonics of Antarctica. In R. J. Adie, ed., *Antarctic geology.* Amsterdam: North-Holland, pp. 692–700.

Wade, F. A. 1974. Geological surveys of Marie Byrd Land and the central Queen Maud Range. *Antarctic Journal of the United States,* 9: 241–42.

Wade, F. A., and Cathey, C. A. 1986. Geology of the basement complex, western Queen Maud Mountains, Antarctica. In M. D. Turner and J. F. Splettstoesser, eds., *Geology of the central Transantarctic Mountains.* Antarctic Research Series, vol. 36. Washington, DC: American Geophysical Union, pp. 429–453.

Wade, F. A., Yeats, V. L., Everett, J. R., Greenlee, D. W., LaPrade, K. E., and Shenk, J. C. 1965a. *The geology of the central Queen Maud Range, Transantarctic Mountains, Antarctica.* Antarctic Research Report Series, 65-1. Lubbock: Texas Technological College.

1965b. Geology of the central portion of the Queen Maud Range, Transantarctic Mountains. *Science,* 150: 1808–9.

Walker, N. W., and Goodge, J. W. 1991. Significance of Late Archean–Early Proterozoic U–Pb ages of individual Nimrod Group detrital zircons and Cambrian plutonism in the Miller Range, Transantarctic Mountains. Geological Society of America Abstracts with Programs, vol. 23, p. A306.

Warren, G. 1965. Geology of Antarctica. In T. Hatherton, ed., *Antarctica.* Wellington: Reed, pp. 279–320.

Warren, G. 1969. Geology of the Terra Nova Bay–McMurdo Sound area, Victoria Land. American Geographic Society Antarctic Map Folio Series, folio 12, XIII.

Waters, A. S. 1991. Basement granitoids of the St. Johns Range, southern Victoria Land, Antarctica. *New Zealand Antarctic Record,* 11(1): 29–35.

Weaver, S. D., Bradshaw, J. D., and Laird, M. G. 1984a. Geochemistry of Cambrian volcanics of the Bowers Supergroup and implications for the Early Paleozoic tectonic evolution of northern Victoria Land, Antarctica. *Earth and Planetary Science Letters,* 68: 128–40.

1984b. Lawrence Peaks volcanics, North Victoria Land. *New Zealand Antarctic Record,* 5(3): 18–22.

Webb, A. W., McDougall, I, and Cooper, J. R. 1964. Potassium–argon dates from Vincennes Bay region and Oates Land. In R. J. Adie, ed., *Antarctic geology*. North-Holland: Amsterdam, pp. 597–660.

Webb, P. N., and McKelvey, B. C. 1959. Geological investigations in South Victoria Land, Antarctica. Part 1: Geology of Victoria Dry Valley. *New Zealand Journal of Geology and Geophysics,* 2: 120–36.

Webby, B. D., Vandenberg, A. H. M., Cooper, R. A., Banks, M. R., Burrett, C. F., Henderson, R. A., Clarkson, P. D., Hughes, C. P., Laurie, J., Stait, B., Thomson, M. R. A., and Webers, G. F. 1981. *The Ordovician system in Australia, New Zealand and Antarctica*. Publication no. 6. Ottawa: International Union of Geological Sciences.

Weber, W. 1982. Beitrag zür geologie des Pensacola-Gebirges (Antarktika). *Freiberger Forschungsheft,* 371: 41–96.

Weber, W., and Fedorov, L. V. 1981. Geology of the northern Neptune Range, Pensacola Mountains. *Geodatische und geophysikalische veroffentlichungen,* ser. 1, no. 8: 68–94.

Weihaupt, J. G. 1960. Reconnaissance of a newly discovered area of mountains in Antarctica. *Journal of Geology,* 68: 669–73.

White, A. J. R., and Chappell, B. W. 1977. Ultrametamorphism and granitoid genesis. *Tectonophysics,* 43: 7–22.

1983. *Granitoid types and their distribution in the Lachlan fold belt, south east Australia*. Memoir, 159. Boulder, CO: Geological Society of America, pp. 21–34.

Williams, P. F., Hobbs, B. E., Vernon, R. H., and Anderson, D. E. 1971. The structural and metamorphic geology of basement rocks in the McMurdo Sound Area, Antarctica. *Journal of the Geological Society of Australia,* 18: 127–42.

Williams, P. L., 1969. Petrology of upper Precambrian and Paleozoic sandstones in the Pensacola Mountains, Antarctica. *Journal of Sedimentary Petrology,* 39: 1455–65.

Wodzicki, A., Bradshaw, J. D., and Laird, M. G. 1982. Petrology of the Wilson and Robertson Bay Groups and Bowers Supergroup, northern Victoria Land, Antarctica. In C. Craddock, ed., *Antarctic geoscience*. Madison: University of Wisconsin Press, pp. 549–54.

Wodzicki, A., and Robert, R., Jr. 1984. Geology and geologic history of the Bowers Supergroup, northern Victoria Land. *Antarctic Journal of the United States,* 14: 29–31.

1986. Geology of the Bowers Supergroup, central Bowers Mountains, northern Victoria Land. In E. Stump, ed., *Geological investigations in northern Victoria Land*. Antarctic Research Series, vol. 46. Washington, DC: American Geophysical Union, pp. 39–68.

Woolfe, K. J., Kirk, P. A., and Sherwood, A. M. 1989. Geology of the Knobhead area, southern Victoria Land, Antarctica, 1:50,000. Miscellaneous Geological Map 19. Lower Hutt: New Zealand Geological Survey, Department of Scientific and Industrial Research.

Wright, T. O. 1980. Sedimentology of the Robertson Bay Group, northern Victoria Land, Antarctica. *Antarctic Journal of the United States,* 15: 6–8.

1981. Sedimentology of the Robertson Bay Group, northern Victoria Land, Antarctica. *Geologisches Jahrbuch,* B41: 127–38.

1982. Structural study of the Leap Year Fault, northern Victoria Land. *Antarctic Journal of the United States,* 5: 11–13.

1985a. Late Precambrian and early Paleozoic tectonism and associated sedimentation in northern Victoria Land, Antarctica. *Geological Society of America Bulletin,* 96: 1332–39.

1985b. Thrust tectonics in northern Victoria Land, Antarctica. In *Abstracts, Sixth Gondwana Symposium*. Miscellaneous Publication 231, Institute of Polar Studies. Columbus: Ohio State University, p. 107.

Wright, T. O., and Brodie, C. 1987. The Handler Formation, a new unit of the Robertson Bay Group, northern Victoria Land, Antarctica. In G. D. McKenzie, ed., *Gondwana Six: Structure tectonics, and geophysics*. Geophysical Monograph 40. Washington, DC: American Geophysical Union, pp. 25–29.

Wright, T. O., and Dallmeyer, R. D. 1991. The age of cleavage development in the Ross orogen, northern Victoria Land, Antarctica: Evidence from $^{40}$Ar/$^{39}$Ar whole-rock slate ages. *Journal of Structural Geology,* 13: 677–90.

1992. The age of cleavage development in the Ross orogen, northern Victoria Land, Antarctica: Evidence from $^{40}$Ar/$^{39}$Ar whole-rock slate ages – Reply. *Journal of Structural Geology,* 14: 891–92.

Wright, T. O., and Findlay, R. H. 1984. Relationships between the Robertson Bay Group and the Bowers Supergroup: New progress and complications from the Victory Mountains, North Victoria Land. *Geologisches Jahrbuch,* B60: 105–16.

Wright, T. O., Ross, R. J., Jr., and Repetski, J. E. 1984. Newly discovered youngest Cambrian or oldest Ordovician fossils from the Robertson Bay terrane (formerly Precambrian), northern Victoria Land, Antarctica. *Geology,* 12: 301–5.

Wu, B., and Berg, J. H. 1992. Early Paleozoic lamprophyre dikes of southern Victoria Land: Geology, petrology and geochemistry. In Y. Yoshida, K. Kaminuma, and K. Shiraishi, eds., *Recent progress in Antarctic earth science.* Tokyo: Terra Scientific, pp. 257–64.

Wyborn, D. 1981. Granitoids of North Victoria Land, Antarctica: Field and petrographic observation. *Geologisches Jahrbuch,* B41: 229–49.

Yochelson, E. L., and Stump, E. 1977. Discovery of Early Cambrian fossils at Taylor Nunatak, Antarctica. *Journal of Paleontology,* 51: 872–75.

Young, D. J., and Ryburn, R. J. 1968. The geology of Buckley and Darwin Nunataks, Beardmore Glacier, Ross Dependency, Antarctica. *New Zealand Journal of Geology and Geophysics,* 11: 922–39.

# Index

Cachets from the author's field seasons.

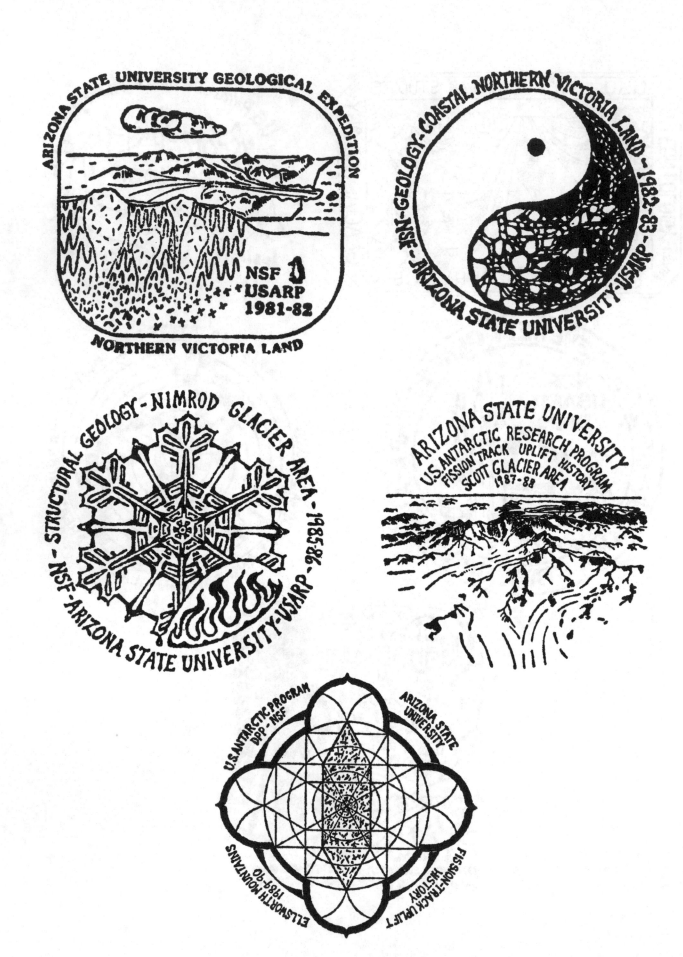